W9-BHR-822

DISCARDED BY
MEMPHIS PUBLIC LIBRARY

The Haynes
Automotive
Brake
Manual

by Bob Henderson
and John H Haynes
Member of the Guild of Motoring Writers

The Haynes Automotive Repair Manual
for maintaining, troubleshooting and
repairing brake systems

ABCDE
FGHIJ
KLMNO
PQRST

Haynes Publishing Group
Sparkford Nr Yeovil
Somerset BA22 7JJ England

Haynes North America, Inc
861 Lawrence Drive
Newbury Park
California 91320 USA

Acknowledgements

We are grateful for the help and cooperation of the Chrysler Corporation, Ford Motor Company, Mazda Motor Company, Nissan Motor Company, Toyota Motor Corporation and Volvo North America Corporation for assistance with technical information and certain illustrations. Kent Reppert II also supplied various illustrations. We also wish to thank Tilton Engineering (Buellton, CA), Wilwood Engineering (Camarillo, CA), Performance Friction (Clover, SC), Neuspeed/Automotive Performance Systems, Inc. (Camarillo, CA) and Jones Brake Technologies, Inc. (St. Thomas, Ontario, Canada) for supplying photos and information on the products shown in Chapter 11. The Lisle Corporation furnished some of the tools and photos shown in Chapter 2.

© **Haynes North America, Inc. 1994**

With permission from J.H. Haynes & Co. Ltd.

A book in the Haynes Automotive Repair Manual Series

Printed in the U.S.A.

All rights reserved. No part of this book may be reproduced or transmitted in any form or by any means, electronic or mechanical, including photocopying, recording or by any information storage or retrieval system, without permission in writing from the copyright holder.

ISBN 1 56392 112 X

Library of Congress Catalog Card Number 94-78658

While every attempt is made to ensure that the information in this manual is correct, no liability can be accepted by the authors or publishers for loss, damage or injury caused by any errors in, or omissions from, the information given.

94-288

Contents

Chapter 5 Disc brakes

Chapter 6 Drum brakes

Chapter 7 Parking brakes

Chapter 8 Brake pedals and brake light switches

Chapter 9 Hydraulic systems and power boosters

Chapter 10 Anti-lock Braking Systems (ABS)

Chapter 11 Modifications

Glossary

Torque specifications

Index

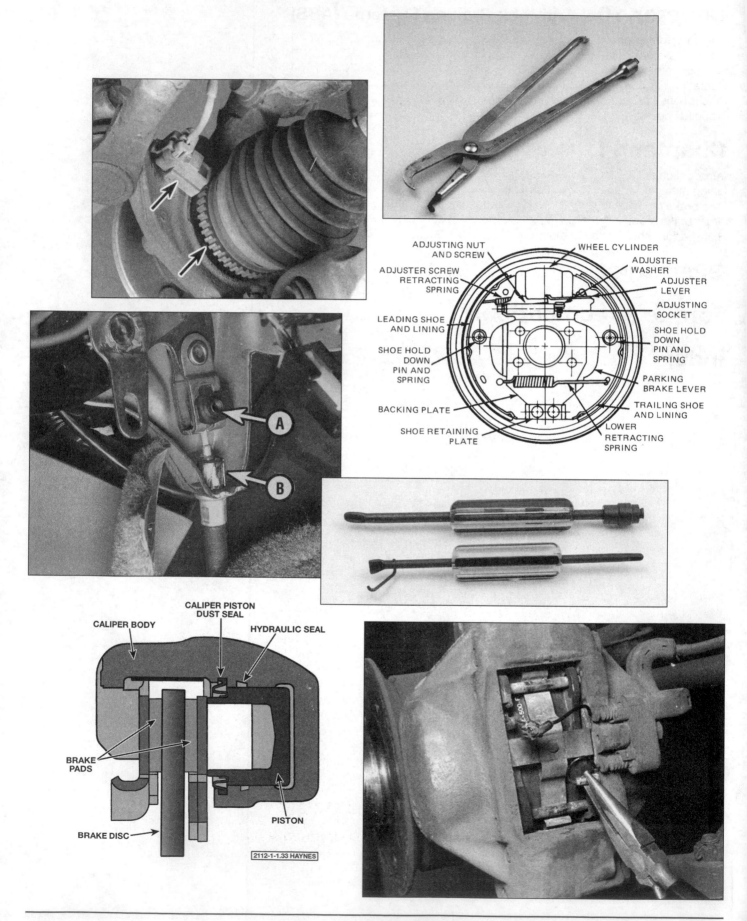

ADJUSTING NUT AND SCREW

WHEEL CYLINDER

ADJUSTER SCREW RETRACTING SPRING

ADJUSTER WASHER

ADJUSTER LEVER

ADJUSTING SOCKET

LEADING SHOE AND LINING

SHOE HOLD DOWN PIN AND SPRING

SHOE HOLD DOWN PIN AND SPRING

PARKING BRAKE LEVER

BACKING PLATE

TRAILING SHOE AND LINING

SHOE RETAINING PLATE

LOWER RETRACTING SPRING

CALIPER BODY

CALIPER PISTON DUST SEAL

HYDRAULIC SEAL

BRAKE PADS

PISTON

BRAKE DISC

2112-1-1.33 HAYNES

1 Introduction

It's a beautiful summer day. You're hurtling down your favorite stretch of asphalt, enjoying the countryside and your favorite CD. You notice, however, an annoying rattle coming from somewhere under the instrument panel. "I'll have to check that out" you say to yourself, as you make a mental note to fix it as soon as you get home. You turn up the stereo.

Suddenly, a very large, slow moving truck pulls onto the road, not more than 50 yards ahead. Instinctively, you check the lane to your left for oncoming traffic and to your dismay, there's lots of it. With nowhere to go, you hit the brakes. *Hard*. Hard enough for the amber "ABS" light on the instrument panel to glow, but that isn't what you're looking at.

You breath a sigh of relief when you realize the truck's liftgate isn't rushing towards you anymore. After a few choice phrases and an exchange of hand gestures, you pass this rolling roadblock and you're back on your way - nothing lost but a little sweat.

A little sweat *and* a minuscule amount of friction material from the brake pads, but you're not thinking about your brakes. They performed flawlessly, and besides, *you* pushed the pedal so *you* get the credit!

When was the last time you *did* think about your vehicle's brakes? The sad fact is that most drivers pay more attention to annoying rattles or scratches in the paint, ignoring their brakes until they start making bad noises, the vehicle pulls to one side, the brake pedal feels mushy or, worse yet, they fail completely. Brakes are probably the most abused and neglected components on a vehicle.

While complete brake failure is not a common occurrence, some forms of brake system trouble are all too common. Most can be avoided with a little attention every few thousand miles. The braking system is the most important system on your vehicle. Your life, the lives of your passengers, pedestrians and other motorists on the road *all* depend upon *your* brakes.

All vehicles will need some form of brake system repair sooner or later. Wear is an inherent feature of all brake systems - your brakes slowly give their lives so you can retain yours. Due to this continual sacrifice, all brake pads and shoes must eventually be retired, to be replaced with the next generation of new lining. Other problems can crop up, too.

But for many, the braking system represents a portion of the vehicle that just shouldn't be tampered with, and because of this, repairs are usually left to someone else. That's where this manual comes in. It has been designed to help just about anyone repair just about any kind of problem in any type of brake system.

Most modern automotive braking systems are actually quite simple to maintain and repair, provided that each step is performed in a careful, deliberate manner. A considerable amount of money can be saved and much knowledge can be gained by taking on these tasks yourself. This will give you a great sense of satisfaction - not only because you saved a bunch of money and your car wasn't tied up for a couple of days in some shop, but because you'll know that the job was done correctly.

This manual unlocks the "mysteries" of the modern automotive braking system. The procedures illustrated throughout the book are general in nature, but with specific differences in component design pointed out where necessary to help you complete each procedure safely and properly. Included are chapters on component and system operating fundamentals, tools and equipment, troubleshooting, maintenance, disc brake pad replacement, drum brake shoe replacement, parking brake systems, brake pedals and brake light switches, hydraulic systems, Anti-lock Brake Systems (ABS) and modifications. At the end of the manual there's a glossary of terms used throughout the book. In short, everything you need to know to successfully maintain and repair your brakes and ensure your safety and that of others around you, as well!

How to use this repair manual

The manual is divided into Chapters. Each chapter is subdivided into sections, some of which consist of consecutively numbered paragraphs (usually referred to as "Steps," since they're normally part of a procedure). If the material is basically informative in nature, rather than a step-by-step procedure, the paragraphs aren't numbered.

The term **(see illustration)** is used in the text to indicate that a photo or drawing has been included to make the information easier to understand (the old cliché "a picture is worth a thousand words" is especially true when it comes to how-to procedures). Also every attempt is made to position illustrations on the same page as the corresponding text to minimize confusion. Some procedures are largely made up of illustrations and captions, with little or no accompanying text. The two types of illustrations used (photographs and line drawings) are referenced by a number preceding the caption. Illustration numbers denote chapter and numerical sequence within the chapter (i.e. 3.4 means Chapter 3, illustration number 4 in order).

The terms "**Note,**" "**Caution,**" and "**Warning**" are used throughout the book with a specific purpose in mind - to attract the reader's attention. A "**Note**" simply provides information required to properly complete a procedure or information which will make the procedure easier to understand. A "**Caution**" outlines a special procedure or special steps which must be taken when completing the procedure where the Caution is found. Failure to pay attention to a Caution can result in damage to the component being repaired or the tools being used. A "**Warning**" is included where personal injury can result if the instructions aren't followed exactly as described.

Even though extreme care has been taken during the preparation of this manual, neither the publisher nor the author can accept responsibility for any errors in, or omissions from, the information given.

The brake system

Hydraulic systems

In the early days of motorized transportation, the entire brake system on a horseless carriage consisted of nothing more than a lever that the driver pulled, causing it to rub against a tire. Although some mechanical advantage could be built into the lever, braking power was largely determined by the strength of the driver. Of course, this was a very inadequate system and could quickly be rendered useless by driving through a little water or mud!

As cars evolved, the brakes at the wheels (drum brakes, which we'll cover in the next Section) were actuated by a linkage or a cable, or a combination of the two. This was usually sufficient for the vehicle weights and the relatively low speeds they were able to achieve. This worked, but cables and linkages were a source of high wear and a definite "weak link" in the brake system chain.

By the late 1920's/early 1930's, most cars were equipped with hydraulically actuated brakes. This greatly increased braking reliability and safety. Since mechanical advantage can easily be designed into a hydraulic system, driving a motor vehicle became a much less tiresome proposition.

Why use fluid in a brake system? Well, for three reasons:

Liquids can't be compressed
Liquids can transmit motion
Liquids can transmit force and can also increase or decrease force

This was discovered long before the automobile was even dreamed of (in 1647), by a French philosopher and mathematician named Blaise Pascal (1623 - 1662). He determined that in a confined fluid, pressure applied externally is transmitted equally in all directions. In a static fluid, force is transmitted at the speed of sound throughout the fluid and acts at right angles on all surfaces in or confining the fluid **(see illustration)**. Of course, this happens because liquids can't be compressed.

Liquids can also transmit motion because of this law of physics. If a piston in a hydraulic cylinder is pushed in a certain amount, a piston in another hydraulic cylinder of equal size, connected by a tube, will move out an equal amount **(see illustration)**.

By the use of different size pistons in a hydraulic system, liquids can increase or decrease force. If a force of 200 pounds is applied to a hydraulic cylinder with an area of one square inch, a pressure of 200 pounds per square inch (psi) will be generated. This pressure of 200 psi will be present in the entire system. If this cylinder is connected by a tube to another cylinder with a piston having a surface area of 20 square inches, a force of 200 psi will push on every square inch of the larger piston. A mechanical advantage of twenty would be generated, and a force of 4000 psi would be pushing on the piston **(see illustration)**. Therefore, a force of 200 pounds could

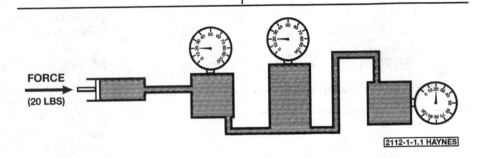

1.1 Pascal's law states that "pressure, when applied to a confined liquid, is transmitted undiminished." In this example, a force of 20 pounds has been exerted on the piston, which has a surface area of one square inch. Notice that each pressure gauge registers 20 pounds per square inch (psi), regardless of the size, shape or location of its chamber.

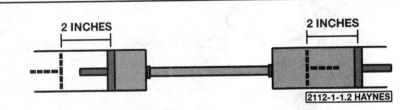

1.2 The piston in the hydraulic chamber at the left has been depressed two inches - the piston in the cylinder on the right has moved out an equal amount. The pistons are of the same surface area and the force transmitted is equal, too.

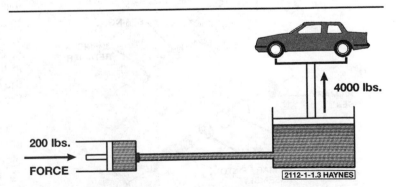

1.3 Here's an example of how liquids can increase force: 200 pounds of effort applied to the piston on the left, which has a surface area of one square inch, will yield an output of 4000 pounds at the piston on the right, which has a surface area of 20 square inches

actually lift or move 4000 pounds! If this particular hydraulic system was operated in reverse, the mechanical advantage would be greatly diminished. A force of 80,000 pounds would have to be applied to the large piston to have 4000 pounds available at the smaller one.

The hydraulic brake system in all motor vehicles is based on these principles. When the driver of a car pushes on the brake pedal, the force exerted on the master cylinder is converted into system pressure and is distributed through the brake lines to the wheel cylinders and/or brake calipers. This action is what applies the brake shoes to the drums or the brake pads to the discs, creating friction, which converts the kinetic energy of the moving vehicle to heat and slows the car down.

The components that make up a modern hydraulic system are only a little more complex than the examples shown in the previous illustrations. Lets take a look at the individual components and the jobs they do.

Master cylinder

The master cylinder converts mechanical input force into system hydraulic pressure. Force exerted by the driver is transmitted through the brake pedal and applied to the master cylinder piston through a pushrod. As the piston moves forward in the cylinder, brake fluid is forced from the master cylinder. Remember, though, that the entire system is filled with fluid, so this motion is transmitted, undiminished, through the brake lines and hoses to the wheel cylinders or calipers. Once all movement is taken up in the wheel cylinders or calipers, pressure in the system begins to rise. Further movement of the components is impossible, but pressure will increase within the system if the driver pushes harder on the brake pedal.

Master cylinders are constructed of either cast iron or aluminum. On most vehicles the master cylinder is mounted on the firewall or power brake booster. Some have integral (non-detachable) fluid reservoirs (see illustration), others have detachable reservoirs made of plastic (see illustration). Some have remote reservoirs that are connected to the master cylinder with hoses (see illustration). A reservoir is necessary to provide additional fluid to compensate for brake lining wear and movement and, in the unfortunate event of a brake fluid leak, a small margin of safety (there will be enough fluid left in the reservoir to enable the driver to pump the brake pedal a few times, hopefully bringing the vehicle to a stop!). Reservoirs must be vented to allow the fluid level to move up and down without creating a vacuum in the system. A rubber diaphragm between the fluid and the cap or cover allows this to happen, while sealing off the brake fluid from outside elements.

Many older vehicles were equipped with single-piston master cylinders (see illustration), but all late model vehicles are equipped with dual-piston master cylinders. The pistons are fitted with seals, or piston cups, which seal tightly against the walls of the cylinder when pressure is developed in the system. This prevents fluid from leaking past

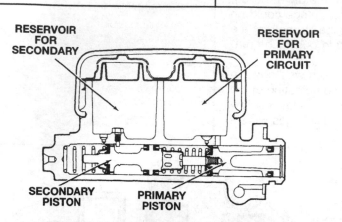

1.4 Cutaway of a typical integral reservoir master cylinder

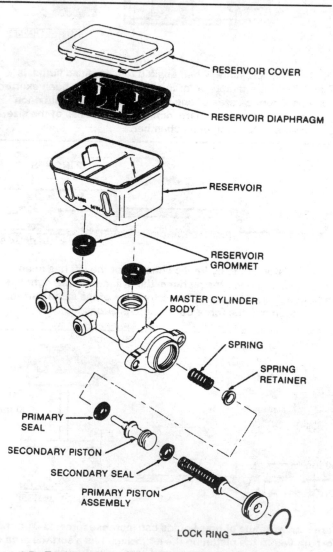

1.5 Exploded view of a typical detachable reservoir master cylinder

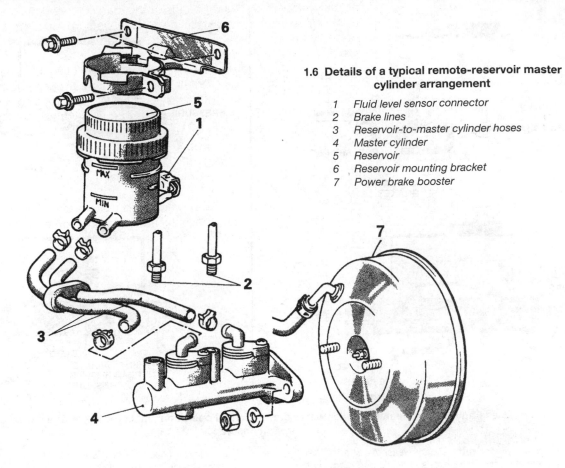

1.6 Details of a typical remote-reservoir master cylinder arrangement

1   Fluid level sensor connector
2   Brake lines
3   Reservoir-to-master cylinder hoses
4   Master cylinder
5   Reservoir
6   Reservoir mounting bracket
7   Power brake booster

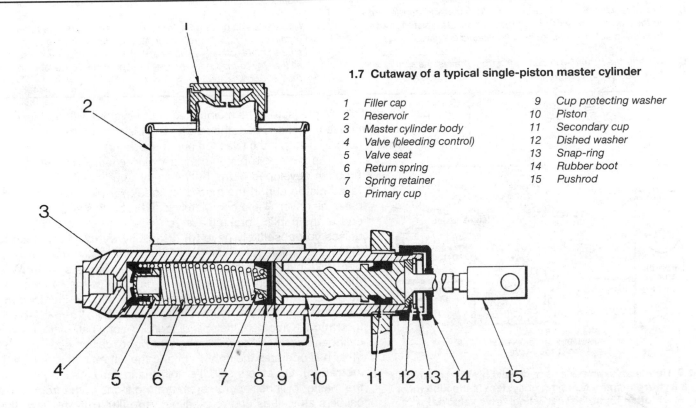

1.7 Cutaway of a typical single-piston master cylinder

1   Filler cap
2   Reservoir
3   Master cylinder body
4   Valve (bleeding control)
5   Valve seat
6   Return spring
7   Spring retainer
8   Primary cup

9   Cup protecting washer
10   Piston
11   Secondary cup
12   Dished washer
13   Snap-ring
14   Rubber boot
15   Pushrod

Haynes Automotive Brake Manual

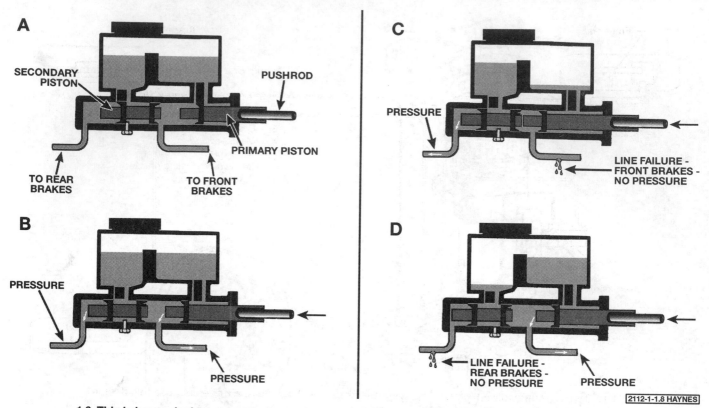

1.8 This is how a dual master cylinder works during normal operation and in the event of a partial brake system failure:

A   The master cylinder in the static (released) position. The chambers in front of the pistons are filled and ready for the application of the brake pedal.

B   Pressure has been applied to the brake pedal and transferred to the master cylinder through the pushrod. The pistons move forward, creating pressure in the hydraulic circuits.

C   Uh-oh - the hydraulic circuit to the front brakes has sprung a leak! Don't fear, though; the primary piston has moved for-

ward and has physically contacted the secondary piston, moving it forward and creating pressure in the rear brake circuit. The fluid will eventually drain from the primary side of the reservoir, but there's still plenty of fluid in the secondary side.

D   Not again! The problem with the front brake circuit has been repaired, but now the rear circuit blew. Thankfully, the pressure from the primary piston has shoved the secondary piston all the way to the front, and plenty of pressure is available to the front brakes.

---

the pistons, and allows pressure to be maintained in the system without bleeding down (decreasing due to leakage past the seals). The pistons are retained in the master cylinder bore by a stop plate (washer) and a snap-ring.

The old single-piston master cylinder worked fine unless a problem developed in the hydraulic system (a leak in a hose or failed piston cup in the master cylinder, for example), in which case the driver would completely lose his brakes and have to resort to the parking brake (have you ever tried to do this? Don't, unless you absolutely have to) or quickly try to find a place to make a fairly safe "crash landing"! On a dual-piston master cylinder brake circuit, each piston operates half of the system. With this design, if there is a problem in one half of the system, the other half of the system will still be functional **(see illustration)**. On some vehicles the system is split from front-to-rear **(see illustration)**. On others it's split diagonally - one half of the master cylinder operates the left front and right rear wheels and the other half operates the right front and the left rear wheels **(see illustration)**. On others still, it's split triangularly - each piston in the master cylinder operates separate pistons in the front brake calipers as well as one rear caliper **(see illustration)**. With this

1.9 Here's an example of a hydraulic brake system with a front/rear split - this type of system is the kind most often found on rear-wheel drive vehicles

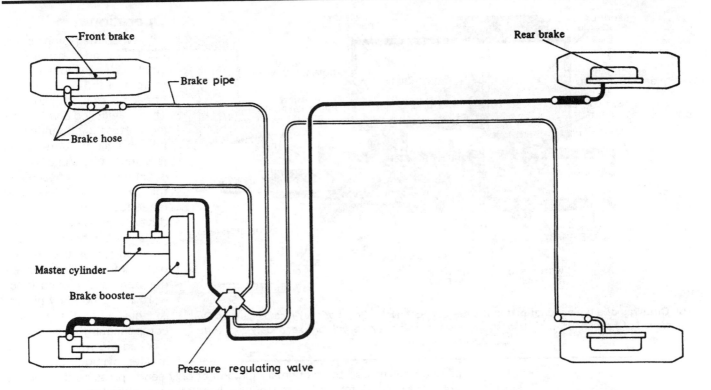

1.10 Typical diagonally split hydraulic system - most small front-wheel drive cars use this design

system, bringing the vehicle to a stop after a partial brake failure is a much less harrowing experience - the vehicle won't veer to one side as much as a diagonally split system, and the rear brakes won't lock-up easily (as would happen if the front brake circuit failed on a front-to-rear split system). In the event of a partial brake failure with any one of these designs, however, increased pedal travel and effort required to stop the vehicle will result.

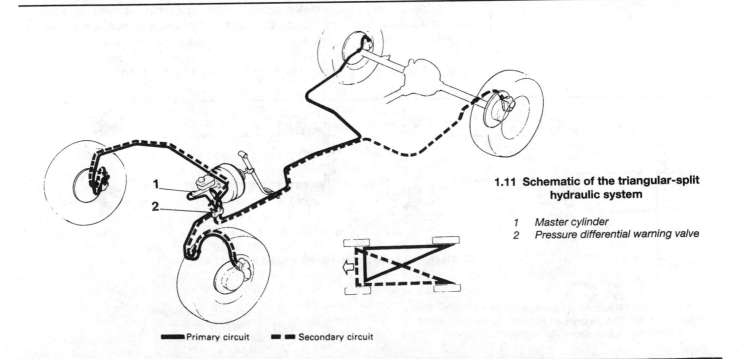

1.11 Schematic of the triangular-split hydraulic system

1   *Master cylinder*
2   *Pressure differential warning valve*

Primary circuit   Secondary circuit

Haynes Automotive Brake Manual

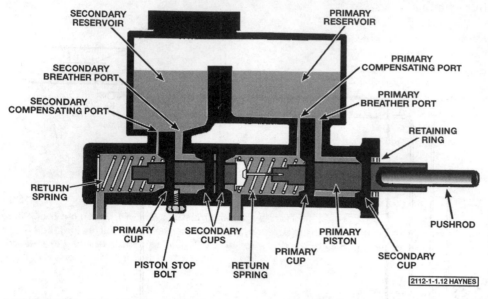

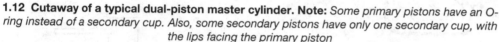

1.12 Cutaway of a typical dual-piston master cylinder. Note: *Some primary pistons have an O-ring instead of a secondary cup. Also, some secondary pistons have only one secondary cup, with the lips facing the primary piston*

Operation

When the pistons in the master cylinder are at rest, small passages located just in front of the primary cups of each piston are exposed. These are called *compensating ports*; they connect the master cylinder bore to the fluid reservoirs and allow the chamber in front of each piston to fill with brake fluid **(see illustration)**. When the driver depresses the brake pedal, the master cylinder pushrod forces the primary piston into the master cylinder bore and moves it toward the secondary piston. At this time, the primary compensating port is blocked off, which seals the master cylinder bore. The pressure developed by the primary piston not only sends pressure through the primary system, it forces the secondary piston forward in the bore, blocking off the secondary compensating port, causing pressure to rise in the secondary system, too.

Behind each compensating port is a *breather port* (sometimes called a *filler port*). This port allows fluid from the reservoir to fill the area behind the piston, preventing air bubbles from traveling around the lips of the primary cups on the pistons when the brakes are released. This fluid also flows past the lips of the primary cups of the pistons when the brakes are released, filling the chambers in front of the pistons, readying them for the next pedal application. A spring fitted to the end of each piston helps return them to their "at rest" positions when the driver releases pressure on the brake pedal.

Now a brief explanation of piston cup terminology. Even though there is a primary piston and a secondary piston in a dual-piston master cylinder, each piston has a primary cup - it's the one that pushes the fluid and creates the pressure **(see illustration)**. Each piston also has a secondary cup. The secondary cup on the secondary piston is the one on the end of the piston closest to the primary piston of the master cylinder. Pressure generated by the primary cup of the pri-

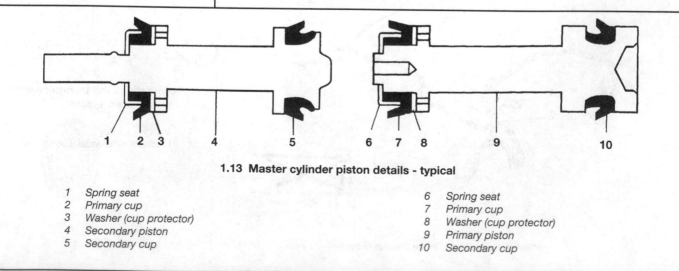

1.13 Master cylinder piston details - typical

1 Spring seat	6 Spring seat
2 Primary cup	7 Primary cup
3 Washer (cup protector)	8 Washer (cup protector)
4 Secondary piston	9 Primary piston
5 Secondary cup	10 Secondary cup

mary piston pushes against the secondary cup of the secondary piston, which forces the secondary piston forward. The secondary cup on the primary piston prevents fluid from leaking out of the master cylinder. Most primary pistons are also equipped with an O-ring at the pushrod end of the piston, which acts as a "wiper ring," preventing contaminants from entering the bore of the master cylinder.

On some (but not all) brake circuits that feed drum brakes, a check valve is installed in the outlet of the master cylinder (**see illustration**). This valve maintains a little residual pressure (about 8 to 16 psi) in the lines and wheel cylinders. This pressure keeps the cups in the wheel cylinders sealed against the bores of the cylinders. In doing so, the cups seal the brake fluid in and don't allow air to infiltrate the system.

When the driver lifts his or her foot from the brake pedal, the pistons begin to return to their "at rest" position. As this happens, brake fluid flows through the bleeder holes in the head of each piston (**see illustration**), forcing the piston cups away from the cylinder wall and allowing fluid to flow into the chamber ahead of each piston. The compensating ports are uncovered when the pistons are fully retracted, and pressure will equalize between the fluid in the cylinder bore and the fluid in the reservoir.

Wheel cylinders

Wheel cylinders are used in drum brake assemblies. They convert the hydraulic pressure created by the master cylinder into mechanical force, which pushes outward on the brake shoes, expanding the shoes against the brake drum.

Wheel cylinders are usually made from cast iron. The most common type is the double-piston wheel cylinder, which uses two pistons (usually aluminum), two rubber cups, cup expanders and a coil spring between the two piston assemblies (**see illustration**). Some designs use pushrods connected to the brake shoes. Other styles use slotted or notched pistons, in which the ends of the brake shoes ride. At each end of the wheel cylinder is a rubber boot, which seals the unit from brake dust and other elements.

Wheel cylinders are drilled and tapped for the brake line connection and bleeder screw. They are retained to the brake backing plate with one or two bolts, or a spring steel clip.

Other types of wheel cylinders include the single-piston cylinder (**see illus-**

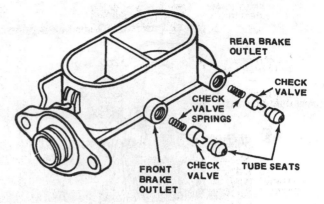

1.14 Here's a typical master cylinder from an old four-wheel drum brake system. Notice the check valves installed in the brake line outlets - they maintain a little residual pressure in the brake system, which keeps the lips of the wheel cylinder seals from leaking fluid or admitting air into the hydraulic system

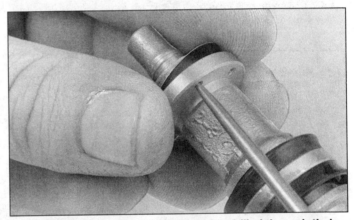

1.15 Most pistons have bleeder holes drilled through their heads, right behind the primary cup. When the brake pedal is released and the pistons retract, fluid rushes through these holes and helps push the lips of the cup off the cylinder wall. This action ensures rapid filling of the pressure chamber in front of the piston.

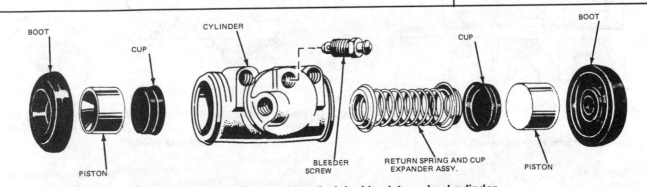

1.16 Exploded view of a typical double-piston wheel cylinder

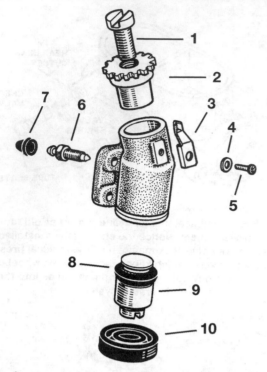

1.17 Here's an exploded view of a typical single-piston wheel cylinder (this one's from a front brake)

1	Trailing shoe seat/adjuster screw	6	Bleeder screw
2	Adjuster star wheel	7	Cover
3	Adjuster detent	8	Cup
4	Washer	9	Piston
5	Screw	10	Dust boot

tration) and the stepped cylinder. The single-piston cylinder is usually used in pairs; each cylinder actuates one brake shoe. Stepped cylinders are essentially the same as double-piston cylinders, except that one half of the cylinder bore is larger than the other. The larger piston is capable of exerting more force than the smaller one. So, stepped cylinders are only used in drum brake designs where it is necessary to apply more pressure to one of the brake shoes than the other.

Calipers

The brake calipers, used on disc brake setups, are basically frames that house the brake pads and the hydraulic cylinders, which convert the system pressure from the master cylinder into mechanical force. This force squeezes the brake pads against the disc, which causes friction and heat dissipation, which in turn slows the vehicle. Calipers will be explained in greater detail under the *Disc brakes* heading.

Pressure differential warning switch

This device **(see illustration)**, used on systems with dual-piston master cylinders, alerts the driver of a partial brake failure by illuminating a warning light on the instrument panel. Most pressure differential warning switches are mounted in the hydraulic system downstream of the master cylinder, but some are mounted in the master cylinder housing **(see illustration)**.

As stated earlier in this chapter, if a portion of a dual-circuit brake system fails, half of the brake system is rendered inoperative, but the other half will still function. The result will be increased effort required to stop the vehicle, and the pedal will travel closer to the floor.

A hydraulically activated switch will alert the driver of this condition. Both circuits of the hydraulic brake system travel through this switch. A small hydraulic

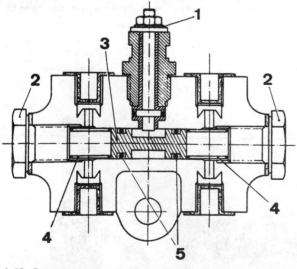

1.18 Cutaway of a pressure differential warning switch

1	Warning switch	4	Springs
2	End plugs	5	Seals
3	Piston		

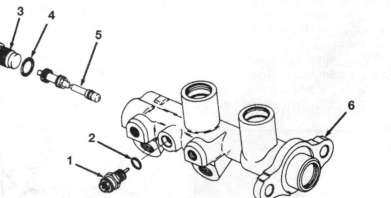

1.19 Here's an example of a pressure differential warning switch housed in the master cylinder (this one's from a mid-80's General Motors vehicle)

1	Warning switch	4	O-ring
2	O-ring	5	Switch piston assembly
3	Plug	6	Master cylinder body

piston is situated between the primary circuit and the secondary circuit. As long as there are no leaks in either circuit, the piston will remain centered in the switch body. But if a brake line in one half of the system springs a leak, which will cause a pressure drop in that half of the system, the piston in the pressure differential warning switch will be forced to the side with less pressure. It takes approximately 80 to 150 pounds of pressure difference to move the piston. This action depresses a plunger on the switch terminal, or allows the plunger on the switch to fall into a cutout in the piston, depending on the design of the switch. In either case, the switch grounds the warning circuit and the light on the dash comes on.

Some pressure differential warning switches are self-centering. This means that after a brake system failure has taken place and the problem has been fixed, it isn't necessary to reset the switch. Other types of switches must be reset manually. This is usually carried out by exerting a force of approximately 450 pounds on the brake pedal, but on some designs it is necessary to have an assistant loosen a bleeder valve on a caliper or wheel cylinder on the half of the brake system that did not have the problem, while the pedal is being depressed. The bleeder valve must be tightened as soon as the light on the instrument panel goes out, or the switch may travel too far, making it necessary to repeat the operation again, but on the other half of the system.

Metering valve

The metering valve, sometimes referred to as the hold-off valve, is used only on vehicles with disc brakes in the front and drum brakes in the rear. It prevents the disc brakes from applying before the drum brakes. This permits a more even, controllable braking action because the front and rear brakes will apply at the same time. This will also ensure that the drum brakes do an equal amount of work during light braking; if a metering valve wasn't used, the front brakes would do *all* the braking during light stops.

This valve, located in the hydraulic circuit to the front brakes **(see illustration)**, prevents brake fluid movement to the calipers until a sufficient amount of pressure has built up in the system - enough to overcome the resistance of the drum brake shoe retracting springs and begin to apply the shoes to the drum. At this designated pressure, the valve opens and allows pressure through to the disc brakes.

Metering valves are usually combined with the pressure differential warning switch and, on models so equipped, the proportioning valve.

Proportioning valve

This valve was once found only on vehicles with drum brakes on the rear wheels, but now most cars with four-wheel disc brakes use them, too. It's located in the hydraulic circuit to the rear wheels and limits the amount of hydraulic pressure to the wheel cylinders or calipers to prevent rear wheel lockup during panic stops **(see illustration)**. The proportioning valve is sometimes remotely mounted (like the metering valve in the previous illustration, or at the rear of the vehicle), but on some brake systems is built right into the master cylinder **(see illustration)**.

On some vehicles, the proportioning valve has a fixed setting. It will regulate system pressure at the same rate, regardless of any other conditions. On other vehicles, espe-

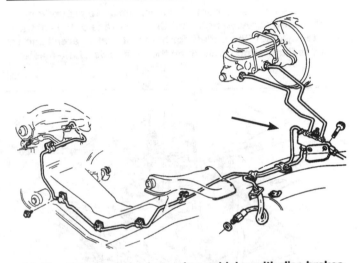

1.20 The metering valve is used on vehicles with disc brakes in the front and drum brakes in the rear. It delays pressure to the front brakes until the rear brakes begin to apply, resulting in a more balanced braking action. It's located in the hydraulic line to the front brakes (arrow). In this example, it's integrated into the combination valve

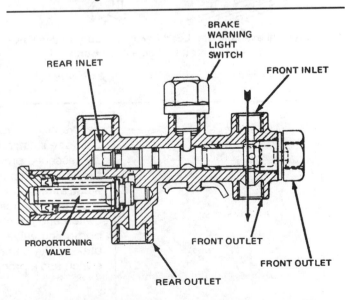

1.21 This proportioning valve is built into the pressure differential warning switch

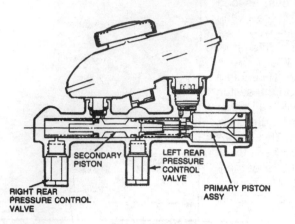

1.22 These pressure control valves, similar to proportioning valves, are screwed right into the rear brake outlets of the master cylinder (the outlets for the front brakes aren't shown in this illustration - they're on the other side of the master cylinder body)

1.23 This variable-rate proportioning valve is mounted to the underside of this vehicle and is connected by a rod to a rear suspension arm. When the rear of the vehicle is loaded, the valve will allow normal pressure to the rear brakes. During hard stops, the rear of the car will tend to rise, shifting weight to the front. The valve will then reduce pressure to the rear brakes, preventing the rear wheel from locking up

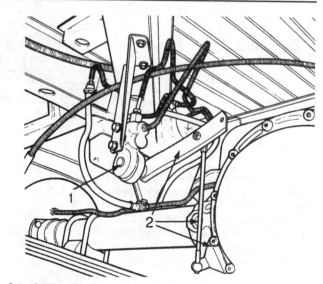

1.24 A typical example of a Load Sensing Proportioning Valve (LSPV) - this one's mounted on a Jeep Comanche

1   LSPV   2   Linkage

cially (but not limited to) pick-up trucks, a variable proportioning valve is used **(see illustrations)**. Sometimes it's called a pressure control valve or load-sensing proportioning valve, but it does the same thing - it limits pressure to the rear brakes just as a normal proportioning valve does, but it's connected to a portion of the rear suspension by a linkage, which allows the valve to monitor ride height and the amount the rear end of the vehicle rises during hard stops. By doing so, the valve can more accurately regulate pressure to the rear brakes. If the vehicle is carrying a heavy load, the valve will allow more pressure to the rear brakes, since the rear of the vehicle won't tend to rise as much during a hard stop as the same vehicle with no load at all (which means there's more weight on the rear wheels, so they won't lock up too easily). It also helps the valve distinguish between a fairly hard stop and an all-out panic stop.

Combination valve

The combination valve combines the pressure differential warning switch, the metering valve and/or the proportioning valve into one unit **(see illustration)**. Each of the separate components works in the same way as described previously, but they're centralized instead of scattered throughout the vehicle. Combination valves are usually located near the master cylinder **(see illustration)**.

Some combination valves only consist of a pressure differential warning switch and a metering valve. This allows the use of a load sensing type of proportioning valve at the rear of the vehicle. On models with four-wheel drum brakes, the combination valve is made up of a pressure differential warning switch and a proportioning valve.

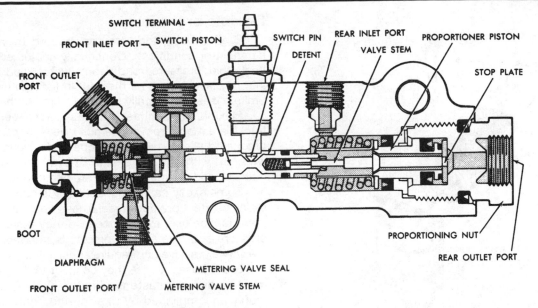

1.25 Cutaway of a typical combination valve - this one incorporates all three functions:

Metering - *Delays pressure application to the front brake calipers until the rear drum brake shoes have overcome the tension of their retracting springs and begun to apply*

Pressure differential warning - *Warns the driver of a failure in the brake hydraulic system*

Proportioning - *Controls, and on some versions ultimately limits, pressure to the rear drum brakes, to prevent the rear wheels from locking up during hard stops*

Brake hoses and lines

Brake lines and hoses are the conduit in which the hydraulic pressure from the master cylinder is transferred to the hydraulic cylinders at the wheel brakes. Non-flexible, steel tubing is used wherever possible. Where the brake lines must span a gap that moves - between the chassis and suspension or steering, for example - flexible brake hoses must be used.

Metal brake lines

Rigid brake lines are constructed from high-quality seamless, double-thickness steel tubing. They are coated (usually with a copper-lead alloy) to prevent corrosion. If it becomes necessary to replace a brake line, only use original-equipment replacement line or brake line of equal quality. Never use copper, aluminum, brass or inferior steel tubing - these materials can't withstand the normal vibration present in a motor vehicle and will fatigue, leading to cracks and subsequent brake failure.

Because of this vibration, the brake lines must be secured to the frame or underbody of the vehicle with clips. Even the high-quality materials used for brake lines will eventually work-harden and break if subjected to vibration for a long enough period of time.

Since relatively high pressures are created in a brake hydraulic system, double-lap flares are used on the ends of the lines where they connect to a brake hose or other component. Even though flaring tools and blank brake lines are readily available, don't be tempted to fabricate your own brake lines. Brake lines, with the flare nuts installed, the fittings flared and the lines pre-bent to the proper shapes, are available, which makes brake line replacement much easier and safer. It isn't easy to make a good, clean, double-lap flare in double-wall steel tubing!

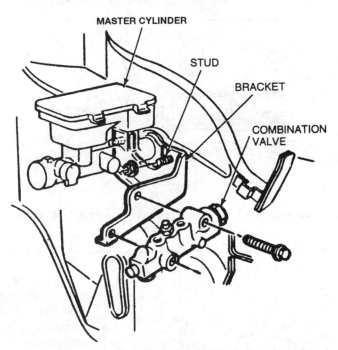

1.26 On many vehicles, the combination valve is mounted on a bracket adjacent to the master cylinder - on others, it's mounted further down, on the frame rail

Flexible brake hose

Brake hoses, too, are constructed to stand up to the harshest of conditions. Continually subjected to movement, heat, cold, pressure, road debris, water, mud, etc., it's amazing they last as long as they do. In fact, on modern vehicles, brake hoses will typically last the lifetime of the vehicle.

Flexible brake hoses on most passenger cars and light trucks are made from numerous layers, or *plies*, of fabric impregnated with a synthetic rubber compound. They are then wrapped with a tough sheath of synthetic rubber or plastic material, and the metal fittings are crimped to the ends.

Some exotic cars and most all race cars use Teflon-lined, stainless steel braided hose. These lines are great for high-performance use, mainly because they resist swelling under pressure and give a more precise brake pedal feel, but they are very expensive and require special fittings.

Brake hoses, where they connect to the rigid metal lines, must be securely fastened in a bracket. This is most commonly accomplished by passing the hose end through the bracket first, securing the hose end to the bracket with a narrow nut or a spring-steel clip, then attaching the metal line with a flare-nut fitting **(see illustrations)**. On some designs, the fitting on the hose is shaped to fit a similar shape in its mounting bracket; this prevents the hose from twisting. If the hose isn't secured to the bracket, vibration will eventually take its toll on the metal line, causing it to fail.

When replacing brake hoses, it's a good idea to take the old one with you to the parts store. Compare the length of the hoses and the fittings to make sure the replacement hose is exactly the same as the original.

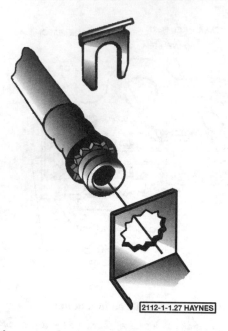

2112-1-1.27 HAYNES

1.27 Here's one method of securing a flexible brake hose to its bracket. The star shape on the hose fitting interlocks with the star shape in the bracket - this will prevent the hose from slipping in its bracket and twisting. The clip will keep the hose fitting from backing out of the bracket, which will prevent the rigid metal line from vibrating.

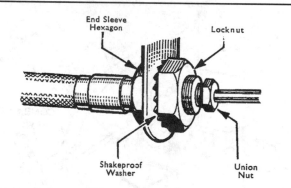

End Sleeve Hexagon

Locknut

Shakeproof Washer

Union Nut

1.28 Instead of a clip, this hose is secured to its bracket with a locknut

Brake fluid

Brake fluid can be considered the "life blood" of the brake hydraulic system. Most vehicles on the road today use brake fluid conforming to DOT 3 or DOT 4 safety standards. DOT stands for Department Of Transportation, the governing body that sets the standards for brake fluid, among other things.

Never introduce any other liquid into the brake hydraulic system. If your vehicle was originally equipped with brake fluid, use only brake fluid in the system. Only a very small percentage of auto manufacturers use anything other than brake fluid in their vehicles. Some high-dollar exotic cars use a type of mineral oil in their brake systems, but brake fluid and mineral oil don't mix.

If any other kind of fluid contaminates your brake system, the entire system must be drained, flushed and refilled with clean brake fluid. But, since it isn't possible to completely remove all fluid from the brake system, this means that every hydraulic component in the system must be removed, disassembled and fitted with new seals. No shortcuts. Mineral oil or petroleum products are not compatible with the material that the brake system seals are made of. These liquids can cause the seals to swell and/or soften - either way, a very dangerous situation would be present.

As you can see, it is important that clean, uncontaminated brake fluid be used in the hydraulic system. It's a unique fluid, with more to its job than you might think. In addition to transmitting pressure, brake fluid must be able to:

a) Maintain a constant viscosity.
b) Be compatible with the lines and seals (i.e. all rubber/synthetic rubber components) in the hydraulic system.
c) Withstand heat - DOT 3 brake fluid has a 460-degree boiling point.

d) Resist freezing.
e) Lubricate the sliding parts in the system.
f) Resist compression.
g) Flow through small orifices with minimum resistance.
h) Resist the formation of sludge or gum deposits, regardless of temperature or pressure.
i) Prevent corrosion of metal parts.
j) Mix with other approved brake fluids.
k) Last, but certainly not least, *last!* Brake fluid must be able to do all of the above even after being in the system for an extended period of time (years).

When adding brake fluid to the brake fluid reservoir, always use brake fluid from a small, sealed brake fluid container. Since brake fluid is hygroscopic (meaning it is able to absorb moisture), fluid stored in an open container, or even in a sealed large container (if it's been sitting on the shelf for a long time) may contain too much moisture. If the fluid is laden with too much moisture, its boiling point is lowered. When the moisture-tainted brake fluid is subjected to the high heat of a modern brake system, the water molecules turn to gas, forming bubbles. This will cause a spongy feeling brake pedal. If the problem is serious enough it can cause complete brake failure. If this happens, the hydraulic system must be purged of all old brake fluid and filled with new fluid.

Use care when handling brake fluid. Here are some important points to remember:
a) Brake fluid is poisonous. If it is accidentally ingested, call a poison control center or seek medical attention immediately.
b) For the above reason, never store or dispose of brake fluid in a beverage container (when bleeding brakes, for example).
c) If you get brake fluid on your skin, wash it off immediately with soap and water. It's a good idea to wear latex gloves when there's a possibility of coming into contact with brake fluid.
d) Wear eye protection whenever bleeding the brakes or working around brake fluid. If you get brake fluid in your eyes, rinse them out with plenty of water, then seek medical attention.
e) Brake fluid will damage paint. Be careful when working on the hydraulic system near painted components. Cover all exposed bodywork in the surrounding area. Immediately rinse off brake fluid spills with lots of water.

Wheel brakes

Now that we know how hydraulic power is generated and transmitted, lets take a look at the different kinds of brakes setups at the wheels - the components that bear the brunt of the braking force and bring your vehicle to a stop.

Regardless of the type of brakes a vehicle may be equipped with, all brakes do the same thing; convert the energy of the moving vehicle into another form of energy - heat. They do this by rubbing parts together, and the resulting friction between the moving parts creates heat. The faster the brakes can create, absorb and dissipate heat, the faster the vehicle will slow down.

There are several factors that determine the efficiency of a brake. These are:
a) *The amount of pressure applied to the brake pads or shoes (more pressure = more friction).*
b) *The total area of the brake lining (greater area = more friction).*
c) *The diameter of the brake disc or drum (larger diameter = more powerful brake - like using a long lever to move a big rock).*
d) *The diameter of the vehicle's tires (a tall tire acts like a long lever turning the brake - the larger the diameter of the tire, the more powerful the brake must be).*
e) *The coefficient of friction between the brake pads and disc or brake shoes and drum.*
f) *The ability of the brake to dissipate heat.*
g) *The coefficient of friction between the tire contact patch and the ground.*

Creating friction is no problem, but dealing with the resulting heat has kept engineers busy for years, constantly searching for ways to improve braking efficiency. Even the brakes of years ago were pretty good at stopping a vehicle from a high rate of speed one or two times before allowing them to cool off, but as the brakes heated up, their ability to dissipate heat decreased, which reduced braking power. This frequently lead to brake fade, a condition that, when mild, requires more pedal effort from the driver to slow the vehicle. When severe, brake fade can lead to a complete loss of braking ability.

While the brakes installed on modern vehicles are certainly not immune to brake fade, they are able to safely handle many more hard stops, one after the other, than the brakes on older vehicles. This is due to improved materials used in the construction of braking components and, to an even greater degree, advances in the design of wheel brakes, enabling them to dispense of heat faster.

Brake fade

Brake fade as a condition that can occur when the brakes have become overheated and lose their ability to create friction and dissipate heat. The driver of a vehicle experiencing complete brake fade finds himself/herself pushing on a hard brake pedal with no stopping ability. Eventually, glazed brake linings occur, which adds to the loss of braking power.

This problem is compounded on a drum brake setup. The drums expand when they get hot, which adds to the distance the shoes must travel before they contact the friction surface. They also lose some of their rigidity and tend to flex outward when the brake shoes start pushing against them.

There's another danger associated with overheated brakes. If the brake fluid is not in top condition (if it's old and has absorbed a lot of moisture), the heat created by the brakes can be conducted to the hydraulic components, resulting in boiling brake fluid. If this happens, the driver is in even more trouble, because even if the brakes cool off enough to allow some degree of operation, the bubbles in the brake fluid will remain - the brake pedal will be soft and mushy and probably travel to the floor, unable to produce any pressure in the hydraulic system. This is because the gas bubbles in the brake fluid are compressible. The only way to get rid of them is to bleed the hydraulic system (at which time all of the old fluid should be purged and replaced with fresh brake fluid).

Brake fade is not only a heat problem - it can be brought on by water, too. While this condition isn't quite as dangerous as heat-induced fade (provided there isn't a parked car directly in front of you as you're trying to stop), it can take a driver by surprise.

The stopping capability of disc brakes isn't affected much by water; the spinning disc throws off the water, and any remaining moisture on the disc or pads is quickly vaporized shortly after the brakes are applied. Drum brakes, however, are very susceptible to water-induced fade.

When water enters a drum brake assembly (usually through the gap between the brake drum and the backing plate) it becomes trapped. It gets thrown outward by the centrifugal force of the spinning drum and acts like a lubricant between the shoes and the drum's friction surface. The problem is made worse when the water mixes with accumulations of brake dust. The resulting mud-like mixture takes longer to burn off than plain water.

If you find yourself experiencing water-induced brake fade, hold your foot down hard on the brake pedal - if there's no immediate change in braking action, release the brakes and apply them again (all the while trying to avoid any obstacles in your path). A couple of repeated applications of the pedal like this should clear up the problem.

There are two basic designs of brakes in use today - disc brakes and drum brakes. There are many different variations of each, but they all operate in a similar manner, using friction to generate heat. The following sections examine the design, components and operation of these brake designs.

Drum brakes

The first wheel brakes fitted to early automobiles were drum brakes. They weren't like the drum brakes on a modern vehicle, though. They worked much like a band in an automatic transmission. The friction material surrounded the outside surface of the drum and clamped tight around the spinning drum when the brakes were applied. It was a very simple arrangement, but had a couple of drawbacks; they couldn't get rid of heat very well and were almost useless if they got wet.

A major breakthrough in braking effectiveness came about when engineers placed the friction surface on the *inside* of the drum. This helped to seal the brake from the elements a little better, and allowed a much greater surface area of the drum to be in contact with the surrounding air, which allowed the brake drum to dissipate the heat it had absorbed much quicker.

The major components of a drum brake assembly are the brake drum, the brake shoes, the backing plate, the wheel cylinder(s), and all the hardware necessary to mount and connect the components **(see illustrations)**. Rear drum brake assemblies also have self-adjusting mechanisms to keep the shoes within a predetermined distance from the drum, even as the lining material wears down.

The following is a description of each major component and its function.

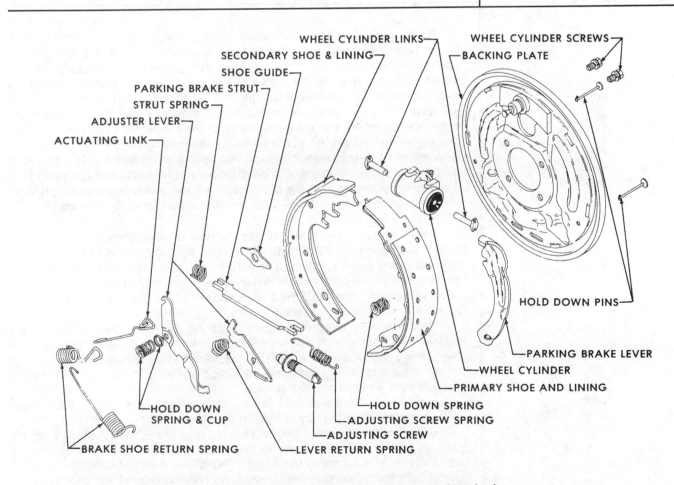

1.29 Exploded view of a typical duo-servo drum brake

Haynes Automotive Brake Manual

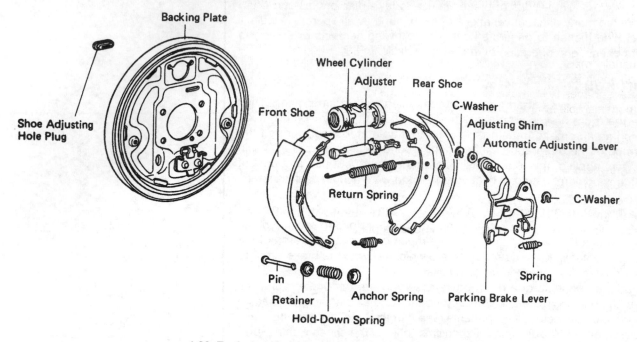

1.30 Exploded view of a typical leading/trailing drum brake

Backing plate - Bolted to the axle housing (rear brake) or steering knuckle (front brake), the backing plate provides a mounting place for the brake shoes and wheel cylinders. It's usually made of heavy-gauge stamped steel. It has raised areas, or lands, for the edges of the brake shoes to ride on. One of the most important parts of the backing plate is the anchor, or on some designs, the anchor pin. The anchor or anchor pin is the immovable rest that the end of the brake shoes rest on, and the point that all braking force for that brake (not heat dissipation, but the portion of the energy of the moving vehicle that the brake is responsible for) is transmitted through. The anchor on some vehicles is riveted to the backing plate. On others, it's bolted to the axle housing or steering knuckle.

Wheel cylinder(s) - This is the component that converts hydraulic pressure into mechanical energy. Wheel cylinders are explained in detail under the *Hydraulics* heading.

Brake shoes - The brake shoe assembly usually consists of two brake shoes. A friction material is fastened to each shoe by means of rivets or by a process called bonding - a form of gluing under a heated drying process. The brake shoes are the components that the wheel cylinder pistons push out on, expanding them to rub against the friction surface of the brake drum. Brake shoes have been designed for maximum rigidity.

On a duo-servo design (we'll get to that) the shoe that faces the front of the vehicle is referred to as the primary shoe. The other one is the secondary shoe. The primary shoe is shorter (or at least the lining is) than the secondary.

On a leading/trailing brake, the front shoe is referred to as the leading shoe; the rear shoe is called the trailing shoe.

In a drum brake setup that has separate wheel cylinders for each brake shoe, both shoes are primary or leading shoes.

Brake lining - The job of the brake lining is to create friction (heat) when it's pressed against the brake drum. As previously stated, the linings are riveted or bonded to the brake shoe. Most brake linings are made of asbestos, held together with special compounds. Some have brass or copper wires impregnated in the lining. Some linings are grooved, slotted or woven to improve cooling. Brake lining wears down in use, and must be changed when it becomes too thin (approximately 1/16-inch from the shoe [bonded linings] or the rivet heads [riveted linings]).

Brake drum - The brake drum spins with the wheel and absorbs and dissipates heat when the brakes are applied and the shoes rub against it. Brake drums are usually made from cast iron, but some are made of aluminum, fitted with cast iron liners for a friction surface. Some drums have cooling vanes cast into their outer surfaces to increase area and aid in cooling.

After a considerable amount of use, brake drums normally will become a little out-of-round. This is due to the intense temperature changes the drum is subjected to. Drums can be machined to get rid of an out-of-round condition or score marks, but all brake drums have a maximum allowable diameter cast or stamped somewhere on the drum's surface. This dimension must not be exceeded, because the drum wouldn't be able to dissipate heat well enough, and could even fatigue and crack.

Attaching hardware and other components - A variety of bits and pieces are required to secure the brake shoes to the backing plate and, on rear drum brakes, actuate the parking brake and ensure that the brake shoes self-adjust during operation. Not all of the following components are found on every drum brake setup.

Hold-down pin, spring and cup - This simple device retains the brake shoe to the backing plate, but allows the brake shoe to move side-to-side a little, if necessary, to better conform to the friction surface of the brake drum. The hold-down pin is inserted from the backside of the backing plate and passes through a hole in the brake shoe. A spring is then placed over the pin and is retained by a cup, or retainer.

Shoe retracting (or return) springs - These springs pull the shoes away from the brake drum after the brake has been released. They're strong enough to overcome the residual pressure in the hydraulic system, so they push the pistons back into the wheel cylinders a little bit, preventing the shoes from dragging against the brake drum.

Parking brake lever and strut - Connected by a cable to the parking brake pedal or lever inside of the vehicle, the parking brake lever inside of a drum brake assembly expands the shoes, with the help of the parking brake strut, out against the drum when the parking brake pedal is depressed or the parking brake lever or rod (inside the vehicle) is lifted. The parking brake lever and strut operate kind of like a "mechanical wheel cylinder."

Self-adjusting mechanism - Self-adjusting mechanisms come in many forms, but they all work to keep the lining on the brake shoes reasonably close to the friction surface of the brake drum, as the lining wears down. On most setups, a small lever, actuated by a cable or linkage, turns an adjuster, or "star" wheel, located between the brake shoes. As the adjuster is turned, the shoes are expanded outward a little bit. Some self-adjusters are activated during normal braking, while others operate when the parking brake is used or the vehicle is driven in reverse and the brakes are applied.

Operation

We've covered the master cylinder and how it creates pressure in the hydraulic system, and how this pressure is converted into mechanical energy by the wheel cylinder. The pistons in the wheel cylinder then push outward on the brake shoes, forcing them against the brake drum, which creates the friction. The heat caused by this friction is then stored in the brake drum and dissipated to the air.

All modern drum brakes are self-energizing, which means the braking force is multiplied by the spinning action of the drum as the brake shoe is applied to it. This self-energizing action is accomplished by the toe (top) of the primary shoe moving out and making contact with the rotating drum. The drum attempts to take the shoe with it - the shoe tends to stick to the drum along its entire span. Of course, it can't begin to spin with the drum, because it's prevented from doing so by the anchor plate or pin.

A *duo-servo* design drum brake takes this self-energizing action one step further (see illustration). Since there is only one anchor pin in this type of brake, the self-energizing force imposed on the primary shoe is transferred to the secondary shoe through a link, or more commonly, the adjusting screw. This forces the bottom of the secondary shoe onto the spinning drum, and the self-energizing process is applied to the secondary shoe to an even greater extent, forcing the top of the secondary shoe into the anchor pin (this is why secondary shoes have more lining material on them). At the same time, the wheel cylinder is also pushing against the secondary shoe. As you can imagine, braking force is greatly multiplied. With this design, the self-energizing action takes place when the brakes are applied when the vehicle is moving forward or backing up - it doesn't matter. Brake designs that have a similar brake shoe/anchor pin layout, but use a single-sided wheel cylinder, are called *uni-servo* brakes (and only have servo-action brakes when going forward).

There's one drawback to the duo-servo design - sometimes they work *too* well, and can lock up the rear wheels. For the most part, duo-servo brakes are used on heavier, rear-wheel drive vehicles. Many other vehicles use self-energizing brakes, but without the duo-servo action. In these designs, the wheel cylinder acts on one end of the brake shoe, but the other end rests against an anchor plate or pin.

One variation of this setup arranges the brake shoes and wheel cylinder in a similar manner to that of the duo-servo design, but instead of the anchor pin at the top and a connecting link between the shoes at the bottom, the bottom of each shoe rests against an anchor (sometimes each shoe has its own anchor pin). This type of brake is commonly referred to as a *leading/trailing* or *double-anchor* design (see illustration 1.30). When the vehicle is traveling forward, only the leading shoe will be self-energized. When the vehicle is braked in reverse, the trailing shoe will be self-energized.

A less common design uses two leading, or primary, shoes. This is called a double-anchor, double-cylinder brake, also called a *leading/leading* brake (see illustration). Each shoe has its own wheel cylinder, and each shoe is self-energizing when the brakes are applied, but only when the vehicle is traveling forward. In reverse there is no self-energizing action at all, and much more muscle power is required to stop the vehicle.

Disc brakes

Once considered an item for sports cars and racers, it didn't take the disc brake long to find a home amongst the most domesticated of vehicles. The disc brake represents the most important advancement in brake design since servo-action.

The disc brake differs from the drum brake in that the friction surface is external, allowing it to cool much more rapidly. There is little deflection or bending in a disc brake assembly, unlike in a drum brake. Friction is created by forcing two brake pads against a rotating disc (see illustration). The disc is pinched between the pads, so the only part that could possibly flex is the caliper, but calipers don't flex very much. The result is a very efficient brake.

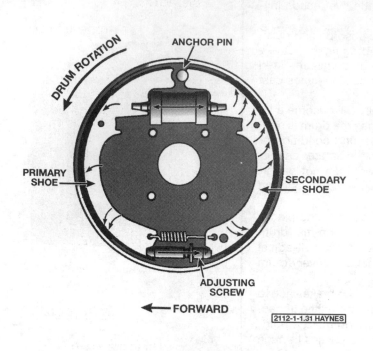

1.31 The self-energizing, or servo-action drum brake uses the rotation of the brake drum to increase the pressure on the brake shoes, making the driver's job easier

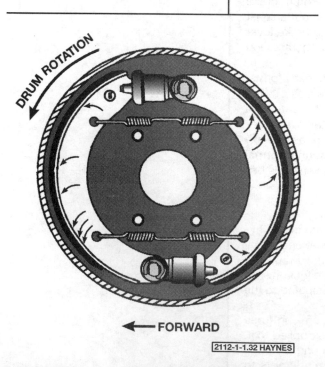

1.32 Double-anchor, double-cylinder design drum brake. This setup makes for a powerful brake, as long as the vehicle is traveling forward

Braking ratio

When the brakes are applied, the natural tendency of the vehicle is to dip forward, placing the greater stopping requirement on the front wheels and less on the rear wheels.

Through increased friction area and a larger cylinder bore on the front brakes (whether they are disc or drum brakes, it doesn't matter), a greater capacity for braking is the result.

Most modern vehicles split the ratio of braking to 60-percent on the front wheels and 40-percent on the rear wheels (this is why most vehicles have disc brakes up front). Under severe braking this ratio can become even more imbalanced, but thanks to the proportioning valve (or pressure control valve), pressure is limited to the rear brakes to keep them from locking up as the rear of the vehicle lightens up.

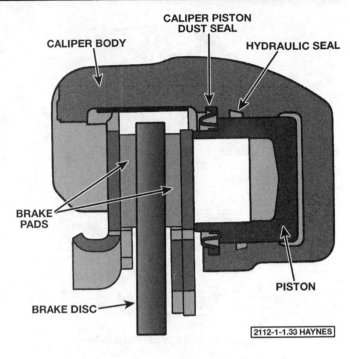

1.33 Cutaway of a typical front disc brake (sliding caliper shown)

Most modern vehicles are equipped with disc brakes on the front wheels, at least. Many are equipped with discs at the rear, too. Since weight transfers to the front of the vehicle under braking, it is important to have more powerful brakes up front - they do most of the work.

Disc brakes have many advantages over drum brakes:

a) Resistance to heat fade - disc brakes dissipate heat more rapidly because more of the braking surface is exposed to air. Heat also makes the discs a little thicker, which have no ill effects on braking (with a drum brake, heat tends to allow the drum to expand and flex).

b) Resistance to water fade - Rotation of the disc tends to throw off the water.

c) Better straight-line stops - because of their clamping action, disc brakes are less likely to pull.

d) Ease of serviceability - disc brake pads, in almost all cases, are easier to replace that drum brake shoes.

e) Lighter weight.

f) Cheaper to manufacture.

The major components of a disc brake assembly are:

The brake disc, or rotor - This part is attached to the hub and turns with the wheel. The thickness and diameter of the disc determine how much heat the brake can absorb and dissipate. Some discs are vented to improve cooling.

The brake pads - These are composed of a steel backing plate to which the brake lining (friction material) is attached. Brake pad linings are attached by rivets or are bonded to the backing plate, just like the brake shoes in a drum brake. The composition of the lining is similar to a drum brake shoe, also.

The brake caliper - This is the component that squeezes the pads against the disc. It is a U-shaped casting that fits over the disc and pads. It contains at least one hydraulic cylinder, which converts the pressure developed by the master cylinder into mechanical energy. Calipers can have up to four pistons, or even more on exotic designs. The caliper is affixed to a mounting bracket that is bolted to the steering knuckle, or fastened to bosses that are integral with the knuckle.

Designs

There are three general designs of disc brakes, determined by caliper arrangement: Fixed caliper, floating calipers and sliding calipers.

Fixed caliper - In this design, the caliper is attached to the steering knuckle

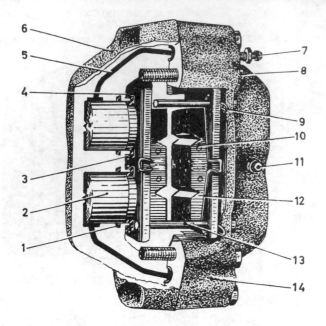

1.34 Cutaway of a typical fixed caliper

1	Piston seal	9	Retaining clip
2	Piston	10	Brake pad
3	Dust boot	11	Bleeder screw
4	Dust boot retaining ring	12	Anti-rattle spring
5	Fluid transfer channel	13	Pad retaining pin
6	Caliper (outer half)	14	Caliper (inner half)
7	Bleeder screw		
8	Bolt (retains caliper halves)		

(or a bracket attached to the knuckle) and is stationary **(see illustrations)**. A minimum of two opposing pistons must be used on this design - one to apply each pad. Some fixed calipers use four pistons (two per pad) and some even use three pistons (a large piston in one half of the caliper and two smaller ones in the other half).

As the brake pedal is applied, pressure from the master cylinder is distributed equally to the pistons, which are forced against the brake pads, which squeeze the disc. The caliper doesn't move. Brake pads are usually retained by a pin (or pins) in this setup.

Most fixed calipers are constructed of two castings containing the hydraulic pistons. The halves are bolted together, some having fluid passages leading from one half to the other. On other designs a fluid transfer tube leads from one side of the caliper to the other to supply the outer piston(s) with hydraulic pressure.

Some fixed calipers are equipped with springs behind the pistons, to keep the pistons from retracting more than just enough to relieve pressure on the disc. This ensures the pads stay right next to the disc (they actually rub against the disc, but with no pressure applied). The result is a firm brake pedal without an excessive amount of travel.

Floating caliper - The floating caliper uses a piston (two in some designs) only in one side of the caliper **(see illustration)**. The caliper is mounted on guide pins that ride in bushings or sleeves, so it is free to move back and forth. As hydraulic pressure builds up behind the caliper piston, the piston pushes on the inner brake pad, forcing it against the disc. As this happens, the inside of the caliper is pushed away from the disc, which simultaneously pulls the outside

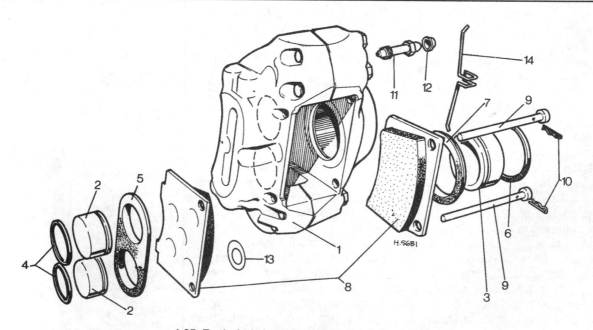

1.35 Exploded view of a three-piston fixed caliper

1	Caliper body	6	Piston seal	11	Bleeder screw
2	Outer pistons	7	Dust boot	12	Dust cap
3	Inner pistons	8	Brake pad	13	Mounting shim
4	Piston seals	9	Pad retaining pin	14	Anti-rattle spring
5	Dust boot	10	Clip		

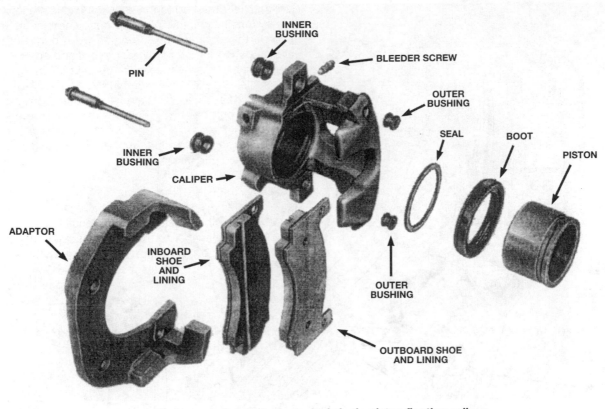

1.36 An exploded view of a typical single-piston floating caliper

of the caliper into the outer brake pad, forcing it against the disc **(see illustration)**. The result is the same powerful clamping action without the complexity of multiple pistons.

Another advantage of the floating caliper is the tendency to absorb pulsations caused by disc runout (wobble). With a fixed caliper, a warped disc will push the pistons back slightly with every rotation of the disc (this is called *knockback*). Under braking, this will be transmitted back through the hydraulic system to the brake pedal, causing it to pulsate under the driver's foot. In a floating caliper setup, the caliper will move back and forth with the pads as the warped disc moves them back and forth, eliminating almost all of the pulsating at the brake pedal.

The outer brake pad in a floating caliper setup is usually affixed to the caliper housing by a retaining spring or tangs. The inner pad is sometimes fastened to the caliper piston or is retained by the caliper mounting bolts. There are exceptions, though. On some designs, both pads ride in the caliper bracket.

In addition, floating calipers are much easier to service than fixed calipers, and are cheaper to manufacture.

Sliding caliper - The sliding caliper is a variation of the floating caliper. Unlike the floating caliper, however, the sliding caliper uses no guide pins; instead, a pair of machined abutments on the caliper adapter position and align the caliper. Retaining clips, springs, keys or "wedge"-type pins keep the caliper on machined guides (sometimes called "ways") on the adapter, allowing lateral movement of

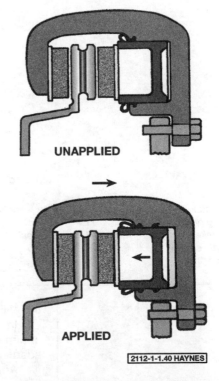

1.37 In a floating caliper, the action of the piston not only pushes the inner brake pad against the disc, it also pulls the outer brake pad against the disc (sliding calipers work like this, too)

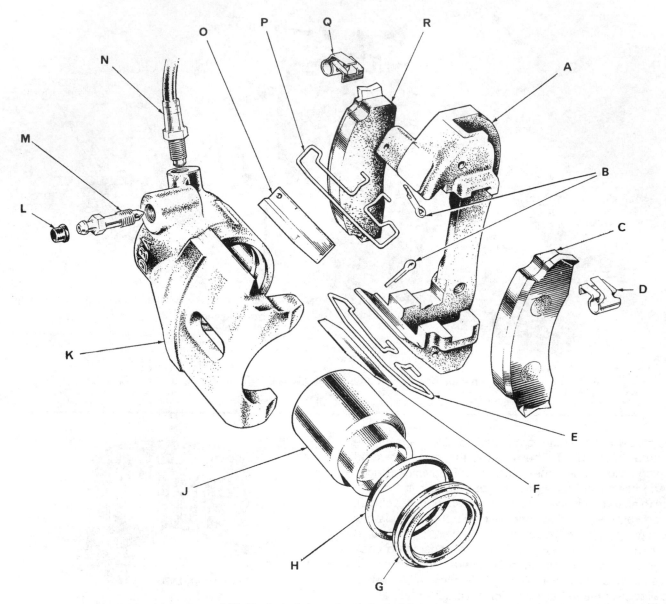

1.38 Exploded view of a typical sliding caliper

A	Caliper mounting bracket	F	Caliper retaining key	K	Caliper housing	P	Tension spring
B	Retaining pins	G	Dust boot	L	Dust cap	Q	Anti-rattle clip
C	Outer brake pad	H	Piston seal	M	Bleeder screw	R	Inner brake pad
D	Anti-rattle clip	I	Not used	N	Brake hose		
E	Tension spring	J	Piston	O	Caliper retaining key		

the caliper **(see illustrations)**.

In a sliding caliper setup, the brake pads sometimes ride on the machined abutments of the caliper adapter. In other designs, one of the pads rides on the adapter and the other pad is affixed to the caliper - either the frame or the piston.

Apart from the way the caliper is mounted and the method of retaining the brake pads, operation of the sliding caliper is the same as the floating caliper.

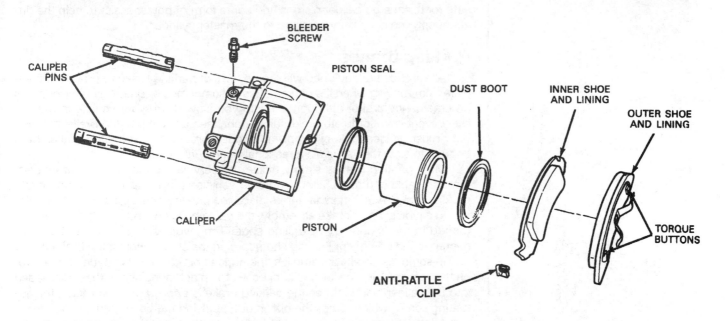

1.39 Here's another kind of sliding caliper - the caliper pins in this Ford setup are spring steel filled with rubber to ensure a tight fit after they're driven into place

Operation

We've actually discussed how the caliper works, but there's an important feature of brake calipers that we haven't covered yet, and that is the self-adjusting capability of the brake caliper.

The piston in a brake caliper fits in its bore with very little clearance (no more than 0.005-inch). To hold the hydraulic fluid in the caliper under pressure, the piston is surrounded by an O-ring near the opening in the piston bore. There's also a dust boot at the top of the piston, and on some calipers the boot seats in a groove just like the piston seal, but serves no purpose in fluid sealing at all. The O-ring seal is much larger than it needs to be for sealing purposes alone. It plays the main role in the self-adjusting feature of disc brakes.

When pressure from the master cylinder builds up behind the caliper piston, the piston is forced out and applies the brake pads. As the piston travels outward, the piston seal is distorted - it bends out with the piston **(see illustration)**. When the brakes are released, pressure behind the piston subsides and the seal pulls the piston back home (remember, we're only talking about a fraction of an inch). As the pads wear down, the piston travels out a little farther, and even though the seal bends out with the piston, it can only go so far. So, the piston slides through the seal enough to take up the slack caused by the thinner brake pads. In doing so, the brake pedal remains firm and doesn't travel to the floor as the brake pads wear down.

Some older multi-piston calipers achieve the same self adjusting action by placing coil springs behind each piston.

As you can see, the disc brake has many advantages over the drum brake, but it has one disadvantage - there's no self energizing action. It could be compared to a drum brake with two trailing shoes. There are two ways to compen-

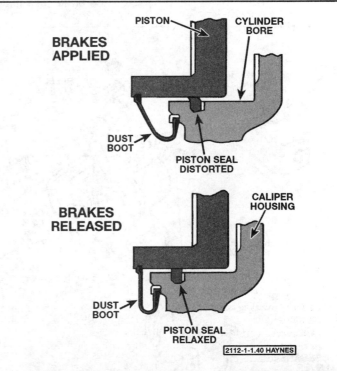

1.40 These drawings illustrate the seal retraction action that takes place after the brakes are released. In the drawing on the left, pressure has pushed the piston from the bore a little. The piston seal has distorted and "followed" the piston. In the drawing on the right, the brakes have been released, and the seal has pulled the piston back into the caliper

sate for this: Large caliper pistons and some form of power assist to help the driver supply the necessary pressure to the master cylinder.

Parking brakes

All vehicles are equipped with some form of parking brake to prevent the vehicle from moving when it's parked. This is sometimes erroneously referred to as an emergency brake. Although it'll help you slow down your vehicle in the unlikely event of a complete brake failure, chances are the parking brake will fade out or just be incapable of supplying enough pressure to apply the brakes hard enough to slow the vehicle in a reasonable amount of time.

Most parking brakes simply utilize the service brakes on the rear wheels, through cable actuation, although some vehicles are equipped with a separate parking brake. A few manufacturers place the parking brake on the front wheels.

In a typical drum brake assembly, the cable pulls on a parking brake lever attached to the secondary or trailing shoe. The lever bears against a strut connected to the primary or leading shoe, forcing the shoes against the brake drum.

In some disc brake assemblies, the parking brake cable is attached to a lever on the rear caliper. The lever is connected to an actuating screw that is threaded into the caliper piston. When the parking brake is applied, the lever turns the actuating screw, which forces the piston out, applying the brake pads.

Another version of this parking brake uses a lever that, when pulled by the parking brake cable, forces a small pushrod into an adjuster spindle threaded into the caliper piston. This pushes the caliper piston from the caliper and applies the brake **(see illustration)**.

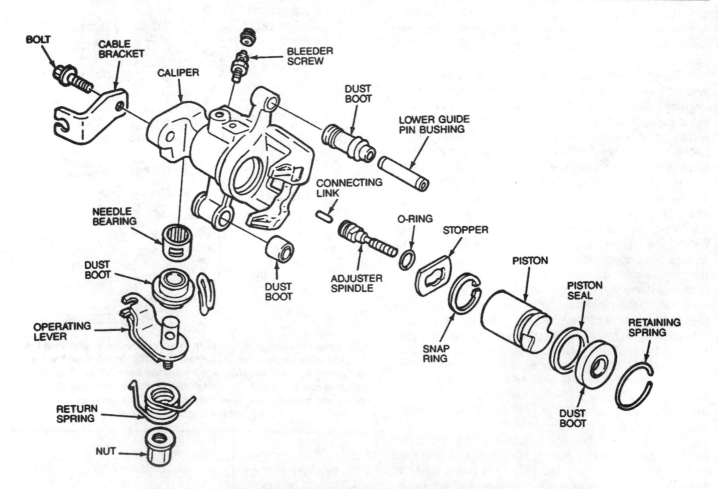

1.41 Exploded view of a typical rear caliper with an adjuster spindle-type parking brake

In other disc brake setups, a separate set of brake pads (much smaller than the regular pads) are clamped against the disc when actuated by the parking brake cable **(see illustration)**. These pads are usually so small that they are good for nothing more than holding the vehicle stationary - they're almost useless for slowing the vehicle down in the event of a total brake failure (thank goodness total brake failures are extremely rare occurrences!).

The other parking brake arrangement on a rear disc brake setup places a pair of parking brake shoes inside the "hat," or inner diameter, of the disc. These work like a miniature drum brake setup **(see illustration)**.

The least common parking brake arrangement placed a drum on the output shaft of the transmission, with the backing plate and brake shoes connected to the transmission housing. This type of arrangement hasn't been used by automobile manufacturers for years.

Power boosters

The power brake booster is a device that helps the driver apply pressure to the master cylinder. On older vehicles with four wheel drum brakes, power assist was not too important because of the self-energizing action inherent to the design of drum brakes. As disc brakes gained popularity, power boosters became much more common. In fact, they are almost a necessity on systems with disc brakes, since so much pressure is required to successfully clamp a spinning disc between the brake pads. Without a power

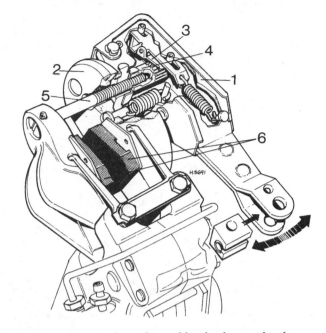

1.42 Here's a cutaway view of a parking brake mechanism on a Jaguar. The small pads are lever operated and self adjusting.

1	Operating lever	4	Adjuster nut
2	Pad carrier assembly	5	Adjuster bolt
3	Pawl assembly	6	Parking brake pads

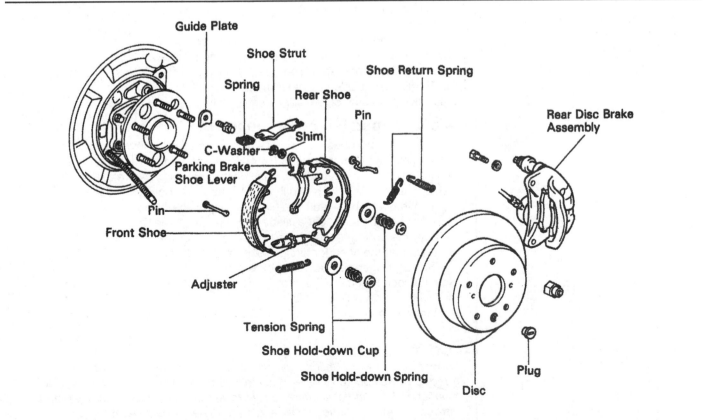

1.43 An exploded view of the "drum-type" parking brake setup on a rear disc brake

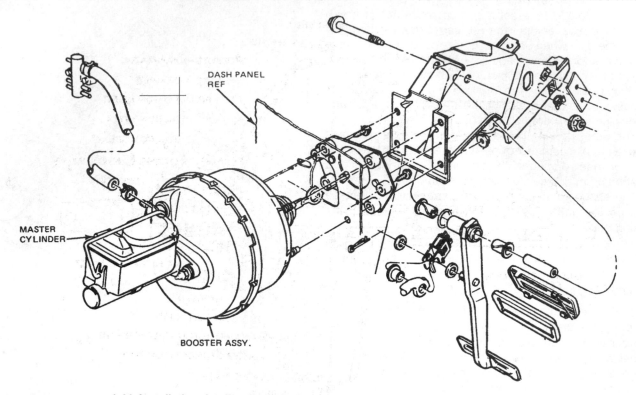

1.44 Installation details of a typical vacuum-operated power brake booster

booster, the cylinders in the brake calipers would have to be bigger, requiring a longer travel master cylinder or more mechanical advantage built into the brake pedal to develop the necessary pressure. Either way, brake pedal travel would be excessive.

There are two general types of power boosters available: Vacuum-operated and hydraulically operated. The most common is the vacuum-operated type, but the hydraulically operated type was used by General Motors quite extensively in the 1980's, and is used by other manufacturers as well, primarily (but not only) in the braking systems of diesel-powered vehicles.

Vacuum-operated power boosters

There are two basic kinds of vacuum-operated power boosters. Vacuum suspended and air-suspended. Both kinds use a diaphragm, or diaphragms, to separate two or four chambers. They each create a pressure differential situation inside to aid in pushing the pushrod into the master cylinder **(see illustrations)**.

In the event of a fault developing inside the booster, enough vacuum is stored to provide sufficient assistance for two or three brake applications and after that, the performance of the hydraulic part of the braking system is only affected in that the need for higher pedal pressures will be noticed.

Vacuum-operated power boosters are constructed of stamped steel housings that lock together. The internals are replaceable but doing so just isn't feasible. When a power booster fails, it's much easier to simply exchange the old unit for a new or rebuilt one, since special tools and expertise are required to overhaul one.

Operation

Note: *The following description details the operation of a typical vacuum suspended power brake booster. The air suspended booster operates in a similar manner, but instead of vacuum being present on both sides of the diaphragm and introducing atmospheric pressure into the brake pedal side of the unit, air is present on both sides of the diaphragm and vacuum is introduced to the master cylinder side of the booster.*

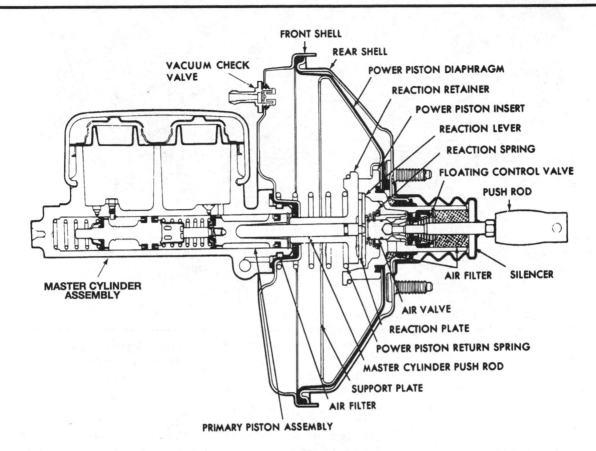

1.45 Cutaway view of a typical vacuum-operated power booster (vacuum-suspended type)

Brakes released - In the "at rest" condition with the engine running, vacuum is present on both sides of the power piston. Air at atmospheric pressure, entering through the filter behind the pushrod, is shut off at the air valve. The floating control valve is held away from the seat in the power piston insert. Any air in the system is drawn through a small passage in the power piston, past the power piston insert valve seat to the insert itself. It then travels through a drilling in the support plate, into the space in front of the power piston then to the intake manifold via a check valve. Vacuum therefore exists on both sides of the power piston which is held against the rear of the housing under spring action.

Brakes applied - When the brake pedal is depressed, the pushrod carries the air valve away from the floating control valve. The floating control valve will follow until it contacts the raised seat in the power piston insert; vacuum is now shut off to the rear power piston and atmospheric pressure (air) enters the housing at the rear of the power piston. The power piston therefore moves forward to operate the floating piston assembly of the hydraulic master cylinder. As pressure increases on the end of the master cylinder piston, the reaction plate is moved off its seat on the lower piston and contacts the reaction levers. These levers swing on their pivots and bear against the end of the air valve operating rod assembly to provide a feedback (approximately 30-percent of the master cylinder load) to the pedal. This enables the driver to "feel" the degree of brake application.

Brake holding - When the desired braking force is achieved, the power piston moves forward until the floating control valve again seats on the air valve. The power piston will now remain stationary until there is a change in applied pedal pressure.

Brakes released - When the pedal pressure is released, the air valve is forced back to contact the power piston under spring action. As it moves, the floating control valve is pushed off its seat on the power piston insert by the air valve. Atmospheric pressure (air) is shut off by the air valve seating on the floating control

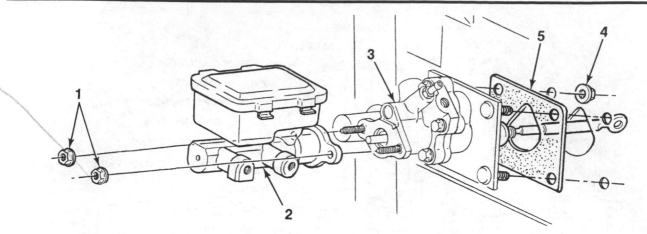

1.46 Mounting details of a General Motors Hydro-boost power brake booster. This unit uses pressure from the power steering pump to assist the driver in applying the brakes.

1	Nuts		4	Nut
2	Master cylinder		5	Gasket
3	Hydro-boost unit			

valve. As the floating control valve lifts from its seat, it opens the rear of the power piston to intake manifold vacuum, and the power piston returns to the rear housing. The hydraulic pressure in the brake system is released as the floating piston assembly returns to the normal position.

Vacuum failure - In the event of vacuum failure, i.e. the engine dies or the vacuum line cracks or the diaphragm inside the booster ruptures, application of the brake pedal moves the pedal pushrod which in turn contacts the master cylinder pushrod and the brakes are applied. This gives a condition as found in a standard, non-power assisted brake system, and a correspondingly higher pedal pressure is required.

Hydraulically operated power boosters

Some power boosters are operated by hydraulic pressure instead of intake manifold vacuum (see illustration). Most of these systems use pressure generated by the power steering pump. When the engine is running, the power steering pump supplies hydraulic pressure to a power flow regulator/accumulator. The regulator/accumulator stores and regulates the pressure to the hydraulic booster unit, which is mounted between the master cylinder and the firewall, just like a vacuum brake booster. When the brake pedal is depressed, a pressure regulating valve opens, admitting power steering fluid under pressure into a chamber. This pressure helps actuate the master cylinder, reducing pedal effort. In the event of a failure in the power steering pump (or if the engine dies), the accumulator stores enough pressure in it for one or two reserve stops. Braking will then revert to non-power assist.

Another kind of hydraulic booster is the General Motors Powermaster unit (see illustration). This power boosting system is integral with the master cylinder. In addition to the master cylinder, it consists of an electro-hydraulic pump, a nitrogen-charged fluid accumulator, a pressure switch and the booster unit.

The accumulator stores brake fluid at pressures of 500 to 700 psi. When pressure drops below this value, the pressure switch signals the pump to activate, which charges up the accumulator. When the driver depresses the brake pedal, pressurized fluid from the accumulator pushes on the power piston in the booster to help apply the master cylinder pistons.

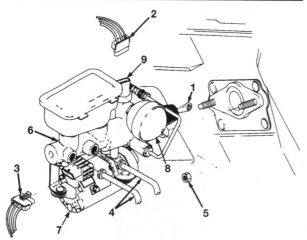

1.47 General Motors Powermaster brake booster assembly

1	Pushrod		6	Master cylinder
2	Electrical connector		7	Electro-hydraulic pump
3	Electrical connector		8	Accumulator
4	Brake lines		9	Pressure switch
5	Nut			

2 Tools and Equipment

A place to work

Establish a place to work. It doesn't have to be particularly large, but it should be clean, safe, well-lit, organized and adequately equipped for the job. True, without a good workshop or garage, you can still service and repair brakes, even if you have to work outside. But major repairs should be carried out in a sheltered area with a roof. Some of the procedures in this book require an environment totally free of dirt, which could cause contamination and subsequent failure if it finds its way into the brake system.

The workshop

The size, shape and location of a shop building is usually dictated by circumstance rather than personal choice. Every do-it-yourselfer dreams of having a spacious, clean, well-lit building specially designed and equipped for working on everything from small engines on lawn and garden equipment to cars and other vehicles. In reality, however, most of us must content ourselves with a garage, basement or shed in the backyard.

Spend some time considering the potential - and drawbacks - of your current facility. Even a well-established workshop can benefit from intelligent design. Lack of space is the most common problem, but you can significantly increase usable space by carefully planning the locations of work and storage areas. One strategy is to look at how others do it. Ask local repair shop owners if you can see their shops. Note how they've arranged their work areas, storage and lighting, then try to scale down their solutions to fit your own shop space, finances and needs.

General workshop requirements

A solid concrete floor is the best surface for a shop area. The floor should be even, smooth and dry. A coat of paint or sealant formulated for concrete surfaces will make oil spills and dirt easier to remove and help cut down on dust - always a problem with concrete.

Paint the walls and ceiling white for maximum reflection. Use gloss or semi-gloss enamel. It's washable and reflective. If your shop has windows, situate workbenches to take advantage of them. Skylights are even better. You can't have too much natural light. Artificial light is also good, but you'll need a lot of it to equal ordinary daylight.

Make sure the building is adequately ventilated. This is critical during the winter months, to prevent condensation problems. It's also a vital safety consideration where solvents, gasoline and other volatile liquids are being used. You should be able to open one or more windows for ventilation. In addition, opening vents in the walls are desirable.

Electricity and lights

Electricity is essential in a shop. It's relatively easy to install if the workshop is part of the house, but it can be difficult and expensive to install if it isn't. Safety should be your primary consideration when dealing with electricity; unless you have a very good working knowledge of electrical installations, have an electrician do any work required to provide power and lights in the shop.

Consider the total electrical requirements of the shop, making allowances for possible later additions of lights and equipment. Don't substitute extension cords for legal and safe permanent wiring. If the wiring isn't adequate, or is substandard, have it upgraded.

Give careful consideration to lights for the workshop. A pair of 150-watt incandescent bulbs, or two 48-inch long, 40-watt fluorescent tubes, suspended approximately 48-inches above the workbench, are the minimum you can get by with. As a general rule, fluorescent lights are probably the best choice. Their light is bright, even, shadow-free and fairly economical, although some people don't care for the bluish tinge they cast on everything. The usual compromise is a good mix of fluorescent and incandescent fixtures.

The position of the lights is important. Don't place a fixture directly above the work area. It will cause shadows, even with fluorescent lights. Attach the light(s) slightly to the rear - or to each side - of the workbench or garage to provide shadow-free lighting. A portable "trouble-light" is very helpful for use when overhead lights are inadequate. If gasoline, solvents or other flammable liquids are present - not an unusual situation in a shop - use special fittings to minimize the risk of fire. And don't use fluorescent lights above machine tools (like a drill press). The flicker produced by alternating current is especially pronounced with this type of light and can make a rotating chuck appear stationary at certain speeds - a very dangerous situation.

Storage and shelves

Set up an organized storage area to avoid losing parts. You'll need storage space for hardware, lubricants and other chemicals, rags, tools and equipment.

If space and finances allow, install metal shelves along the walls. Arrange the shelves so they're widely spaced near the bottom to take large or heavy items. Metal shelf units are pricey, but they make the best use of available space. And the shelf height is adjustable on most units.

Wood shelves **(see illustration)** are sometimes a cheaper storage solution. But they must be built - not just assembled. They must be much heftier than metal shelves to carry the same weight, the shelves can't be adjusted vertically and you can't just disassemble them and take them with you if you move. Wood also absorbs oil and other liquids and is obviously a much greater fire hazard.

Store small parts in plastic drawers or bins mounted on metal racks attached to the wall. They're available from most hardware, home and lumber stores. Bins come in various sizes and usually have slots for labels.

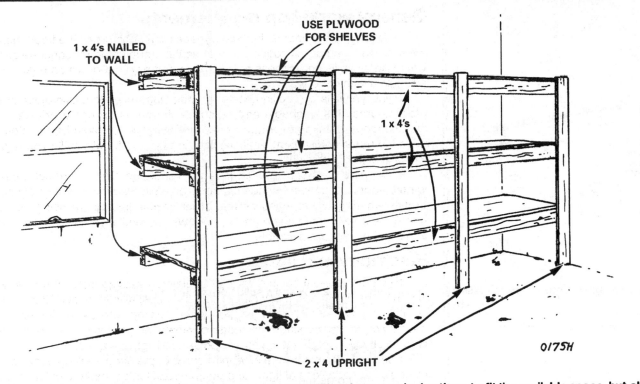

1 x 4's NAILED
TO WALL

USE PLYWOOD
FOR SHELVES

1 x 4's

2 x 4 UPRIGHT

0175H

2.1 Homemade wood shelves are relatively inexpensive to build and you can design them to fit the available space, but all that wood can be a fire hazard

All kinds of containers are useful in a shop. Glass jars are handy for storing fasteners, but they're easily broken. Cardboard boxes are adequate for temporary use, but if they become damp, the bottoms eventually weaken and fall apart if you store oily or heavy parts in them. Plastic containers come in a variety of sizes and colors for easy identification. Egg cartons are excellent organizers for tiny parts. Large ice cream tubs are suitable for keeping small parts together. Get the type with a snap cover. Old metal cake pans, bread pans and muffin tins also make good storage containers for small parts.

Workbenches

A workbench is essential - it provides a place to lay out parts and tools during repair procedures, and it's a lot more comfortable than working on a floor or the driveway. The workbench should be as large and sturdy as space and finances allow. If cost is no object, buy industrial steel benches. They're more expensive than home-built benches, but they're very strong, they're easy to assemble, and - if you move - they can be disassembled quickly and you can take them with you. They're also available in various lengths, so you can buy the exact size to fill the space along a wall.

If steel benches aren't in the budget, fabricate a bench frame from slotted angle-iron or Douglas fir (use 2 x 6's rather than 2 x 4's) **(see illustration)**. Cut the pieces of the frame to the required size and bolt them together with carriage bolts. A 30 or 36 by 80-inch, solid-core door with hardboard surfaces makes a good bench top. And you can flip it over when one side is worn out.

An even cheaper - and quicker - solution? Assemble a bench by attaching the bench top frame pieces to the wall with angled braces and use the wall studs as part of the framework.

Regardless of the type of frame you decide to use for the workbench, be sure to position the bench top at a comfortable working height and make sure everything is level. Shelves installed below the bench will make it more rigid and provide useful storage space.

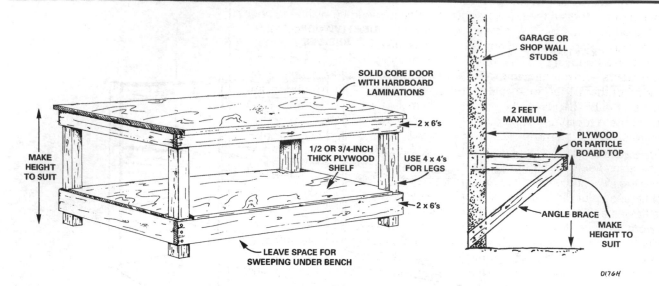

2.2 You can build a sturdy, inexpensive workbench with 4 x 4s, 2 x 6s and a solid core door with hardboard laminations - or build a bench using the wall as an integral member as shown

Tools and equipment

For some home mechanics, the idea of using the correct tool is completely foreign. They'll cheerfully tackle the most complex procedures with only a set of cheap open-end wrenches of the wrong type, a single screwdriver with a worn tip, a large hammer and an adjustable wrench. Though they often get away with it, this cavalier approach is stupid and dangerous. It can result in relatively minor annoyances like stripped fasteners, or cause catastrophic consequences like brake failure. It can also result in serious injury.

A complete assortment of good tools is a given for anyone who plans to work on cars. If you don't already have most of the tools listed below, the initial investment may seem high, but compared to the spiraling costs of routine maintenance and repairs, it's a deal. Besides, you can use a lot of the tools around the house for other types of mechanical repairs. While some of the tools we'll describe aren't necessary to complete most brake repair operations, they are representative of the kinds of tools you would expect to find in a well-equipped shop.

Buying tools

There are two ways to buy tools. The easiest and quickest way is to simply buy an entire set. Tool sets are often priced substantially below the cost of the same individually priced tools - and sometimes they even come with a tool box. When purchasing such sets, you often wind up with some tools you don't need or want. But if low price and convenience are your concerns, this might be the way to go. Keep in mind that you're going to keep a quality set of tools a long time (maybe the rest of your life), so check the tools carefully; don't skimp too much on price, either. Buying tools individually is usually a more expensive and time-consuming way to go, but you're more likely to wind up with the tools you need and want. You can also select each tool on its relative merits for the way you use it.

You can get most of the hand tools on our list from the tool department of any large department store or hardware store chain that sells hand tools. Blackhawk, Craftsman, Lisle, KD, Proto and SK are fairly inexpensive, good-quality choices. Specialty tools are available from mechanics' tool companies such as Snap-on, Mac, Matco, Cornwall, Kent-Moore, OTC, etc. These companies also supply the other tools you need, but they'll probably be more expensive.

Also consider buying second-hand tools from garage sales or used tool outlets. You may have limited choice in sizes, but you can usually determine from the condition of the tools if they're worth buying. You can end up with a number of unwanted or duplicate tools, but it's a cheap way of putting a basic tool kit together, and you can always sell off any surplus tools later.

Until you're a good judge of the quality levels of tools, avoid mail order firms (excepting Sears and other name-brand suppliers), flea markets and swap meets. Some of them offer good value for the money, but many sell cheap, imported tools of dubious quality. Like other consumer products counterfeited in the Far East, these tools run the gamut from acceptable to unusable.

If you're unsure about how much use a tool will get, the following approach may help. For example, if you need a set of combination wrenches but aren't sure which sizes you'll end up using most, buy a cheap or medium-priced set (make sure the jaws fit the fastener sizes marked on them). After some use over a period of time, carefully examine each tool in the set to assess its condition. If all the tools fit well and are undamaged, don't bother buying a better set. If one or two are worn, replace them with high-quality items - this way you'll end up with top-quality tools where they're needed most and the cheaper ones are sufficient for occasional use. On rare occasions you may conclude the whole set is poor quality. If so, buy a better set, if necessary, and remember never to buy that brand again.

In summary, try to avoid cheap tools, especially when you're purchasing high-use items like screwdrivers, wrenches and sockets. Cheap tools don't last long. Their initial cost plus the additional expense of replacing them will exceed the initial cost of better-quality tools.

Hand tools

Note: *The information that follows is for early-model engines with only Standard fastener sizes. On some late-model engines, you'll need Metric wrenches, sockets and Allen wrenches. Generally, manufacturers began integrating metric fasteners into their vehicles around 1975.*

A list of general-purpose hand tools you should have in your shop

Adjustable wrench - 10-inch
Allen wrench set (1/8 to 3/8-inch or 4 mm to 10 mm)
Ball peen hammer - 12 oz (any steel hammer will do)
Brake bleeding kit
Brass hammer
Brushes (various sizes, for cleaning small parts)
Combination (slip-joint) pliers - 6-inch
Center punch
Cold chisels - 1/4 and 1/2-inch
Combination wrench set (1/4 to 1-inch or 7mm to 19mm)
Dial indicator and base
Extensions - 1-, 6-, 10- and 12-inch
E-Z out (screw extractor) set
Feeler gauge set
Files (assorted)
Flare-nut wrenches
Floor jack
Gasket scraper
Hacksaw and assortment of blades
Impact screwdriver and bits
Locking pliers
Micrometer(s) (a one-inch micrometer is suitable for most work)
Phillips screwdriver (no. 2 x 6-inch)
Phillips screwdriver (no. 3 x 8-inch)
Phillips screwdriver (stubby - no. 2)
Pin punches (1/16, 1/8, 3/16-inch)
Pliers - lineman's

Pliers - needle-nose
Pliers - snap-ring (internal and external)
Pliers - vise-grip
Pliers - diagonal cutters
Ratchet
Scraper (made from flattened copper tubing)
Scribe
Socket set (6-point sockets are preferred, but some fasteners require the use of 12-point sockets)
Soft-face hammer (plastic/rubber)
Spark plug socket (with rubber insert)
Spark plug gap adjusting tool
Standard screwdriver (1/4-inch x 6-inch)
Standard screwdriver (5/16-inch x 6-inch)
Standard screwdriver (3/8-inch x 10-inch)
Standard screwdriver (5/16-inch - stubby)
Steel ruler - 6-inch
Tap and die set
Thread gauge
Torque wrench (same size drive as sockets)
Torx socket(s)
Universal joint
Vacuum gauge/pump (hand-held)
Wire brush (large)
Wire cutter pliers

What to look for when buying hand tools and general purpose tools

Wrenches and sockets

Wrenches vary widely in quality. One indication of their cost is their quality: The more they cost, the better they are. Buy the best wrenches you can afford. You'll use them a lot.

Start with a set containing wrenches from 1/4 to 1-inch in size. The size, stamped on the wrench **(see illustration)**, indicates the distance across the nut or bolt head, or the distance between the wrench jaws - not the diameter of the threads on the fastener - in inches. For example, a 1/4-inch bolt usually has a 7/16-inch hex head - the size of the wrench required to loosen or tighten it. However, the relationship between thread diameter and hex size doesn't always hold true. In some instances, an unusually small hex may be used to discourage over-tightening or because space around the fastener head is limited. Conversely, some fasteners have a disproportionately large hex-head.

Wrenches are similar in appearance, so their quality level can be difficult to judge just by looking at them. There are bargains to be had, just as there are overpriced tools with well-known brand names. On the other hand, you may buy what looks like a reasonable value set of wrenches only to find they fit badly or are made from poor-quality steel.

With a little experience, it's possible to judge the quality of a tool by looking at it. Often, you may have come across the brand name before and have a good idea of the quality. Close examination of the tool can often reveal some hints as to its quality. Prestige tools are usually polished and chrome-plated over their entire surface, with the working faces ground to size. The polished finish is largely cosmetic, but it does make them easy to keep clean. Ground jaws normally indicate the tool will fit well on fasteners.

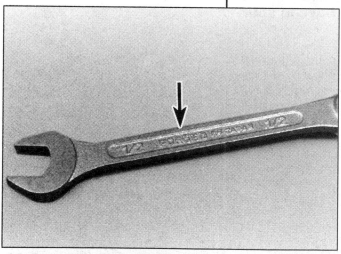

2.3 One quick way to determine whether you're looking at a quality wrench is to read the information printed on the handle - if it says "chrome vanadium" or "forged", it's made out of the right material

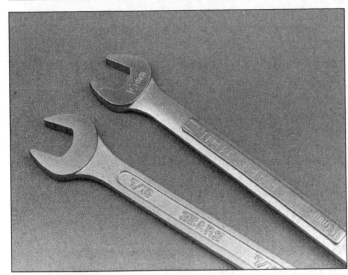

2.4 The size stamped on a wrench indicates the distance across the nut or bolt head (or the distance between the wrench jaws) in inches, not the diameter of the threads on the fastener

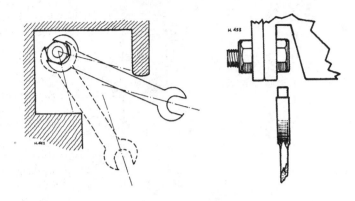

2.5 Open-end wrenches can do several things other wrenches can't - for example, they can be used on bolt heads with limited clearance (above) and they can be used in tight spots where there's little room to turn a wrench by flipping the offset jaw over every few degrees of rotation

A side-by-side comparison of a high-quality wrench with a cheap equivalent is an eye opener. The better tool will be made from a good-quality material, often a forged/chrome-vanadium steel alloy **(see illustration)**. This, together with careful design, allows the tool to be kept as small and compact as possible. If, by comparison, the cheap tool is thicker and heavier, especially around the jaws, it's usually because the extra material is needed to compensate for its lower quality. If the tool fits properly, this isn't necessarily bad - it is, after all, cheaper - but in situations where it's necessary to work in a confined area, the cheaper tool may be too bulky to fit.

Open-end wrenches

Because of its versatility, the open-end wrench is the most common type of wrench. It has a jaw on either end, connected by a flat handle section. The jaws either vary by a size, or overlap sizes between consecutive wrenches in a set. This allows one wrench to be used to hold a bolt head while a similar-size nut is removed. A typical fractional size wrench set might have the following jaw sizes: 1/4 x 5/16, 3/8 x 7/16, 1/2 x 9/16, 9/16 x 5/8 and so on.

Typically, the jaw end is set at an angle to the handle, a feature which makes them very useful in confined spaces; by turning the nut or bolt as far as the obstruction allows, then turning the wrench over so the jaw faces in the other direction, it's possible to move the fastener a fraction of a turn at a time **(see illustration)**. The handle length is generally determined by the size of the jaw and is calculated to allow a nut or bolt to be tightened sufficiently by hand with minimal risk of breakage or thread damage (though this doesn't apply to soft materials like brass or aluminum).

Common open-end wrenches are usually sold in sets and it's rarely worth buying them individually unless it's to replace a lost or broken tool from a set. Single tools invariably cost more, so check the sizes you're most likely to need regularly and buy the best set of wrenches you can afford in that range of sizes. If money is limited, remember that you'll use open-end wrenches more than any other type - it's a good idea to buy a good set and cut corners elsewhere.

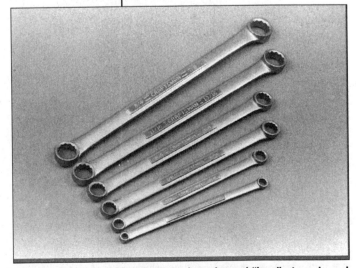

2.6 Box-end wrenches have a ring-shaped "box" at each end - when space permits, they offer the best combination of "grip" and strength

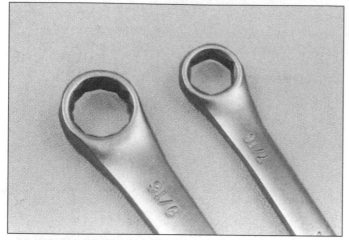

2.7 Box-end wrenches are available in 12 (left) and 6-point (right) openings; even though the 12-point design offers twice as many wrench positions, buy the 6-point first - it's less likely to strip off the corners of a nut or bolt head

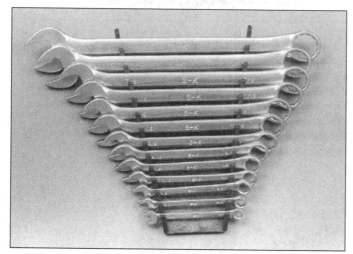

2.8 Buy a set of combination wrenches from 1/4 to 1-inch or 7mm to 19mm

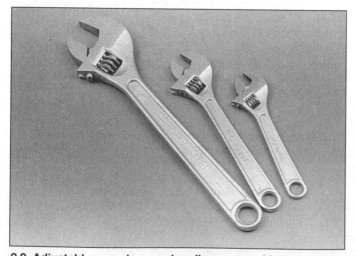

2.9 Adjustable wrenches can handle a range of fastener sizes - they're not as good as single-size wrenches but they're handy for loosening and tightening those odd-sized fasteners for which you haven't yet bought the correct wrench

Box-end wrenches

Box-end wrenches **(see illustration)** have ring-shaped ends with a 6-point (hex) or 12-point (double hex) opening **(see illustration)**. This allows the tool to fit on the fastener hex at 15 (12-point) or 30-degree (6-point) intervals. Normally, each tool has two ends of different sizes, allowing an overlapping range of sizes in a set, as described for open-end wrenches.

Although available as flat tools, the handle is usually offset at each end to allow it to clear obstructions near the fastener, which is normally an advantage. In addition to normal length wrenches, it's also possible to buy long handle types to allow more leverage (very useful when trying to loosen rusted or seized nuts). It is, however, easy to shear off fasteners if not careful, and sometimes the extra length impairs access.

As with open-end wrenches, box-ends are available in varying quality, again often indicated by finish and the amount of metal around the ring ends. While the same criteria should be applied when selecting a set of box-end wrenches, if your budget is limited, go for better-quality open-end wrenches and a slightly cheaper set of box-ends.

Combination wrenches

These wrenches **(see illustration)** combine a box-end and open-end of the same size in one tool and offer many of the advantages of both. Like the others, they're widely available in sets and as such are probably a better choice than box-ends only. They're generally compact, short-handled tools and are well suited for tight spaces where access is limited.

Adjustable wrenches

Adjustable wrenches **(see illustration)** come in several sizes. Each size can handle a range of fastener sizes. Adjustable wrenches aren't as effective as one-size tools and it's easy to damage fasteners with them. However, they can be an invaluable addition to any tool kit - if they're used with discretion. **Note:** *If you attach the wrench to the fastener with the movable jaw pointing in the direction of wrench rotation* **(see illustration)**, *an adjustable wrench will be less*

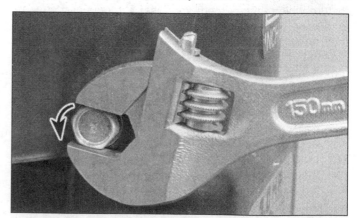

2.10 When you use an adjustable wrench, make sure the movable jaw points in the direction the wrench is being turned (arrow) so the wrench doesn't distort and slip off the fastener head

likely to slip and damage the fastener head.

The most common adjustable wrench is the open-end type with a set of parallel jaws that can be set to fit the head of a fastener. Most are controlled by a threaded spindle, though there are various cam and spring-loaded versions available. Don't buy large tools of this type; you'll rarely be able to find enough clearance to use them.

Flare nut wrenches

These wrenches, sometimes called line wrenches, are used for loosening and tightening hydraulic line fittings (tube nuts). Construction is similar to a six-point box end wrench, but a portion of one of the flats is cut out to allow the wrench to pass over a line or hose **(see illustration)**. This design offers much more surface area of the wrench to be in contact with the flats on the fitting, which will prevent the fittings from being rounded off because the load is distributed over as much area as possible. They are also thicker than most open or box end wrenches, which adds to the surface area in contact with the tube nut. **Caution:** *When loosening a tube nut on a fitting connected to a flexible hose, always use a backup wrench to hold the larger, female fitting stationary. This will prevent the hydraulic line from twisting.*

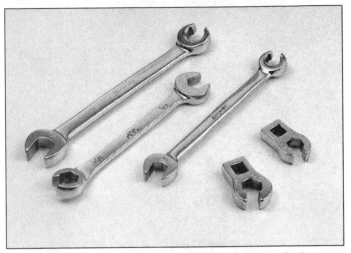

2.11 Flare nut wrenches should always be used when loosening or tightening line fittings (tube nuts) - the two odd-shaped items on the right are flare-nut *crows feet*, which are useful for loosening hard-to-reach fittings when used with a long extension

Ratchet and socket sets

Ratcheting socket wrenches **(see illustration)** are highly versatile. Besides the sockets themselves, many other interchangeable accessories - extensions, U-drives, step-down adapters, screwdriver bits, Allen bits, crow's feet, etc. - are available. Buy six-point sockets - they're less likely to slip and strip the corners off bolts and nuts. **Note:** *Some brake system fasteners (caliper mounting bolts on some vehicles, for example) require the use of 12-point sockets.* Don't buy sockets with extra-thick walls - they might be stronger but they can be hard to use on recessed fasteners or fasteners in tight quarters.

A 3/8-inch drive set is adequate for most work. Sometimes a 1/2-inch drive set is required when working with large fasteners. Although the larger drive is bulky and more expensive, it has the capacity of accepting a very wide range of large sockets. Later, you may want to consider a 1/4-inch drive for little stuff like ignition and carburetor work.

Interchangeable sockets consist of a forged-steel alloy cylinder with a hex or double-hex formed inside one end. The other end is formed into the square drive recess that engages over the corresponding square end of various socket drive tools.

Sockets are available in 1/4, 3/8, 1/2 and 3/4-inch drive sizes. A 3/8-inch drive set is most useful for engine repairs, although 1/4-inch drive sockets and accessories may occasionally be needed.

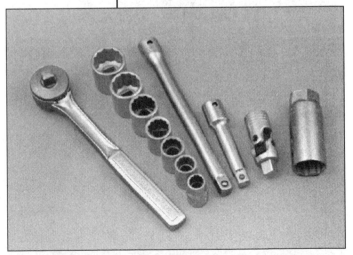

2.12 A typical ratchet and socket set includes a ratchet, a set of sockets, a long and a short extension, a universal joint and a spark plug socket

The most economical way to buy sockets is in a set. As always, quality will govern the cost of the tools. Once again, the "buy the best" approach is usually advised when selecting sockets. While this is a good idea, since the end result is a set of quality tools that should last a lifetime, the cost is so high it's difficult to justify the expense for home use.

As far as accessories go, you'll need a ratchet, at least one extension (buy a three or six-inch size), a spark plug socket and maybe a T-handle or breaker bar. Other desirable, though less essential items, are a speeder handle, a U-joint, extensions of various other lengths and adapters from one drive size to another

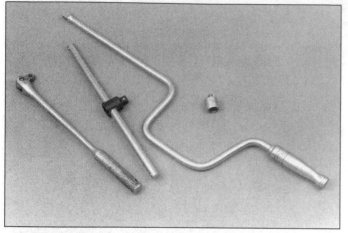

2.13 Lots of other accessories are available for ratchets; From left to right, a breaker bar, a sliding T-handle, a speed handle and a 3/8-to-1/4-inch adapter

2.14 Deep sockets enable you to loosen or tighten an elongated fastener, or to get at a nut with a long bolt protruding from it

(see illustration). Some of the sets you find may combine drive sizes; they're well worth having if you find the right set at a good price, but avoid being dazzled by the number of pieces.

Above all, be sure to completely ignore any label that reads "86-piece Socket Set," which refers to the number of pieces, not to the number of sockets (sometimes even the metal box and plastic insert are counted in the total!).

Apart from well-known and respected brand names, you'll have to take a chance on the quality of the set you buy. If you know someone who has a set that has held up well, try to find the same brand, if possible. Take a pocketful of nuts and bolts with you and check the fit in some of the sockets. Check the operation of the ratchet. Good ones operate smoothly and crisply in small steps; cheap ones are coarse and stiff - a good basis for guessing the quality of the rest of the pieces.

One of the best things about a socket set is the built-in facility for expansion. Once you have a basic set, you can purchase extra sockets when necessary and replace worn or damaged tools. There are special deep sockets for reaching recessed fasteners or to allow the socket to fit over a projecting bolt or stud **(see illustration)**. You can also buy screwdriver, Allen and Torx bits to fit various drive tools (they can be very handy in some applications) **(see illustration)**. Most socket sets include a special deep socket for 14 millimeter spark plugs. They have rubber inserts to protect the spark plug porcelain insulator and hold the plug in the socket to avoid burned fingers.

Torque wrenches

Torque wrenches **(see illustration)** are essential for tightening critical fasteners. A fastener that's not tight enough may eventually escape from its hole. An

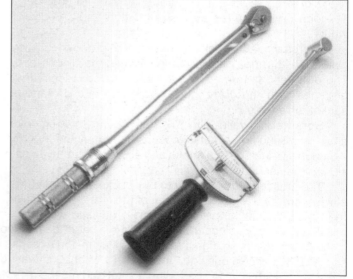

2.15 Standard and Phillips bits, Allen-head and Torx drivers will expand the versatility of your ratchet and extensions even further

2.16 Torque wrenches (click-type on left, beam-type on right) are the only way to accurately tighten critical fasteners like connecting rod bolts, cylinder head bolts, etc.

overtightened fastener, on the other hand, could break under stress.

There are two types of torque wrenches - the "beam" type, which indicates torque loads by deflecting a flexible shaft and the "click" type (see illustrations), which emits an audible click when the torque resistance reaches the specified resistance.

Torque wrenches are available in a variety of drive sizes and torque ranges for particular applications. For most work, 0 to 100 ft-lbs should be adequate. Keep in mind that "click" types are usually more accurate (and more expensive).

Impact drivers

The impact driver (see illustration) belongs with the screwdrivers, but it's mentioned here since it can also be used with sockets (impact drivers normally are 3/8-inch square drive). As explained later, an impact driver works by converting a hammer blow on the end of its handle into a sharp twisting movement. While this is a great way to jar a seized fastener loose, the loads imposed on the socket are excessive. Use sockets only with discretion and expect to have to replace damaged ones on occasion.

Using wrenches and sockets

Although you may think the proper use of tools is self-evident, it's worth some thought. After all, when did you last see instructions for use supplied with a set of wrenches?

Which wrench?

Before you start working, figure out the best tool for the job; in this instance the best wrench for a hex-head fastener. Sit down with a few nuts and bolts and look at how various tools fit the bolt heads.

A golden rule is to choose a tool that contacts the largest area of the hex-head. This distributes the load as evenly as possible and lessens the risk of damage. The shape most closely resembling the bolt head or nut is another hex, so a 6-point socket or box-end wrench is usually the best choice (see illustration). Many sockets and box-end wrenches have double hex (12-point) openings. If

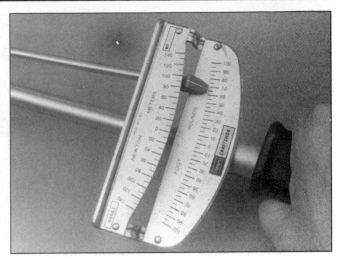

2.17 The deflecting beam-type torque wrench is inexpensive and simple to use - just tighten the fastener until the pointer points to the specified torque setting

2.18 "Click" type torque wrenches can be set to "give" at a pre-set torque, which makes them very accurate and easy to use

2.19 The impact driver converts a sharp blow into a twisting motion - this is a handy addition to your socket arsenal for those fasteners that won't let go - you can use it with any bit that fits a 38-inch drive ratchet

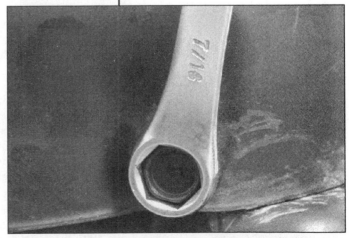

2.20 Try to use a six-point box wrench (or socket) whenever possible - it's shape matches that of the fastener, which means maximum grip and minimum slip

2.21 Sometimes a six-point tool just doesn't offer you any grip when you get the wrench at the angle it needs to be in to loosen or tighten a fastener - when this happens, pull out the 12-point sockets or wrenches - but remember; they're much more likely to strip the corners off a fastener

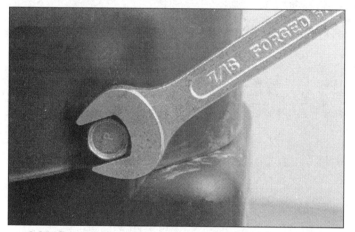

2.22 Open-end wrenches contact only two sides of the fastener and the jaws tend to open up when you put some muscle on the wrench handle - that's why they should only be used as a last resort

you slip a 12-point box-end wrench over a nut, look at how and where the two are in contact. The corners of the nut engage in every other point of the wrench. When the wrench is turned, pressure is applied evenly on each of the six corners **(see illustration)**. This is fine unless the fastener head was previously rounded off. If so, the corners will be damaged and the wrench will slip. If you encounter a damaged bolt head or nut, always use a 6-point wrench or socket if possible. If you don't have one of the right size, choose a wrench that fits securely and proceed with care.

If you slip an open-end wrench over a hex-head fastener, you'll see the tool is in contact on two faces only **(see illustration)**. This is acceptable provided the tool and fastener are both in good condition. The need for a snug fit between the wrench and nut or bolt explains the recommendation to buy good-quality open-end wrenches. If the wrench jaws, the bolt head or both are damaged, the wrench will probably slip, rounding off and distorting the head. In some applications, an open-end wrench is the only possible choice due to limited access, but always check the fit of the wrench on the fastener before attempting to loosen it; if it's hard to get at with a wrench, think how hard it will be to remove after the head is damaged.

The last choice is an adjustable wrench or self-locking plier/wrench (Vise-Grips). Use these tools only when all else has failed. In some cases, a self-locking wrench may be able to grip a damaged head that no wrench could deal with, but be careful not to make matters worse by damaging it further.

Bearing in mind the remarks about the correct choice of tool in the first place, there are several things worth noting about the actual use of the tool. First, make sure the wrench head is clean and undamaged. If the fastener is rusted or coated with paint, the wrench won't fit correctly. Clean off the head and, if it's rusted, apply some penetrating oil. Leave it to soak in for a while before attempting removal.

It may seem obvious, but take a close look at the fastener to be removed before using a wrench. On many mass-produced machines, one end of a fastener may be fixed or captive, which speeds up initial assembly and usually makes removal easier. If a nut is installed on a stud or a bolt threads into a captive nut or tapped hole, you may have only one fastener to deal with. If, on the other hand, you have a separate nut and bolt, you must hold the bolt head while the nut is removed. In some areas this can be difficult, particularly where engine mounts are involved. In this type of situation you may need an assistant to hold the bolt head with a wrench while you remove the nut from the other side. If this isn't possible, you'll have to try to position a box-end wrench so it wedges against some other component to prevent it from turning.

Be on the lookout for left-hand threads. They aren't common, but are sometimes used on the ends of rotating shafts to make sure the nut doesn't come loose during operation. If you can see the shaft end, the thread type can be checked visually. If you're unsure, place your thumbnail in the threads and see which way you have to turn your hand so your nail "unscrews" from the shaft. If you have to turn your hand counterclockwise, it's a conventional right-hand thread.

Beware of the upside-down fastener syndrome. If you're loosening a fastener from the under side of a something, it's easy to get confused about which way to turn it. What seems like counterclockwise to you is actually clockwise (from the fastener's point of view). Even after years of experience, this can still catch you once in a while.

In most cases, a fastener can be removed simply by placing the wrench on the nut or bolt head and turning it. Occasionally, though, the condition or location of the fastener may make things more difficult. Make sure the wrench is square on the head. You may need to reposition the tool or try another type to obtain a snug fit. Make sure the engine you're working on is secure and can't move when you turn the wrench. If necessary, get someone to help steady it for you. Position yourself so you can get maximum leverage on the wrench.

If possible, locate the wrench so you can pull the end towards you. If you have to push on the tool, remember that it may slip, or the fastener may move suddenly. For this reason, don't curl your fingers around the handle or you may crush or bruise them when the fastener moves; keep your hand flat, pushing on the wrench with the heel of your thumb. If the tool digs into your hand, place a rag between it and your hand or wear a heavy glove.

If the fastener doesn't move with normal hand pressure, stop and try to figure out why before the fastener or wrench is damaged or you hurt yourself. Stuck fasteners may require penetrating oil, heat or an impact driver or air tool.

Using sockets to remove hex-head fasteners is less likely to result in damage than if a wrench is used. Make sure the socket fits snugly over the fastener head, then attach an extension, if needed, and the ratchet or breaker bar. Theoretically, a ratchet shouldn't be used for loosening a fastener or for final tightening because the ratchet mechanism may be overloaded and could slip. In some instances, the location of the fastener may mean you have no choice but to use a ratchet, in which case you'll have to be extra careful.

Never use extensions where they aren't needed. Whether or not an extension is used, always support the drive end of the breaker bar with one hand while turning it with the other. Once the fastener is loose, the ratchet can be used to speed up removal.

Pliers

Some tool manufacturers make 25 or 30 different types of pliers. You only need a fraction of this selection **(see illustration)**. Get a good pair of slip-joint pliers for general use. A pair of needle-nose models is handy for reaching into hard-to-get-at places. A set of diagonal wire cutters (dikes) is essential for electrical work and pulling out cotter pins. Vise-Grips are adjustable, locking pliers that grip a fastener firmly - and won't let go - when locked into place. Parallel-jaw, adjustable pliers have angled jaws that remain parallel at any degree of opening. They're also referred to as Channel-lock (the original manufacturer) pliers, arc-joint pliers and water pump pliers. Whatever you call them, they're terrific for gripping things with a lot of force.

Slip-joint pliers have two open positions; a figure eight-shaped, elongated slot in one handle slips back-and-forth on a pivot pin on the other handle to change them. Good-quality pliers have jaws made of tempered steel and there's usually a wire-cutter at the base of the jaws. The primary uses of slip-joint pliers are for holding objects, bending and cutting throttle wires and crimping and bending metal parts, not loosening nuts and bolts.

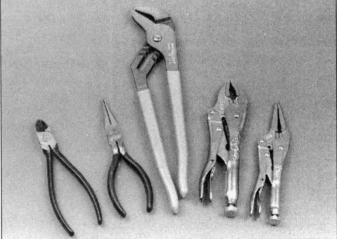

2.23 A typical assortment of the types of pliers you need to have in your box - from the left; diagonal cutters (dikes), needle-nose pliers, Channel-lock pliers, Vise-grip pliers, needle-nose Vise-grip pliers

Arc-joint or "Channel-lock" pliers have parallel jaws you can open to various widths by engaging different tongues and grooves, or channels, near the pivot pin. Since the tool expands to fit many size objects, it has countless uses for engine and equipment maintenance. Channel-lock pliers come in various sizes. The medium size is adequate for general work; small and large sizes are nice to have as your budget permits. You'll use all three sizes frequently.

Vise-Grips (a brand name) come in various sizes; the medium size with curved jaws is best for all-around work. However, buy a large and small one if

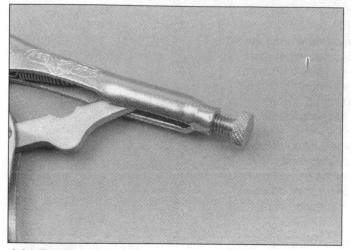

2.24 To adjust the jaws on a pair of vise-grips, grasp the part you want to hold with the jaws, tighten them down by turning the knurled knob on the end of one handle and snap the handles together - if you tightened the knob all the way down, you'll probably have to open it up (back it off) a little before you can close the handles

possible, since they're often used in pairs. Although this tool falls somewhere between an adjustable wrench, a pair of pliers and a portable vise, it can be invaluable for loosening and tightening fasteners - it's the only pliers that should be used for this purpose.

The jaw opening is set by turning a knurled knob at the end of one handle. The jaws are placed over the head of the fastener and the handles are squeezed together, locking the tool onto the fastener (see illustration). The design of the tool allows extreme pressure to be applied at the jaws and a variety of jaw designs enable the tool to grip firmly even on damaged heads (see illustration). Vise-Grips are great for removing fasteners that've been rounded off by badly-fitting wrenches.

As the name suggests, needle-nose pliers have long, thin jaws designed for reaching into holes and other restricted areas. Most needle-nose, or long-nose, pliers also have wire cutters at the base of the jaws.

Look for these qualities when buying pliers: Smooth operating handles and jaws, jaws that match up and grip evenly when the handles are closed, a nice finish and the word "forged" somewhere on the tool.

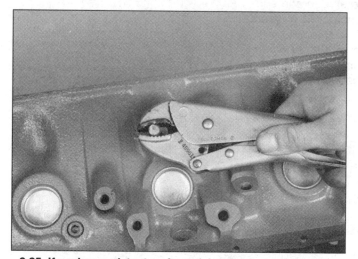

2.25 If you're persistent and careful, most fasteners can be removed with vise-grips

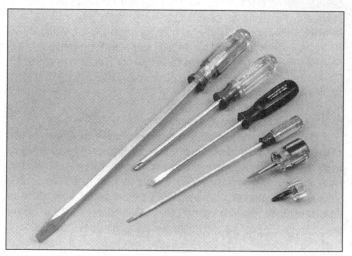

2.26 Screwdrivers come in a myriad of lengths, sizes and styles

Screwdrivers

Screwdrivers (see illustration) come in a wide variety of sizes and price ranges. Anything from Craftsman on up is fine. But don't buy screwdriver sets for ten bucks at discount tool stores. Even if they look exactly like more expensive brands, the metal tips and shafts are made with inferior alloys and aren't properly heat treated. They usually bend the first time you apply some serious torque.

A screwdriver consists of a steel blade or shank with a drive tip formed at one end. The most common tips are standard (also called straight slot and flat-blade) and Phillips. The other end has a handle attached to it. Traditionally, handles were made from wood and secured to the shank, which had raised tangs to prevent it from turning in the handle. Most screwdrivers now come with plastic handles, which are generally more durable than wood.

The design and size of handles and blades vary considerably. Some handles are specially shaped to fit the human hand and provide a better grip. The shank may be either round or square and some have a hex-shaped bolster under the handle to accept a wrench to provide more leverage when trying to turn a stubborn screw. The shank diameter, tip size and overall length vary too.

If access is restricted, a number of special screwdrivers are designed to fit into confined spaces. The "stubby" screwdriver has a specially shortened handle and blade. There are also offset screwdrivers and special screwdriver bits that attach to a ratchet or extension.

The important thing to remember when buying screwdrivers is that they really do come in sizes designed to fit different size fasteners. The slot in any screw has definite dimensions - length, width and depth. Like a bolt head or a nut, the screw slot must be driven by a tool that uses all of the available bearing surface and doesn't slip. Don't use a big wide blade on a small screw and don't try to turn a large screw slot with a tiny, narrow blade. The same principles apply to Allen heads, Phillips heads, Torx heads, etc. Don't even think of using a slotted screwdriver on one of these heads! And don't use your screwdrivers as levers, chisels or punches! This kind of abuse turns them into very bad screwdrivers. It's also dangerous.

Standard screwdrivers

These are used to remove and install conventional slotted screws and are available in a wide range of sizes denoting the width of the tip and the length of the shank (for example: a 3/8 x 10-inch screwdriver is 3/8-inch wide at the tip and the shank is 10-inches long). You should have a variety of screwdrivers so screws of various sizes can be dealt with without damaging them. The blade end must be the same width and thickness as the screw slot to work properly, without slipping. When selecting standard screwdrivers, choose good-quality tools, preferably with chrome moly, forged steel shanks. The tip of the shank should be ground to a parallel, flat profile (hollow ground) and not to a taper or wedge shape, which will tend to twist out of the slot when pressure is applied **(see illustration)**.

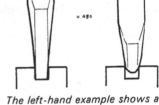

Misuse of a screwdriver – the blade shown is both too narrow and too thin and will probably slip or break off

The left-hand example shows a snug-fitting tip. The right-hand drawing shows a damaged tip which will twist out of the slot when pressure is applied

2.27 Standard screwdrivers - wrong size (left), correct fit in screw slot (center) and worn tip (right)

All screwdrivers wear in use, but standard types can be reground to shape a number of times. When reshaping a tip, start by grinding the very end flat at right angles to the shank. Make sure the tip fits snugly in the slot of a screw of the appropriate size and keep the sides of the tip parallel. Remove only a small amount of metal at a time to avoid overheating the tip and destroying the temper of the steel.

Phillips screwdrivers

Phillips screws are sometimes installed during initial assembly with air tools and are next to impossible to remove later without ruining the heads, particularly if the wrong size screwdriver is used. And don't use other types of cross-head screwdrivers (Torx, Posi-drive, etc.) on Phillips screws - they won't work.

The only way to ensure the screwdrivers you buy will fit properly, is to take a couple of screws with you to make sure the fit between the screwdriver and fastener is snug. If the fit is good, you should be able to angle the blade down almost vertically without the screw slipping off the tip. Use only screwdrivers that fit exactly - anything else is guaranteed to chew out the screw head instantly.

The idea behind all cross-head screw designs is to make the screw and screwdriver blade self-aligning. Provided you aim the blade at the center of the screw head, it'll engage correctly, unlike conventional slotted screws, which need careful alignment. This makes the screws suitable for machine installation on an assembly line (which explains why they're sometimes so tight and difficult to remove). The drawback with these screws is the driving tangs on the screwdriver tip are very small and must fit very precisely in the screw head. If this isn't the case, the huge loads imposed on small flats of the screw slot simply tear the metal away, at which point the screw ceases to be removable by normal methods. The problem is made worse by the normally soft material chosen for screws.

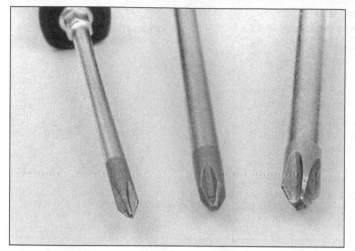

2.28 The tip size on a Phillips screwdriver is indicated by a number from 1 to 4, with 1 the smallest (left - No. 1: center - Nos. 2; eight - No. 3)

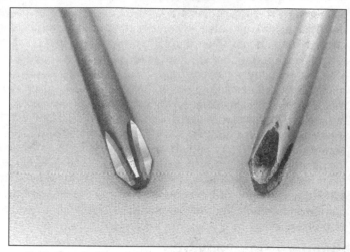

2.29 New (left) and worn (right) Phillips screwdriver tips

To deal with these screws on a regular basis, you'll need high-quality screwdrivers with various size tips so you'll be sure to have the right one when you need it. Phillips screwdrivers are sized by the tip number and length of the shank (for example: a number 2 x 6-inch Phillips screwdriver has a number 2 tip - to fit screws of only that size recess - and the shank is 6-inches long). Tip sizes 1, 2 and 3 should be adequate for engine repair work **(see illustration)**. If the tips get worn or damaged, buy new screwdrivers so the tools don't destroy the screws they're used on **(see illustration)**.

Here's a tip that may come in handy when using Phillips screwdrivers - if the screw is extremely tight and the tip tends to back out of the recess rather than turn the screw, apply a small amount of valve lapping compound to the screwdriver tip so it will grip the screw better.

Hammers

Resorting to a hammer should always be the last resort. When nothing else will do the job, a medium-size ball peen hammer, a heavy rubber mallet and a heavy soft-brass hammer **(see illustration)** are often the only way to loosen or install a part.

A ball-peen hammer has a head with a conventional cylindrical face at one end and a rounded ball end at the other and is a general-purpose tool found in almost any type of shop. It has a shorter neck than a claw hammer and the face is tempered for striking punches and chisels. A fairly large hammer is preferable to a small one. Although it's possible to find small ones, you won't need them very often and it's much easier to control the blows from a heavier head. As a general rule, a single 12 or 16-ounce hammer will work for most jobs, though occasionally larger or smaller ones may be useful.

A soft-face hammer is used where a steel hammer could cause damage to the component or other tools being used. A steel hammer head might crack an aluminum part, but a rubber or plastic hammer can be used with more confidence. Soft-face hammers are available with interchangeable heads (usually one made of rubber and another made of relatively hard plastic). When the heads are worn out, new ones can be installed. If finances are really limited, you can get by without a soft-face hammer by placing a small hardwood block between the component and a steel hammer head to prevent damage.

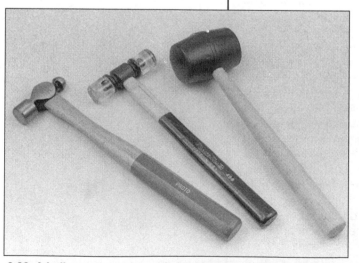

2.30 A ball-peen hammer, soft-face hammer and rubber mallet (left-to-right) will be needed for various tasks (any steel hammer can be used in place of the ball peen hammer)

Hammers should be used with common sense; the head should strike the desired object squarely and with the right amount of force. For many jobs, little effort is needed - simply allow the weight of the head to do the work, using the length of the swing to control the amount of force applied. With practice, a hammer can be used with surprising finesse, but it'll take a while to achieve. Initial mistakes include striking the object at an angle, in which case the hammer head may glance off to one side, or hitting the edge of the object. Either one can result in damage to the part or to your thumb, if it gets in the way, so be careful. Hold the hammer handle near the end, not near the head, and grip it firmly but not too tightly.

Check the condition of your hammers on a regular basis. The danger of a loose head coming off is self-evident, but check the head for chips and cracks too. If damage is noted, buy a new hammer - the head may chip in use and the resulting fragments can be extremely dangerous. It goes without saying that eye protection is essential whenever a hammer is used.

Punches and chisels

Punches and chisels **(see illustration)** are used along with a hammer for various purposes in the shop. Drift punches are often simply a length of round steel bar used to drive a component or fastener out of a bore. A typical use would be for removing or installing a bearing or bushing. A drift of the same diameter as the bearing outer race is placed against the bearing and tapped with a hammer to knock it in or out of the bore. Most manufacturers offer special drifts for the various bearings in a particular engine. While they're useful to a busy dealer service department, they are prohibitively expensive for the do-it-your-selfer who may only need to use them once. In such cases, it's better to improvise. For bearing removal and installation, it's usually possible to use a socket of the appropriate diameter to tap the bearing in or out; an unorthodox use for a socket, but it works.

Smaller diameter drift punches can be purchased or fabricated from steel bar stock. In some cases, you'll need to drive out items like caliper retaining pins. Here, it's essential to

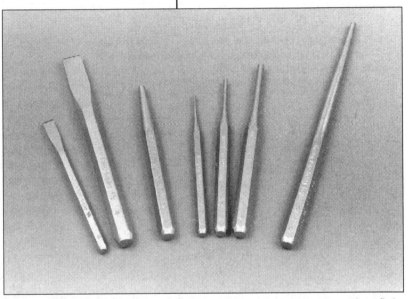

2.31 Cold chisels, center-punches, pin punches and line-up punches (left-to-right) will be needed sooner or later for many jobs

avoid damaging the pin or caliper, so the drift must be made from a soft material. Brass or copper is the usual choice for such jobs; the drift may be damaged in use, but the pin and surrounding components will be protected.

Punches are available in various shapes and sizes and a set of assorted types will be very useful. One of the most basic is the center punch, a small cylindrical punch with the end ground to a point. It'll be needed whenever a hole is drilled. The center of the hole is located first and the punch is used to make a small indentation at the intended point. The indentation acts as a guide for the drill bit so the hole ends up in the right place. Without a punch mark the drill bit will wander and you'll find it impossible to drill with any real accuracy. You can also buy automatic center punches. They're spring loaded and are pressed against the surface to be marked, without the need to use a hammer.

Pin punches are intended for removing items like roll pins (semi-hard, hollow pins that fit tightly in their holes). Pin punches have other uses, however. You may occasionally have to remove rivets or bolts by cutting off the heads and driving out the shanks with a pin punch. They're also very handy for aligning holes in components while bolts or screws are inserted.

Of the various sizes and types of metal-cutting chisels available, a simple cold chisel is essential in any mechanic's workshop. One about 6-inches long

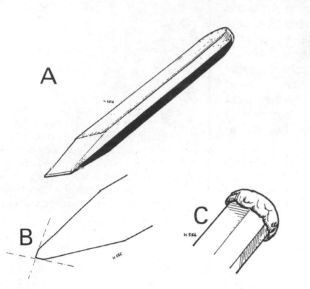

2.32 A typical general purpose cold chisel - note the angle of the cutting edge (A), which should be checked and resharpened on a regular basis; the mushroomed head (B) is dangerous and should be filed to restore it to its original shape

with a 1/2-inch wide blade should be adequate. The cutting edge is ground to about 80-degrees **(see illustration)**, while the rest of the tip is ground to a shallower angle away from the edge. The primary use of the cold chisel is rough metal cutting - this can be anything from sheet metal work to cutting off the heads of seized or rusted bolts or splitting nuts. A cold chisel is also useful for turning out screws or bolts with messed-up heads.

All of the tools described in this section should be good quality items. They're not particularly expensive, so it's not really worth trying to save money on them. More significantly, there's a risk that with cheap tools, fragments may break off in use - a potentially dangerous situation.

Even with good-quality tools, the heads and working ends will inevitably get worn or damaged, so it's a good idea to maintain all such tools on a regular basis. Using a file or bench grinder, remove all burrs and mushroomed edges from around the head. This is an important task because the build-up of material around the head can fly off when it's struck with a hammer and is potentially dangerous. Make sure the tool retains its original profile at the working end, again, filing or grinding off all burrs. In the case of cold chisels, the cutting edge will usually have to be reground quite often because the material in the tool isn't usually much harder than materials typically being cut. Make sure the edge is reasonably sharp, but don't make the tip angle greater than it was originally; it'll just wear down faster if you do.

The techniques for using these tools vary according to the job to be done and are best learned by experience. The one common denominator is the fact they're all normally struck with a hammer. It follows that eye protection should be worn. Always make sure the working end of the tool is in contact with the part being punched or cut. If it isn't, the tool will bounce off the surface and damage may result.

Hacksaws

A hacksaw **(see illustration)** consists of a handle and frame supporting a flexible steel blade under tension. Blades are available in various lengths and most hacksaws can be adjusted to accommodate the different sizes. The most common blade length is 10-inches.

Most hacksaw frames are adequate. There's little difference between brands. Pick one that's rigid and allows easy blade changing and repositioning.

The type of blade to use, indicated by the number of teeth per inch, (TPI) **(see illustration)**, is determined by the material being cut. The rule of thumb is to make sure at least three teeth are in contact with the metal being cut at any one time **(see illustration)**. In practice, this means a fine blade for cutting thin sheet materials, while a coarser blade can be used for faster cutting through thicker items such as bolts or bar stock. When cutting thin materials, angle the saw so the blade cuts at a shallow angle. More teeth are in contact and there's less chance of the blade binding and breaking, or teeth breaking.

When you buy blades, choose a reputable brand. Cheap, unbranded blades may be perfectly acceptable, but you can't tell by looking at them. Poor quality blades will be insufficiently hardened on the teeth edge and will dull quickly. Most rep-

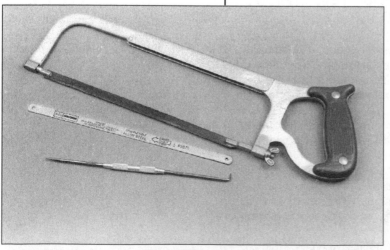

2.33 Hacksaws are handy for little cutting jobs like sheet metal and rusted fasteners

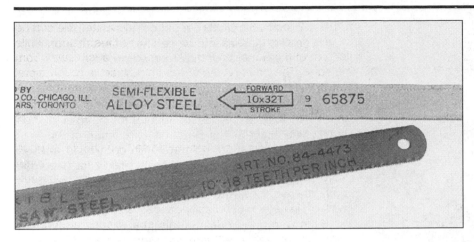

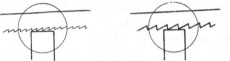

2.34 Hacksaw blades are marked with the number of teeth per inch (TPI - use a relatively course blade for aluminum and thicker items such as bolts or bar stock; use a finer blade for materials like thin sheet steel

When cutting thin materials, check that at least three teeth are in contact with the workpiece at any time. Too coarse a blade will result in a poor cut and may break the blade. If you do not have the correct blade, cut at a shallow angle to the material

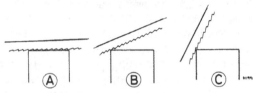

The correct cutting angle is important. If it is too shallow (A) the blade will wander. The angle shown at (B) is correct when starting the cut, and may be reduced slightly once under way. In (C) the angle is too steep and the blade will be inclined to jump out of the cut

2.35 Correct procedure for use of a hacksaw

2.36 Good quality hacksaw blades are marked like this

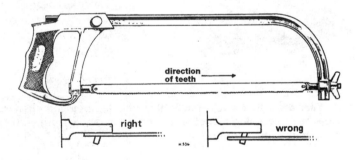

2.37 Correct installation of a hacksaw blade - the teeth must point away from the handle and butt against the locating lugs

utable brands will be marked "Flexible High Speed Steel" or a similar term, to indicate the type of material used **(see illustration)**. It is possible to buy "unbreakable" blades (only the teeth are hardened, leaving the rest of the blade less brittle).

Sometimes, a full-size hacksaw is too big to allow access to a frozen nut or bolt. On most saws, you can overcome this problem by turning the blade 90-degrees. Occasionally you may have to position the saw around an obstacle and then install the blade on the other side of it. Where space is really restricted, you may have to use a handle that clamps onto a saw blade at one end. This allows access when a hacksaw frame would not work at all and has another advantage in that you can make use of broken off hacksaw blades instead of throwing them away. Note that because only one end of the blade is supported, and it's not held under tension, it's difficult to control and less efficient when cutting.

Before using a hacksaw, make sure the blade is suitable for the material being cut and installed correctly in the frame **(see illustration)**. Whatever it is you're cutting must be securely supported so it can't move around. The saw cuts on the forward stroke, so the teeth must point away from the handle. This might seem obvious, but it's easy to install the blade backwards by mistake and ruin the teeth on the first few strokes. Make sure the blade is tensioned adequately or it'll

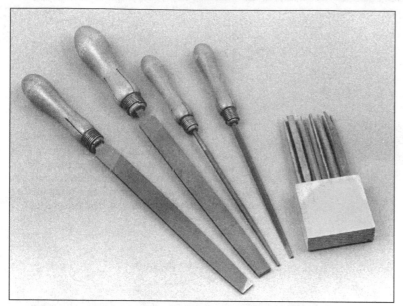

2.38 Get a good assortment of files - they're handy for deburring, marking parts, removing rust, filing the heads off rivets, restoring threads and fabricating small parts

2.39 Files are either single-cut (left) or double-cut (right) - generally speaking, use a single-cut file to produce a very smooth surface; use a double-cut file to remove large amounts of material quickly

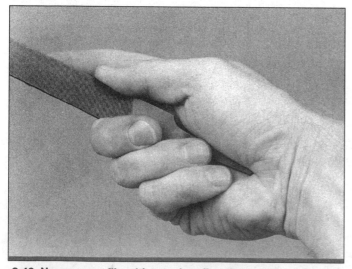

2.40 Never use a file without a handle - the tang is sharp and could puncture your hand

distort and chatter in the cut and may break. Wear safety glasses and be careful not to cut yourself on the saw blade or the sharp edge of the cut.

Files

Files **(see illustration)** come in a wide variety of sizes and types for specific jobs, but all of them are used for the same basic function of removing small amounts of metal in a controlled fashion. Files are used by mechanics mainly for deburring, marking parts, removing rust, filing the heads off rivets, restoring threads and fabricating small parts.

File shapes commonly available include flat, half-round, round, square and triangular. Each shape comes in a range of sizes (lengths) and cuts ranging from rough to smooth. The file face is covered with rows of diagonal ridges which form the cutting teeth. They may be aligned in one direction only (single cut) or in two directions to form a diamond-shaped pattern (double-cut) **(see illustration)**. The spacing of the teeth determines the file coarseness, again, ranging from rough to smooth in five basic grades: Rough, coarse, bastard, second-cut and smooth.

You'll want to build up a set of files by purchasing tools of the required shape and cut as they're needed. A good starting point would be flat, half-round, round and triangular files (at least one each - bastard or second-cut types). In addition, you'll have to buy one or more file handles (files are usually sold without handles, which are purchased separately and pushed over the tapered tang of the file when in use) **(see illustration)**. You may need to buy more than one size handle to fit the various files in your tool box, but don't attempt to get by without them. A file tang is fairly sharp and you almost certainly will end up stabbing yourself in the palm of the hand if you use a file without a handle and it catches in the workpiece during use. Adjustable handles are also available for use with files of various sizes, eliminating the need for several handles **(see illustration)**.

Exceptions to the need for a handle are fine swiss pattern files, which have a rounded handle instead of a tang. These small files are usually sold in sets with a number of different shapes. Originally intended for very fine work, they can be very handy for use in inaccessible areas. Swiss files are normally the best choice if piston ring ends require filing to obtain the correct end gap.

The correct procedure for using files is fairly easy to master. As with a hacksaw, the work should be clamped securely in a vise, if needed, to prevent it from moving around while being worked on. Hold the file by the handle, using your free hand at the file end to guide it and keep it flat in relation to the surface being filed. Use smooth cutting strokes and be careful not to rock the file as it passes over the surface. Also, don't slide it diagonally across the surface or the teeth will make grooves in the workpiece. Don't drag a file back across the workpiece at the end of the stroke - lift it

2.52 Although it's not absolutely necessary, an air compressor can make many jobs easier and produce better results, especially when air powered tools are available to use with it

2.53 Another indispensable piece of equipment is the bench grinder (with a wire wheel mounted on one arbor) - make sure it's securely bolted down and never use it with the rests or eye shields removed

Electric drills

Countersinking bolt holes, enlarging oil passages, honing cylinder bores, removing rusted or broken off fasteners, enlarging holes and fabricating small parts - an electric drill **(see illustration)** is indispensable for your shop. A 3/8-inch chuck (drill bit holder) will handle most jobs. Collect several different wire brushes to use in the drill and make sure you have a complete set of sharp metal drill bits **(see illustration)**. Cordless drills are extremely versatile because they don't force you to work near an outlet. They're also handy to have around for a variety of non-mechanical jobs.

Twist drills

Drilling operations are done with twist drills, either in a hand drill or a drill press. Twist drills (or drill bits, as they're

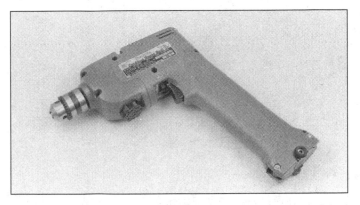

2.54 Electric drills can be cordless (above) or 115-volt, AC-powered (below)

2.55 Get a set of good quality drill bits for drilling holes and wire brushes of various sizes for cleaning up metal parts - make sure the bits are designed for drilling in metal!

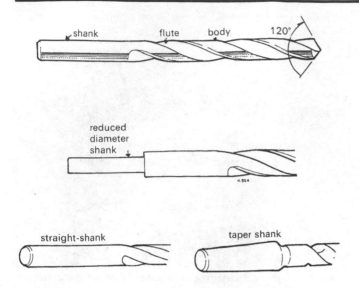

2.56 A typical drill bit (top), a reduced shank bit (center), and a tapered shank bit (bottom right)

2.57 Drill bits in the range most commonly used are available in fractional sizes (left) and number sizes (right) so almost any size hole can be drilled

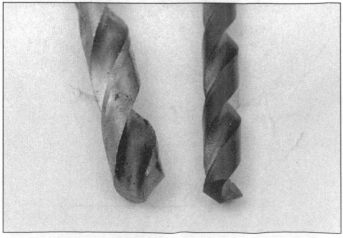

2.58 If a bit gets dull (left), discard it or resharpen it so it looks like the bit on the right

often called) consist of a round shank with spiral flutes formed into the upper two-thirds to clear the waste produced while drilling, keep the drill centered in the hole and finish the sides of the hole.

The lower portion of the shank is left plain and used to hold the drill in the chuck. In this section, we will discuss only normal parallel shank drills **(see illustration)**. There is another type of bit with the plain end formed into a special size taper designed to fit directly into a corresponding socket in a heavy-duty drill press. These drills are known as Morse Taper drills and are used primarily in machine shops.

At the cutting end of the drill, two edges are ground to form a conical point. They're generally angled at about 60-degrees from the drill axis, but they can be reground to other angles for specific applications. For general use the standard angle is correct - this is how the drills are supplied.

When buying drills, purchase a good-quality set (sizes 1/16 to 3/8-inch). Make sure the drills are marked "High Speed Steel" or "HSS". This indicates they're hard enough to withstand continual use in metal; many cheaper, unmarked drills are suitable only for use in wood or other soft materials. Buying a set ensures the right size bit will be available when it's needed.

Twist drill sizes

Twist drills are available in a vast array of sizes, most of which you'll never need. There are three basic drill sizing systems: Fractional, number and letter **(see illustration)** (we won't get involved with the fourth system, which is metric sizes).

Fractional sizes start at 1/64-inch and increase in increments of 1/64-inch. Number drills range in descending order from 80 (0.0135-inch), the smallest, to 1 (0.2280-inch), the largest. Letter sizes start with A (0.234-inch), the smallest, and go through Z (0.413-inch), the largest.

This bewildering range of sizes means it's possible to drill an accurate hole of almost any size within reason. In practice, you'll be limited by the size of chuck on your drill (normally 3/8 or 1/2-inch). In addition, very few stores stock the entire range of possible sizes, so you'll have to shop around for the nearest available size to the one you require.

Sharpening twist drills

Like any tool with a cutting edge, twist drills will eventually get dull **(see illustration)**. How often they'll need sharpening depends to some extent on whether they're used correctly. A dull twist drill will soon make itself known. A good indication of the condition of the cutting edges is to watch the waste emerging from the hole being drilled. If the tip is in good condition, two even spirals of waste metal will be produced; if this fails to happen or the tip gets hot, it's safe to assume that sharpening is required.

With smaller size drills - under about 1/8-inch - it's easier and more economical to throw the worn drill away and buy another one. With larger (more expensive) sizes, sharpening is a better bet. When sharpening twist drills, the included angle of the cutting edge must be maintained at the original 120-degrees and the small chisel edge at the tip must be retained. With some practice, sharpening can be slightly and pull it back to prevent damage to the teeth.

Files don't require maintenance in the usual sense, but they should be kept clean and free of metal filings. Steel is a reasonably easy material to work with, but softer metals like aluminum tend to clog the file teeth very quickly, which will result in scratches in the workpiece. This can be avoided by rubbing the file face with chalk before using it. General cleaning is carried out with a file card or a fine wire brush. If kept clean, files will last a long time - when they do eventually dull, they must be replaced; there is no satisfactory way of sharpening a worn file.

Taps and dies

Taps

Tap and die sets **(see illustration)** are available in inch and metric sizes. Taps are used to cut internal threads and clean or restore damaged threads. A tap consists of a fluted shank with a drive square at one end. It's threaded along part of its length - the cutting edges are formed where the flutes intersect the threads **(see illustration)**. Taps are made from hardened steel so they will cut threads in materials softer than what they're made of.

Taps come in three different types: Taper, plug and bottoming. The only real difference is the length of the chamfer on the cutting end of the tap. Taper taps are chamfered for the first 6 or 8 threads, which makes them easy to start but prevents them from cutting threads close to the bottom of a hole. Plug taps are chamfered up about 3 to 5 threads, which makes them a good all around tap because they're relatively easy to start and will cut nearly to the bottom of a hole. Bottoming taps, as the name implies, have a very short chamfer (1-1/2 to 3 threads) and will cut as close to the bottom of a blind hole as practical. However, to do this, the threads should be started with a plug or taper tap.

Although cheap tap and die sets are available, the quality is usually very low and they can actually do more harm than good when used on threaded holes in aluminum parts. The alternative is to buy high-quality taps if and when you need them, even though they aren't cheap, especially if you need to buy two or more thread pitches in a given size. Despite this, it's the best option - you'll probably only need taps on rare occasions, so a full set isn't absolutely necessary.

Taps are normally used by hand (they can be used in machine tools, but primarily for manufacturing purposes). The square drive end of the tap is held in a tap wrench (an adjustable T-handle). For smaller sizes, a T-handled chuck can be used. The tapping process starts by drilling a hole of the correct diameter. For each tap size, there's a corresponding twist drill that will produce a hole of the correct size. Note how the tapered section progressively decreases across the ridge. Plug taps are normally needed for finishing tapped holes in blind bores.

This is important; too large a hole will leave the finished thread with the tops missing, producing a weak

2.41 Adjustable handles that will work with many different size files are also available

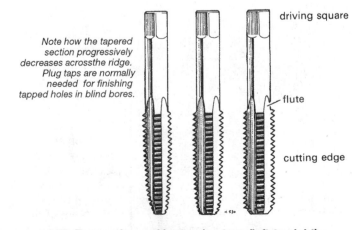

2.42 Tap and dies sets are available in inch and metric sizes - taps are used for cutting internal threads and cleaning and restoring damaged threads; dies are used for cutting, cleaning and restoring external threads

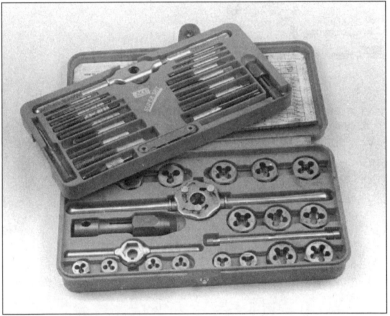

Note how the tapered section progressively decreases across the ridge. Plug taps are normally needed for finishing tapped holes in blind bores.

driving square

flute

cutting edge

2.43 Taper, plug and bottoming taps (left-to-right)

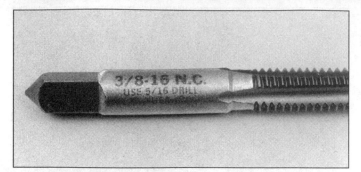

2.44 If you need to drill and tap a hole, the drill bit size to use for a given bolt (tap) size is marked on the tap

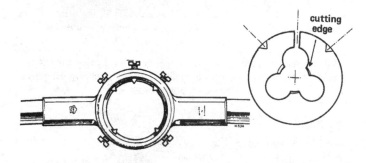

2.45 A die (right) is used for cutting external threads (this one is a split-type/adjustable die) and is held in a tool called a die stock (left)

2.46 Hex-shaped dies are especially handy for mechanic's work because they can be turned with a wrench

and unreliable grip. Conversely, too small a hole will place excessive loads on the hard and brittle shank of the tap, which can break it off in the hole. Removing a broken off tap from a hole is no fun! The correct tap drill size is normally marked on the tap itself or the container it comes in **(see illustration)**.

Dies

Dies are used to cut, clean or restore external threads. Most dies are made from a hex-shaped or cylindrical piece of hardened steel with a threaded hole in the center. The threaded hole is overlapped by three or four cutouts, which equate to the flutes on taps and allow metal waste to escape during the threading process. Dies are held in a T-handled holder (called a die stock) **(see illustration)**. Some dies are split at one point, allowing them to be adjusted slightly (opened and closed) for fine control of thread clearances.

Dies aren't needed as often as taps, for the simple reason it's normally easier to install a new bolt than to salvage one. However, it's often helpful to be able to extend the threads of a bolt or clean up damaged threads with a die. Hex-shaped dies are particularly useful for mechanic's work, since they can be turned with a wrench **(see illustration)** and are usually less expensive than adjustable ones.

The procedure for cutting threads with a die is broadly similar to that described above for taps. When using an adjustable die, the initial cut is made with the die fully opened, the adjustment screw being used to reduce the diameter on successive cuts until the finished size is reached. As with taps, a cutting lubricant should be used, and the die must be backed off every few turns to clear swarf from the cutouts.

Pullers

You may need a general-purpose puller for some brake work. Pullers can remove seized or corroded parts and push driveaxles from wheel hubs. Universal two- and three-legged pullers are widely available in numerous designs and sizes.

The typical puller consists of a central boss with two or three pivoting arms attached. The outer ends of the arms are hooked jaws which grab the part you want to pull off **(see illustration)**. You can reverse the arms on most pullers to use the puller on internal openings when necessary. The central boss is threaded to accept a puller bolt, which does the work. You can also get hydraulic versions of these tools which are capable of more pressure, but they're expensive.

You can adapt pullers by purchasing, or fabricating, special jaws for specific jobs. If you decide to make your own jaws, keep in mind that the pulling force should be concentrated as close to the center of the component as possible to avoid damaging it.

Before you use a puller, assemble it and check it to make sure it doesn't snag on anything and the loads on the part to be removed are distributed evenly. If you're dealing with a part held on a shaft by a nut, loosen the nut but don't remove it. Leaving the nut on helps prevent distortion of the shaft end under pressure from the puller bolt and stops the part from flying off the shaft when it comes loose.

Tighten a puller gradually until the assembly is under moderate pressure, then try to jar the component loose by striking the puller bolt a few times with a hammer. If this doesn't work, tighten the bolt a little further and repeat the process. If this approach doesn't work, stop and reconsider. At some point you must make a decision whether to continue applying pressure in this manner. Sometimes, you can apply penetrating oil around the joint and leave it overnight, with the puller in place and tightened securely. By the next day, the taper has separated and the problem has resolved itself.

If nothing else works, try heating the area surrounding the troublesome part with a propane or gas welding torch (we don't, however, recommend messing around with welding equipment if you're not already experienced in its use). Apply the heat to the hub area of the component you wish to remove. Keep the flame moving to avoid uneven heating and the risk of distortion. Keep pressure applied with the puller and make sure that you're able to deal with the resulting hot component and the puller jaws if it does come free. Be very careful to keep the flame away from aluminum parts.

If all reasonable attempts to remove a part fail, don't be afraid to give up. It's cheaper to quit now than to damage components. Either buy or borrow the correct tool, or take the component (or vehicle) to a dealer service department or other repair shop to have the part removed for you.

2.47 A two or three-jaw puller will come in handy for many tasks in the shop and can also be used for working on other types of equipment

Drawbolt extractors

The simple drawbolt extractor is easy to make up and valuable in every workshop. There are no commercially available tools of this type; you simply make a tool to suit a particular application. You can use a drawbolt extractor to pull out stubborn piston pins and to remove bearings and bushings.

To make a drawbolt extractor, you'll need an assortment of threaded rods in various sizes (available at hardware stores), and nuts to fit them. You'll also need assorted washers, spacers and tubing. For things like piston pins, you'll usually need a longer piece of tube.

Some typical drawbolt uses are shown in the accompanying line drawings **(see illustration)**. They also reveal the order of assembly of the various pieces. The same arrangement, minus the tubular spacer section, can usually be used to install a new bushing or piston pin. Using the tool is quite simple. Just make sure you get the bush or pin square to the bore when you install it. Lubricate the part being pressed into place, where appropriate.

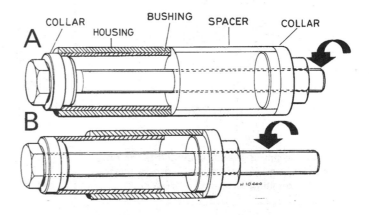

2.48 Typical drawbolt uses - in A, the nut is tightened to pull the collar and bushing into the large spacer; in B, the spacer is left out and the drawbolt is repositioned to install the new bushing

Pullers for use in blind bores

Bushings or bearings installed in "blind holes" often require special pullers. Some bearings can be removed without a puller if you heat the engine or component evenly in an oven and tap it face down on a clean wooden surface to dislodge the bearing. Wear heavy gloves to protect yourself when handling the heated components. If you need a puller to do the job, get a slide-hammer with interchangeable tips. Slide hammers range from universal two or three-jaw puller arrangements to special bearing pullers. Bearing pullers are hardened steel tubes with a flange around the bottom edge. The tube is split at several places, which allows a wedge to expand the tool once it's in place. The tool fits inside the bearing inner race and is tightened so the flange or lip is locked under the edge of the race.

The slide-hammer consists of a steel shaft with a stop at its upper end. The

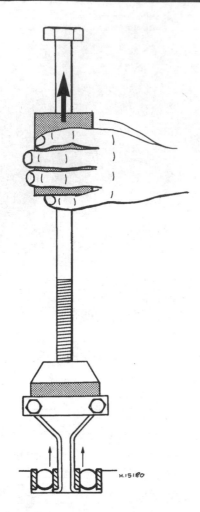

2.49 A slide hammer with special attachments can be used for removing bearings and bushings from blind holes

shaft carries a sliding weight which slides along the shaft until it strikes the stop. This allows the tool holding the bearing to yank it out of the bore **(see illustration)**.

Bench vise

The bench vise **(see illustration)** is an essential tool in a sho Buy the best quality vise you can afford. A good vise is expensive, b the quality of its materials and workmanship are worth the ext money. Size is also important - bigger vises are usually more versati Make sure the jaws open at least four inches. Get a set of soft jaws fit the vise as well - you'll need them to grip engine parts that could damaged by the hardened vise jaws **(see illustration)**.

Power tools

Really, the only power tool you absolutely need is an electric d But if you have an air compressor and electricity, there's a wide rai of pneumatic and electric hand tools to make all sorts of jobs ea and faster.

Air compressor

An air compressor **(see illustration)** makes most jobs easier a faster. Drying off parts after cleaning them with solvent, blowing c passages in a block or head, running power tools - the list is endles Once you buy a compressor, you'll wonder how you ever got alor without it. Air tools really speed up tedious procedures like removi and installing wheel lug nuts.

Bench-mounted grinder

A bench grinder **(see illustration)** is also handy. With a wire wh on one end and a grinding wheel on the other, it's great for cleaning fasteners, sharpening tools and removing rust. Make sure the grind is fastened securely to the bench or stand, always wear eye protecti when operating it and never grind aluminum parts on the grindir wheel.

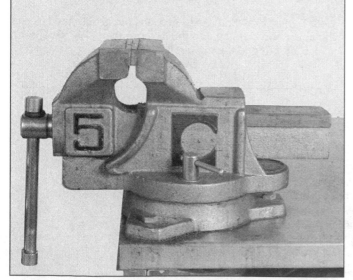

2.50 A bench vise is one of the most useful pieces of equipment you can have in the shop - bigger is usually better with vises, so get a vise with jaws that open at least four inches

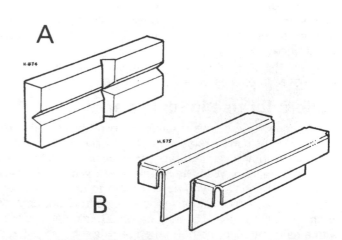

2.51 Sometimes, the parts you have to jig up in the vise are delicate, or made of soft materials - to avoid damaging them, get a pair of fiberglass or plastic "soft jaws" (A) or fabricate your own with 1/8-inch thick aluminum sheet (B)

done freehand on a bench grinder, but it should be noted that it's very easy to make mistakes. For most home mechanics, a sharpening jig that mounts next to the grinding wheel should be used so the drill is clamped at the correct angle **(see illustration)**.

Drilling equipment

Tools to hold and turn drill bits range from simple, inexpensive hand-operated or electric drills to sophisticated and expensive drill presses. Ideally, all drilling should be done on a drill press with the workpiece clamped solidly in a vise. These machines are expensive and take up a lot of bench or floor space, so they're out of the question for many do-it-yourselfers.

The best tool for the home shop is an electric drill with a 3/8-inch chuck. Both cordless and AC drills (that run off household current) are available. If you're purchasing one for the first time, look for a well-known, reputable brand name and variable speed as minimum requirements. A 1/4-inch chuck, single-speed drill will work, but it's worth paying a little more for the larger, variable speed type.

All drills require a key to lock the bit in the chuck. When removing or installing a bit, make sure the cord is unplugged to avoid accidents. Initially, tighten the chuck by hand, checking to see if the bit is centered correctly. This is especially important when using small drill bits which can get caught between the jaws. Once the chuck is hand tight, use the key to tighten it securely - remember to remove the key afterwards!

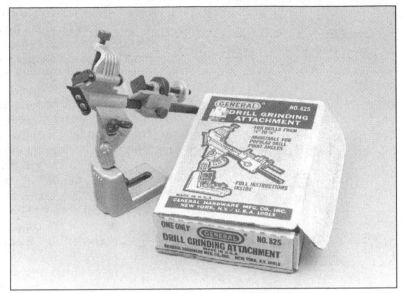

2.59 Inexpensive drill bit sharpening jigs designed to be used with a bench grinder are widely available

Drilling and finishing holes

Preparation for drilling

If possible, make sure the part you intend to drill in is securely clamped in a vise. If it's impossible to get the work to a vise, make sure it's stable and secure. Twist drills often dig in during drilling - this can be dangerous, particularly if the work suddenly starts spinning on the end of the drill. Make sure the work supported securely.

Start by locating the center of the hole you're drilling. Use a center punch to make an indentation for the drill bit so it won't wander. If you're drilling out a broken-off bolt, be sure to position the punch in the exact center of the bolt **(see illustration)**.

If you're drilling a large hole (above 1/4-inch), you may want to make a pilot hole first. As the name suggests, it will guide the larger drill bit and minimize drill bit wandering. Before actually drilling a hole, make sure the area immediately behind the bit is clear of anything you don't want drilled.

Drilling

When drilling steel, especially with smaller bits, no lubrication is needed. If a large bit is involved, oil can be used to ensure a clean cut and prevent overheating of the drill tip. When drilling aluminum, which tends to smear around the cutting edges and clog the drill bit flutes, use kerosene as a lubricant.

Wear safety goggles or a face shield and assume a

2.60 Before you drill a hole, use a centerpunch to make an indentation for the drill bit so it won't wander

comfortable, stable stance so you can control the pressure on the drill easily. Position the drill tip in the punch mark and make sure, if you're drilling by hand, the bit is perpendicular to the surface of the workpiece. Start drilling without applying much pressure until you're sure the hole is positioned correctly. If the hole starts off center, it can be very difficult to correct. You can try angling the bit slightly so the hole center moves in the opposite direction, but this must be done before the flutes of the bit have entered the hole. It's at the starting point that a variable-speed drill is invaluable; the low speed allows fine adjustments to be made before it's too late. Continue drilling until the desired hole depth is reached or until the drill tip emerges at the other side of the workpiece.

Cutting speed and pressure are important - as a general rule, the larger the diameter of the drill bit, the slower the drilling speed should be. With a single-speed drill, there's little that can be done to control it, but two-speed or variable speed drills can be controlled. If the drilling speed is too high, the cutting edges of the bit will tend to overheat and dull. Pressure should be varied during drilling. Start with light pressure until the drill tip has located properly in the work. Gradually increase pressure so the bit cuts evenly. If the tip is sharp and the pressure correct, two distinct spirals of metal will emerge from the bit flutes. If the pressure is too light, the bit won't cut properly, while excessive pressure will overheat the tip.

Decrease pressure as the bit breaks through the workpiece. If this isn't done, the bit may jam in the hole; if you're using a hand-held drill, it could be jerked out of your hands, especially when using larger size bits.

Once a pilot hole has been made, install the larger bit in the chuck and enlarge the hole. The second bit will follow the pilot hole - there's no need to attempt to guide it (if you do, the bit may break off). It is important, however, to hold the drill at the correct angle.

After the hole has been drilled to the correct size, remove the burrs left around the edges of the hole. This can be done with a small round file, or by chamfering the opening with a larger bit or a countersink **(see illustration)**. Use a drill bit that's several sizes larger than the hole and simply twist it around each opening by hand until any rough edges are removed.

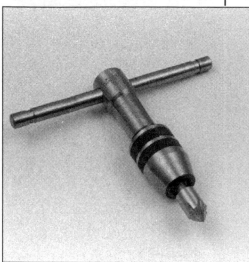

2.61 Use a large drill bit or a countersink mounted in a tap wrench to remove burrs from a hole after drilling or enlarging it

Enlarging and reshaping holes

The biggest practical size for bits used in a hand drill is about 1/2-inch. This is partly determined by the capacity of the chuck (although it's possible to buy larger drills with stepped shanks). The real limit is the difficulty of controlling large bits by hand; drills over 1/2-inch tend to be too much to handle in anything other than a drill press. If you have to make a larger hole, or if a shape other than round is involved, different techniques are required.

If a hole simply must be enlarged slightly, a round file is probably the best tool to use. If the hole must be very large, a hole saw will be needed, but they can only be used in sheet metal.

Large or irregular-shaped holes can also be made in sheet metal and other thin materials by drilling a series of small holes very close together. In this case the desired hole size and shape must be marked with a scribe. The next step depends on the size bit to be used; the idea is to drill a series of almost touching holes just inside the outline of the large hole. Center punch each location, then drill the small holes. A cold chisel can then be used to knock out the waste material at the center of the hole, which can then be filed to size. This is a time consuming process, but it's the only practical approach for the home shop. Success is dependent on accuracy when marking the hole shape and using the center punch.

High-speed grinders

A good die grinder **(see illustration)** will deburr blocks, radius piston domes and chamfer oil holes ten times as fast as you can do any of these jobs by hand. Used in conjunction with an abrasive disc, it can be used to de-glaze brake discs.

2.62 A good die grinder will deburr blocks, radius piston domes, chamfer oil holes and do a lot of other little jobs what would be tedious if done manually

2.63 Buy at least one fire extinguisher before you open shop - make sure it's rated for flammable liquid fires and KNOW HOW TO USE IT!

Safety items that should be in every shop

Fire extinguishers

Buy at least one fire extinguisher **(see illustration)** before doing any maintenance or repair procedures. Make sure it's rated for flammable liquid fires. Familiarize yourself with its use as soon as you buy it - don't wait until you need it to figure out how to use it. And be sure to have it checked and recharged at regular intervals. Refer to the safety tips at the end of this chapter for more information about the hazards of gasoline and other flammable liquids.

Gloves

If you're handling hot parts or metal parts with sharp edges, wear a pair of industrial work gloves to protect yourself from burns, cuts and splinters **(see illustration)**. Wear a pair of heavy duty rubber gloves (to protect your hands when you wash parts in solvent or brake cleaner.

Safety glasses or goggles

Never work on a bench or high-speed grinder without safety glasses **(see illustration)**. Don't take a chance on

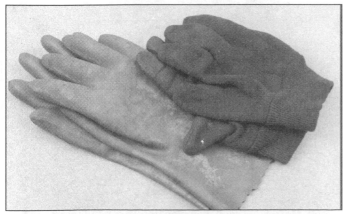

2.64 Get a pair of heavy work gloves for handling hot or sharp-edged objects and a pair of rubber gloves for washing parts with solvent or brake cleaner

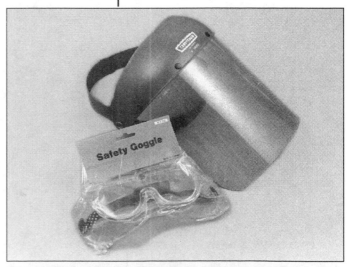

2.65 One of the most important items you'll need in the shop is a face shield or safety goggles, especially when you're hitting metal parts with a hammer, washing parts, bleeding brakes or grinding something on the bench grinder

2.66 Don't begin work on your brakes until you're wearing a filtering mask like this

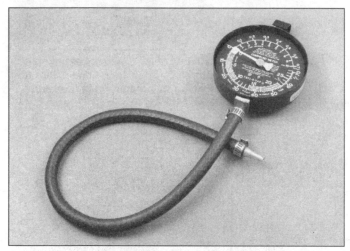

2.67 The vacuum gauge indicates intake manifold vacuum, in inches of mercury (in-Hg)

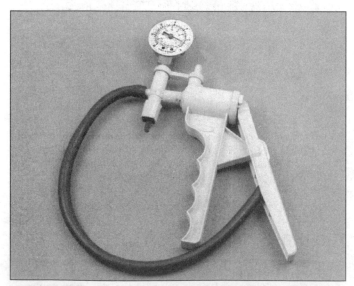

2.68 The vacuum/pressure pump can create a vacuum in a circuit, or pressurize it, to simulate the actual operating conditions

getting a metal sliver in your eye. It's also a good idea to wear safety glasses when you're washing parts and while bleeding brakes.

Filtering mask

The linings of most brake pads and shoes contain asbestos, which is extremely hazardous to your health. The dust deposited all over your brakes (and wheels) is made up of a high percentage of asbestos fibers. Be sure to always wear a filtering mask **(see illustration)** when working on or around your brakes - it'll greatly reduce the risk of inhaling asbestos fibers.

Special brake-related tools

Vacuum gauge

The vacuum gauge **(see illustration)** indicates intake manifold vacuum, in inches of mercury (in-Hg). Power brake boosters depend on a healthy vacuum to operate properly. A hard pedal on a vehicle equipped with a vacuum-actuated power booster would normally indicate a faulty booster, but before replacing anything it would be a good idea to check the intake manifold vacuum with a gauge like this.

Vacuum/pressure pump

The hand-operated vacuum/pressure pump **(see illustration)** can create a vacuum, or build up pressure, in a circuit to check components that are vacuum or pressure operated, such as power brake boosters.

Torx Bits

Many owners become surprised, sometimes even frustrated, when they tackle a service or repair procedure and

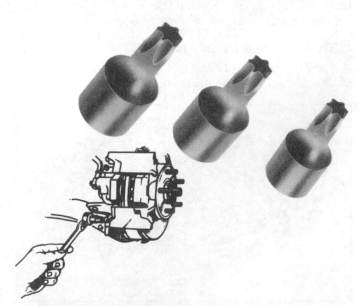

2.69 Many late model vehicles use Torx head fasteners to secure the brake calipers - always use the proper size bit when loosening or tightening these bolts (photo courtesy of the Lisle Corp.)

run into a fastener with a little six-pointed, star-shaped recess in its head. These are called Torx head bolts and are becoming increasingly popular with automobile manufacturers, especially for mounting brake calipers **(see illustration)**. It's very important to use the proper Torx bit on these fasteners. Never try to use an Allen wrench on a Torx fastener - you'll strip the head out and really create a problem!

Brake pad spreader

When replacing brake pads, the caliper piston(s) must be pushed back into the caliper to make room for the new pads. On some caliper designs (sliding calipers) this can be done before the caliper is removed, using an ordinary C-clamp. On other kinds of calipers (especially multiple-piston calipers) a brake pad spreader is very helpful in accomplishing this task **(see illustration)**. You just unbolt the caliper, slide it off the brake disc, insert this tool between the pads and turn the screw - the piston(s) will be forced to the bottom of the bore(s).

Universal disc brake caliper tool

This kit **(see illustration)** combines two kinds of caliper piston depressors. One serves the same purpose as the tool described above. The other is for use on rear disc brake calipers with threaded parking brake actuators in the caliper pistons. On these calipers the pistons can't be pushed into their bores - they have to be rotated back in.

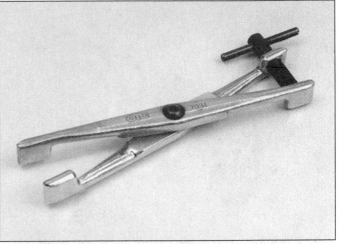

2.70 A pad spreader like this is used for pushing the caliper piston(s) back into the caliper bore(s) to make room for the new brake pads

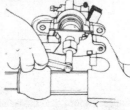

2.71 This universal brake caliper tool works on many different kinds of calipers and greatly facilitates brake pad replacement, especially on rear disc brakes (photo courtesy of the Lisle Corp.)

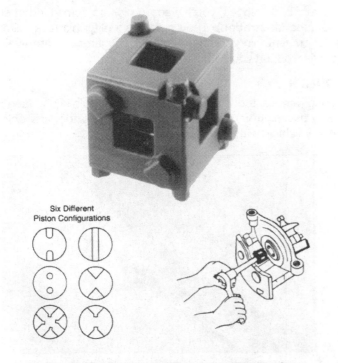

Six Different Piston Configurations

2.72 This six-in-one rear caliper piston retractor is an economical alternative to the universal disc brake caliper tool (shown in illustration 2.71) - it also simplifies the task of turning the piston on actuator-screw rear calipers (photo courtesy of the Lisle Corp.)

Rear disc brake piston tool

Here's another tool for turning the rear brake caliper pistons back into their bores **(see illustration)**. This one offers six different lug arrangements - one to fit almost any vehicle using this design rear caliper.

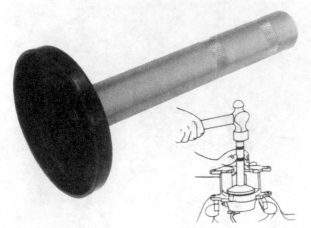

2.73 This tool will help you cleanly install dust boots that have a steel case (photo courtesy of the Lisle Corp.)

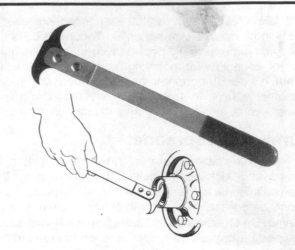

2.74 A seal removal tool like this will facilitate the removal of stubborn seals (photo courtesy of the Lisle Corp.)

Brake caliper dust boot installer

A dust boot installer **(see illustration)** isn't an absolute necessity for overhauling brake calipers, but it eases the dust boot installation procedure and reduces the chance of damaging the boot or caliper. It's for use only on dust boots with a rigid outer diameter.

Seal removal tool

If you remove your brake discs or drums for machining, and they are integral with the hubs, you'll have to clean and repack the wheel bearings. On some kinds of hubs the seals are difficult to remove, in which case a seal removal tool **(see illustration)** will come in handy.

Brake spring tools

If you're replacing drum brake shoes, you'll want to pick up some kind of spring removal/installation tool **(see illustration)**. They greatly simplify this task, eliminate frustration and can even prevent injury. Many different configurations are available. Be sure to get one that will work on your vehicle.

Brake spring pliers

Another kind of spring removal tool combines three spring tools in one **(see illustration)**. This is probably the most versatile type of brake spring tool available and will work on just about any vehicle with drum brakes.

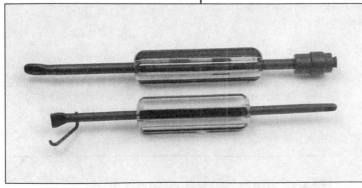

2.75 The only way to correctly and safely remove or install brake shoe retractor springs is to use a spring tool

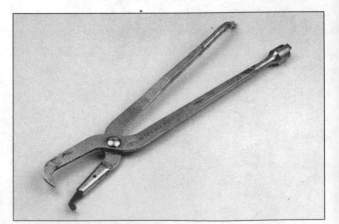

2.76 Brake spring pliers like this are able to handle just about any spring setup you're likely to encounter and offer lots of leverage for removing and installing those really strong springs

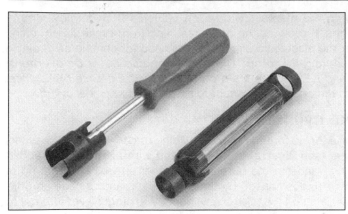

2.77 Pliers can be used to remove brake shoe hold-down springs, but if you don't want to fumble around and waste time, do it the right way and use a hold-down spring tool

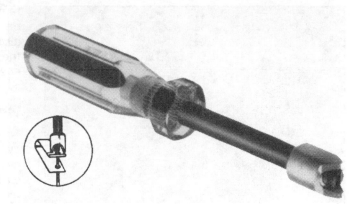

2.78 If the drum brakes on the vehicle you're working on use hold-down clips instead of springs, use a tool like this (photo courtesy of the Lisle Corp.)

Hold-down spring tool

This tool, available in different sizes and designs, eases the removal of the brake shoe hold-down springs and cups **(see illustration)**. Just set the tool on the cup, push down and turn 90-degrees and ease up, and the cup and spring are removed.

Hold-down clip tool

Similar to the hold-down spring tool, this device simplifies the removal of brake shoe hold-down clips, used mostly on imported cars and trucks **(see illustration)**.

Brake shoe adjuster tool

These tools are available in a few different shapes to work on many different types of drum brakes, but they're all used for the same thing - to reach through the hole in the backing plate (or drum on some models) and turn the adjuster (or "star") wheel to expand or retract the brake shoes **(see illustration)**.

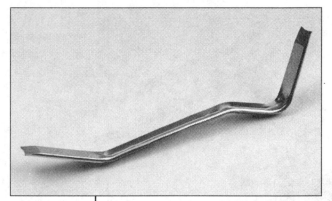

2.79 Here's the correct tool to use for adjusting drum brakes (a screwdriver will work, but the way this tool is angled makes turning the star wheel much easier)

Wheel cylinder clip expander

This is a tool not everyone will need - only people who want to easily remove the wheel cylinders on General Motors vehicles that use clips to retain the cylinder to the backing plate **(see illustration)**.

Hose pinchers

Hose pinchers, or clamps **(see illustration)**, are very useful tools, provided they are used correctly. Placed over a hose and tightened, they will prevent fluid from flowing through the hose. They're great for using on fluid feed hoses when

2.80 This tool spreads the ears of the wheel cylinder retaining clips on some GM vehicles

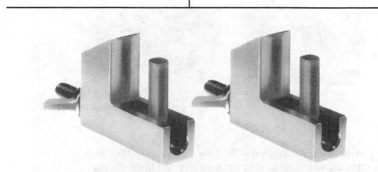

2.81 Hose pinchers are useful for clamping off the hoses to the remote brake fluid reservoir (on vehicles so equipped) when removing the master cylinder (photo courtesy of the Lisle Corp.)

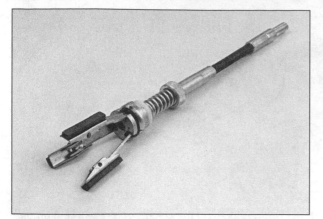

2.82 You'll need a brake cylinder hone like this if you intend to rebuild cast iron master or wheel cylinders

removing the master cylinder on vehicles that employ remote reservoirs. It reduces the mess resulting from brake fluid running all over the place, and eliminates the need for completely draining the reservoir. **Warning:** *These tools should never be overtightened, and NEVER use them on high-pressure flexible brake lines, such as the ones that connect to the calipers or wheel cylinders.*

Brake cylinder hone

This tool is for use when rebuilding master cylinders or wheel cylinders **(see illustration)**. Lubricated with plenty of brake fluid and run in-and-out the bore at the proper speed, it will get rid of glaze and ridges caused by the piston cups and provide an optimum sealing surface for the new cups. **Warning:** *Never attempt to hone aluminum cylinders!*

Brake bleeder wrenches

These special wrenches **(see illustration)** fit perfectly on bleeder valves and are offset so as not to place undue strain on fragile bleeders. They are also quite long for their opening sizes to provide plenty of leverage.

Vacuum pump brake bleeding kit

When it's time to bleed the brakes and there aren't any friends around to pump the brake pedal for you, this device **(see illustration)** will allow you to finish the job. Just hook it up to the bleeder valve (follow the tool manufacturer's instructions), open the bleeder valve and operate the pump. All air will be drawn from the portion of the system you're bleeding.

One-man brake bleeder

This tool **(see illustration)** will also let you complete the brake bleeding procedure by yourself. It's basically a container and hose with a one-way check valve in it. This allows you to open the bleeder valve, pump the pedal to purge the system of air without any air or old fluid being drawn back into the system.

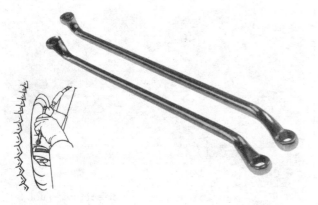

2.83 These long bleeder wrenches have six-point box ends - the perfect combination for loosening bleeder valves that are rusty and stuck (photo courtesy of the Lisle Corp.)

2.84 This brake bleeding kit will allow you to bleed the brake system by yourself (photo courtesy of the Lisle Corp.)

2.85 The one-man brake bleeder helps you achieve the same objective as the vacuum pump bleeder kit, but it's considerably less expensive (the only drawback is that you'll have to get out from underneath the car, hop into the driver's seat, pump the pedal a few times, then get out and tighten the bleeder valve) (photo courtesy of the Lisle Corp.)

Micrometers

The most accurate way to measure the thickness of a brake disc is with a micrometer. When doing a brake job or just checking your brakes, you'll have to confirm that the discs haven't worn down to their minimum thickness, or that there's enough material left to allow machining. Most discs are less than one-inch in diameter. If this is the case with your vehicle, your tool of choice should be the trusty one-inch outside micrometer **(see illustration)**.

Insist on accuracy to within one ten-thousandths of an inch (0.0001-inch) when you shop for a micrometer. You'll probably never need that kind of precision, but the extra decimal place will help you decide which way to round off a close measurement.

High-quality micrometers have a range of one inch. If you plan to work on a wide variety of vehicles, or if your vehicle is equipped with large, ventilated brake discs, you'll need a one-to-two inch micrometer.

Standard micrometers will work, but a special brake disc micrometer with a pointed anvil is recommended. It will take into account any deep scratches or grooves that a normal micrometer wouldn't be able to measure.

2.86 The one-inch micrometer is an essential precision measuring device for determining the thickness of brake discs. If your discs are thicker than one inch, you'll need a one-to-two inch micrometer

How to read a micrometer

The outside micrometer is without a doubt the most widely used precision measuring tool. It can be used to make a variety of highly accurate measurements without much possibility of error through misreading, a problem associated with other measuring instruments, such as vernier calipers.

Like any slide caliper, the outside micrometer uses the "double contact" of its spindle and anvil **(see illustration)** touching the object to be measured to determine that object's dimensions. Unlike a caliper, however, the micrometer also features a unique precision screw adjustment which can be read with a great deal more accuracy than calipers.

Why is this screw adjustment so accurate? Because years ago toolmakers discovered that a screw with 40 precision machined threads to the inch will advance one-fortieth (0.025) of an inch with each complete turn. The screw threads on the spindle revolve inside a fixed nut concealed by a sleeve.

On a one-inch micrometer, this sleeve is engraved longitudinally with exactly 40 lines to the inch, to correspond with the number of threads on the spindle. Every fourth line is made longer and is numbered one-tenth inch, two-tenths, etc. The other lines are often staggered to make them easier to read.

The thimble (the barrel which moves up and down the sleeve as it rotates) is divided into 25 divisions around the circumference of its beveled edge and is numbered from zero to 25. Close the micrometer spindle till it touches the anvil: You should see nothing but the zero line on the sleeve next to the beveled edge of the thimble. And the zero line of the thimble should be aligned with the horizontal (or axial) line on the sleeve. Remember: Each full revolution of the spindle from zero to zero advances or retracts the spindle one-fortieth or 0.025-inch. Therefore, if you rotate the thimble from zero on the beveled edge to the first

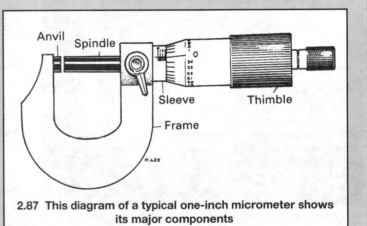

2.87 This diagram of a typical one-inch micrometer shows its major components

Anvil · Spindle · Sleeve · Thimble · Frame

graduation, you will move the spindle 1/25th of 1/40th, or 1/25th of 25/1000ths, which equals 1/1000th, or 0.001-inch.

Remember: Each numbered graduation on the sleeve represents 0.1-inch, each of the other sleeve graduations represents 0.025-inch and each graduation on the thimble represents 0.001-inch. Remember those three and you're halfway there.

For example: Suppose the 4 line is visible on the sleeve. This represents 0.400-inch. Then suppose there are an additional three lines (the short ones without numbers) showing. These marks are worth 0.025-inch each, or 0.075-inch. Finally, there are also two marks on the beveled edge of the thimble beyond the zero mark, each good for 0.001-inch, or a total of 0.002-inch. Add it all up and you get 0.400 plus 0.075 plus 0.002, which equals 0.477-inch.

Some beginners use a "dollars, quarters and cents" analogy to simplify reading a micrometer. Add up the bucks and change, then put a decimal point instead of a dollar sign in front of the sum!

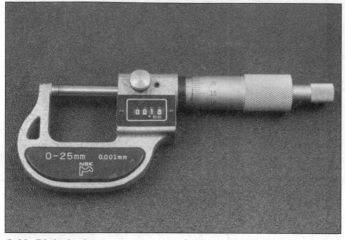

2.88 Digital micrometers are easier to read than conventional micrometers, are just as accurate and are finally starting to become affordable

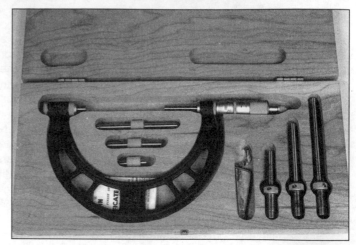

2.89 Avoid micrometer "sets" with interchangeable anvils - they're awkward to use when measuring little parts and changing the anvils is a hassle

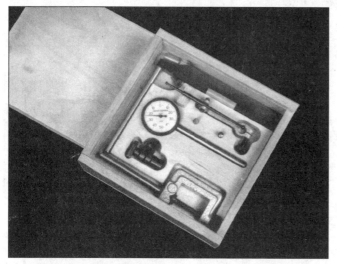

2.90 The dial indicator is the only way to measure brake disc runout

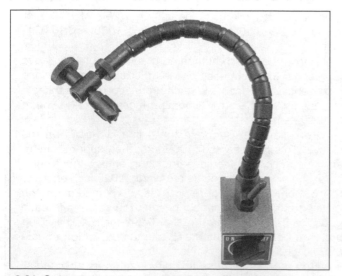

2.91 Get an adjustable, flexible fixture like this one, and a magnetic base, to ensure maximum versatility from your dial indicator

Digital micrometers **(see illustration)** are easier to read than conventional micrometers, are just as accurate and are finally starting to become affordable. If you're uncomfortable reading a conventional micrometer (see sidebar), then get a digital.

Unless you're not going to use them very often, stay away from micrometers with interchangeable anvils **(see illustration)**. In theory, one of these beauties can do the work of five or six single-range micrometers. The trouble is, they're awkward to use when measuring little parts, and changing the anvils is a hassle.

Dial indicators

The dial indicator **(see illustration)** is another measuring mainstay. You'll use this tool for measuring brake disc runout (warpage). Make sure the dial indicator you buy is graduated in 0.001-inch increments.

Buy a dial indicator set that includes a flexible fixture and a magnetic stand **(see illustration)**. If the model you buy doesn't have a magnetic base, buy one separately. Make sure the magnet is plenty strong. If a weak magnet comes loose and the dial indicator takes a tumble on a concrete floor, you can kiss it good-bye. Make sure the arm that attaches the dial indicator to the flexible fixture is sturdy and the locking clamps are easy to operate.

Calipers

Vernier calipers **(see illustration)** aren't quite as accurate as a micrometer, but they're handy for quick measurements and they're relatively inexpensive. Most calipers have inside and outside jaws, so you can measure the inside diameter of a hole, or the outside diameter of a part.

Better-quality calipers have a dust shield over the geared rack that turns the dial to prevent small metal particles from jamming the mechanism. Make sure there's no play in the moveable jaw. To check, put a thin piece of metal between the jaws and measure its thickness with the metal close to the rack, then out near the tips of the jaws. Compare your two measurements. If they vary by more than 0.001-inch, look at another caliper - the jaw mechanism is deflecting.

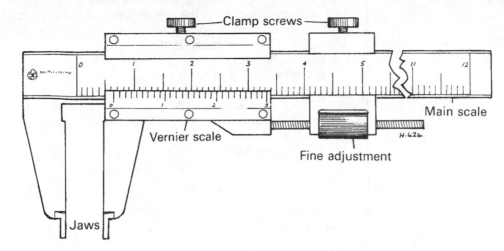

2.92 Vernier calipers aren't quite as accurate as micrometers, but they're handy for quick measurements and relatively inexpensive, and because they've got jaws that can measure internal and external dimensions, they're versatile

If your eyes are going bad, or already are bad, vernier calipers can be difficult to read. Dial calipers **(see illustration)** are a better choice. Dial calipers combine the measuring capabilities of vernier calipers with the convenience of dial indicators. Because they're much easier to read quickly than vernier calipers, they're ideal for taking quick measurements when absolute accuracy isn't necessary. Like conventional vernier calipers, they have both inside and outside jaws which allow you to quickly determine the diameter of a hole or a part. Get a six-inch dial caliper, graduated in 0.001-inch increments.

The latest calipers **(see illustration)** have a digital LCD display that indicates both inch and metric dimensions. If you can afford one of these, it's the hot setup.

How to read a vernier caliper

On the lower half of the main beam, each inch is divided into ten numbered increments, or tenths (0.100-inch, 0.200-inch, etc.). Each tenth is divided into four increments of 0.025-inch each. The vernier scale has 25 increments, each representing a thousandth (0.001) of an inch.

First read the number of inches, then read the number of tenths. Add to this

2.93 Dial calipers are a lot easier to read than conventional vernier calipers, particularly if your eyesight isn't as good as it used to be!

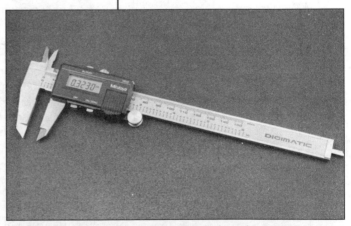

2.94 The latest calipers have a digital readout that is even easier to read than a dial caliper - another advantage of digital calipers is that they have a small microchip that allows them to convert instantaneously from inch to metric dimensions

0.025-inch for each additional graduation. Using the English vernier scale, determine which graduation of the vernier lines up exactly with a graduation on the main beam. This vernier graduation is the number of thousandths which are to be added to the previous readings.

For example, let's say:

1) The number of inches is zero, or 0.000-inch;

2) The number of tenths is 4, or 0.400-inch;

3) The number of 0.025's is 2, or 0.050-inch; and

4) The vernier graduation which lines up with a graduation on the main beam is 15, or 0.015-inch.

5) Add them up:
	0.000
	0.400
	0.050
	0.015

6) And you get: 0.46-inch

That's all there is to it!

Storage and care of tools

Good tools are expensive, so treat them well. After you're through with your tools, wipe off any dirt, grease or metal chips and put them away. Don't leave tools lying around in the work area. General purpose hand tools - screwdrivers, pliers, wrenches and sockets - can be hung on a wall panel or stored in a tool box. Store precision measuring instruments, gauges, meters, etc. in a tool box to protect them from dust, dirt, metal chips and humidity.

Fasteners

Fasteners - nuts, bolts, studs and screws - hold parts together. Keep the following things in mind when working with fasteners: All threaded fasteners should be clean and straight, with good threads and unrounded corners on the hex head (where the wrench fits). Make it a habit to replace all damaged nuts and bolts with new ones. Almost all fasteners have a locking device of some type, either a lockwasher, locknut, locking tab or thread adhesive. Don't reuse special locknuts with nylon or fiber inserts. Once they're removed, they lose their locking ability. Install new locknuts.

Flat washers and lockwashers, when removed from an assembly, should always be replaced exactly as removed. Replace any damaged washers with new ones. Never use a lockwasher on any soft metal surface (such as aluminum), thin sheet metal or plastic.

Apply penetrant to rusted nuts and bolts to loosen them up and prevent breakage. Some mechanics use turpentine in a spout-type oil can, which works quite well. After applying the rust penetrant, let it work for a few minutes before trying to loosen the nut or bolt. Badly rusted fasteners may have to be chiseled or sawed off or removed with a special nut breaker, available at tool stores.

If a bolt or stud breaks off in an assembly, it can be drilled and removed with a special tool commonly available for this purpose. Most automotive machine shops can perform this task, as well as other repair procedures, such as the repair of threaded holes that have been stripped out.

Fastener Sizes

For a number of reasons, automobile manufacturers are making wider and wider use of metric fasteners. Therefore, it's important to be able to tell the difference between standard (sometimes called USS or SAE) and metric hardware, since they cannot be interchanged.

All bolts, whether standard or metric, are sized in accordance with their diameter, thread pitch and length **(see illustration).** For example, a standard 1/2-13 x 1 bolt is 1/2 inch in diameter, has 13 threads per inch and is 1 inch long. An

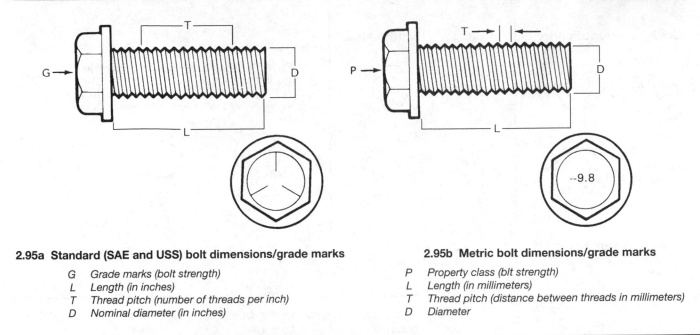

2.95a Standard (SAE and USS) bolt dimensions/grade marks

G   Grade marks (bolt strength)
L   Length (in inches)
T   Thread pitch (number of threads per inch)
D   Nominal diameter (in inches)

2.95b Metric bolt dimensions/grade marks

P   Property class (blt strength)
L   Length (in millimeters)
T   Thread pitch (distance between threads in millimeters)
D   Diameter

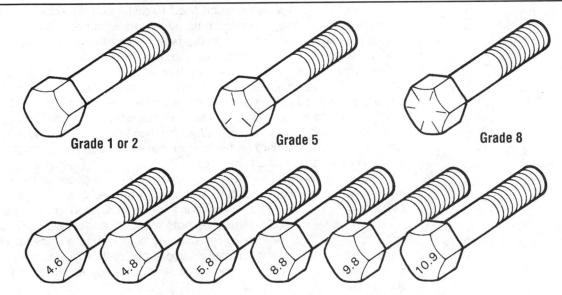

2.96 Bolt strength markings (top - standard/ SAE; bottom - metric)

M12-1.75x25 metric bolt is 12mm in diameter, has a thread pitch of 1.75 mm (the distance between threads) and is 25 mm long. The two bolts are nearly identical, and easily confused, but they are not interchangeable.

In addition to the differences in diameter, thread pitch and length, metric and standard bolts can also be distinguished by examining the bolt heads. The distance across the flats on a standard bolt head is measured in inches; the same dimension on a metric bolt or nut is sized in millimeters. So don't use a standard wrench on a metric bolt, or vice versa.

Most standard bolts also have slashes radiating out from the center of the head **(see illustration)** to denote the grade or strength of the bolt, which is an indication of the amount of torque that can be applied to it. The greater the number of slashes, the greater the strength of the bolt. Grades 0 through 5 are commonly used on automobiles. Metric bolts have a property class (grade) number, rather than a slash, molded into their heads to indicate bolt strength. In this case, the

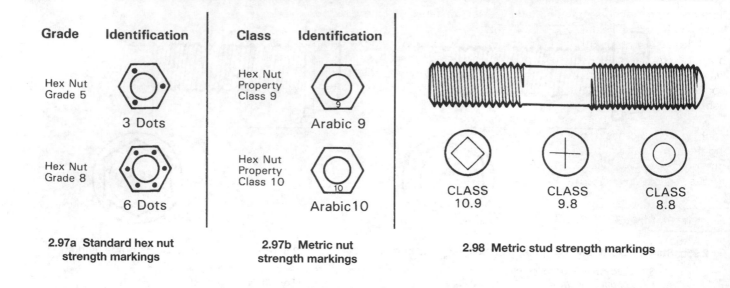

Grade	Identification
Hex Nut Grade 5	3 Dots
Hex Nut Grade 8	6 Dots

2.97a Standard hex nut strength markings

Class	Identification
Hex Nut Property Class 9	Arabic 9
Hex Nut Property Class 10	Arabic 10

2.97b Metric nut strength markings

CLASS 10.9   CLASS 9.8   CLASS 8.8

2.98 Metric stud strength markings

higher the number, the stronger the bolt. Property class numbers 8.8, 9.8 and 10.9 are commonly used on automobiles.

Strength markings can also be used to distinguish standard hex nuts from metric hex nuts. Many standard nuts have dots stamped into one side, while metric nuts are marked with a number **(see illustrations).** The greater the number of dots, or the higher the number, the greater the strength of the nut.

Metric studs are also marked on their ends **(see illustration)** according to property class (grade). Larger studs are numbered (the same as metric bolts), while smaller studs carry a geometric code to denote grade.

It should be noted that many fasteners, especially Grades 0 through 2, have no distinguishing marks on them. When such is the case, the only way to determine whether it's standard or metric is to measure the thread pitch or compare it to a known fastener of the same size.

Standard fasteners are often referred to as SAE, as opposed to metric. However, it should be noted that SAE technically refers to a non-metric fine thread fastener only. Coarse thread non-metric fasteners are referred to as US sizes.

Since fasteners of the same size (both standard and metric) may have different strength ratings, be sure to reinstall any bolts, studs or nuts removed from your vehicle in their original locations. Also, when replacing a fastener with a new one, make sure that the new one has a strength rating equal to or greater than the original.

Tightening sequences and procedures

Most threaded fasteners should be tightened to a specific torque value **(see charts on opposite page). Warning:** *These are general torque specifications for conventional fasteners. Some fasteners used in the brake system (most notably the disc brake caliper bolts or guide pins) are specifically designed for their purposes and would not fall into the categories listed below.* Torque is the twisting force applied to a threaded component such as a nut or bolt. Overtightening the fastener can weaken it and cause it to break, while undertightening can cause it to eventually come loose. Bolts, screws and studs, depending on the material they are made of and their thread diameters, have specific torque values, many of which are noted in the Specifications at the beginning of each Chapter. Be sure to follow the torque recommendations closely. For fasteners not assigned a specific torque, a general torque value chart is presented here as a guide. These torque values are for dry (unlubricated) fasteners threaded into steel or cast iron (not aluminum). As was previously mentioned, the size and grade of a fastener determine the amount of torque that can safely be applied to it. The figures listed here are approximate for Grade 2 and Grade 3 fasteners. Higher grades can tolerate higher torque values.

If fasteners are laid out in a pattern - such as cylinder head bolts, oil pan bolts, differential cover bolts, etc. - loosen and tighten them in sequence to avoid warping the component. Where it matters, we'll show you this sequence. If a specific pattern isn't that important, the following rule-of thumb guide will prevent warping.

First, install the bolts or nuts finger-tight. Then tighten them one full turn each, in a criss-cross or diagonal pattern. Then return to the first one and, following the same pattern, tighten them all one-half turn. Finally, tighten each of them one-quarter turn at a time until each fastener has been tightened to the proper torque. To loosen and remove the fasteners, reverse this procedure.

Metric thread sizes	Ft-lbs	Nm
M-6	6 to 9	9 to 12
M-8	14 to 21	19 to 28
M-10	28 to 40	38 to 54
M-12	50 to 71	68 to 96
M-14	80 to 140	109 to 154

Pipe thread sizes		
1/8	5 to 8	7 to 10
1/4	12 to 18	17 to 24
3/8	22 to 33	30 to 44
1/2	25 to 35	34 to 47

U.S. thread sizes		
1/4 – 20	6 to 9	9 to 12
5/16 – 18	12 to 18	17 to 24
5/16 – 24	14 to 20	19 to 27
3/8 – 16	22 to 32	30 to 43
3/8 – 24	27 to 38	37 to 51
7/16 – 14	40 to 55	55 to 74
7/16 – 20	40 to 60	55 to 81
1/2 – 13	55 to 80	75 to 108

How to remove broken fasteners

Sooner or later, you're going to break off a bolt inside its threaded hole. There are several ways to remove it. Before you buy an expensive extractor set, try some of the following cheaper methods first.

First, regardless of which of the following methods you use, be sure to use penetrating oil. Penetrating oil is a special light oil with excellent penetrating power for freeing dirty and rusty fasteners. But it also works well on tightly torqued broken fasteners.

If enough of the fastener protrudes from its hole and if it isn't torqued down too tightly - you can often remove it with vise-grips or a small pipe wrench. If that doesn't work, or if the fastener doesn't provide sufficient

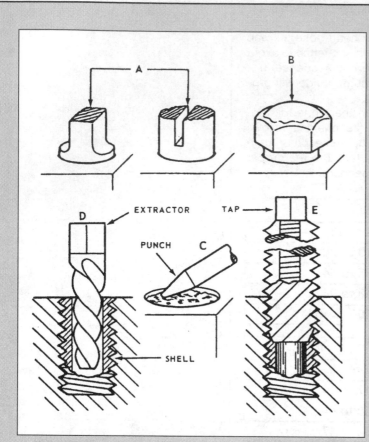

2.99 There are several ways to remove a broken fastener

A	File it flat or slot it
B	Weld on a nut
C	Use a punch to unscrew it
D	Use a screw extractor (like an E-Z-Out)
E	Use a tap to remove the shell

purchase for pliers or a wrench, try filing it down to take a wrench, or cut a slot in it to accept a screwdriver **(see illustration)**. If you still can't get it off - and you know how to weld - try welding a flat piece of steel, or a nut, to the top of the broken fastener. If the fastener is broken off flush with - or below - the top of its hole, try tapping it out with a small, sharp punch. If that doesn't work, try drilling out the broken fastener with a bit only slightly smaller than the inside diameter of the hole. For example, if the hole is 1/2-inch in diameter, use a 15/32-inch drill bit. This leaves a shell which you can pick out with a sharp chisel.

If THAT doesn't work, you'll have to resort to some form of screw extractor, such as E-Z-Out **(see illustration)**. Screw extractors are sold in sets which can remove anything from 1/4-inch to 1-inch bolts or studs. Most extractors are fluted and tapered high-grade steel. To use a screw extractor, drill a hole slightly smaller than the O.D. of the extractor you're going to use (Extractor sets include the manufacturer's recommendations for what size drill bit to use with each extractor size). Then screw in the extractor **(see illustration)** and back it - and the broken fastener - out. Extractors are reverse-threaded, so they won't unscrew when you back them out.

A word to the wise: Even though an E-Z-Out will usually save your bacon, it can cause even more grief if you're careless or sloppy. Drilling the hole for the extractor off-center, or using too small, or too big, a bit for the size of the fastener you're removing will only make things worse. So be careful!

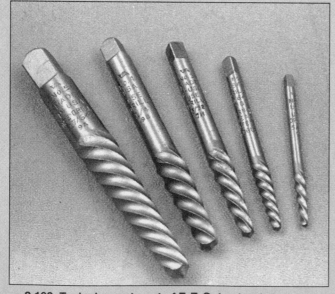

2.100 Typical assortment of E-Z-Out extractors

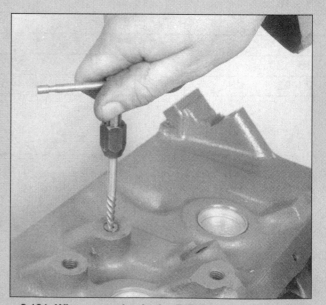

2.101 When screwing in the E-Z-Out, make sure it's centered properly

How to repair broken threads

Warning: *Never attempt to repair the threads of the caliper mounting bolt holes, torque plate mounting bolt holes, stripped-out holes in two-piece calipers, wheel cylinder mounting bolt holes or any other critical brake system component. Instead, replace the part with a new one.*

Sometimes, the internal threads of a nut or bolt hole can become stripped, usually from overtightening. Stripping threads is an all-too common occurrence, especially when working with aluminum parts, because aluminum is so soft that it easily strips out. Overtightened spark plugs are another common cause of stripped threads.

Usually, external or internal threads are only partially stripped. After they've been cleaned up with a tap or die, they'll still work. Sometimes, however, threads are badly damaged. When this happens, you've got three choices:

1) Drill and tap the hole to the next suitable oversize and install a larger diameter bolt, screw or stud.

2) Drill and tap the hole to accept a threaded plug, then drill and tap the plug to the original screw size. You can also buy a plug already threaded to the original size. Then you simply drill a hole to the specified size, then run the threaded plug into the hole with a bolt and jam nut. Once the plug is fully seated, remove the jam nut and bolt.

3) The third method uses a patented thread repair

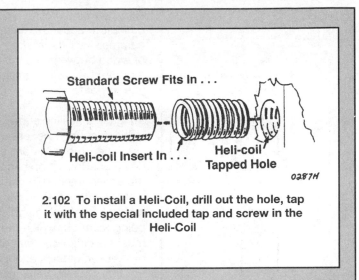

2.102 To install a Heli-Coil, drill out the hole, tap it with the special included tap and screw in the Heli-Coil

kit like Heli-Coil or Slimsert. These easy-to-use kits are designed to repair damaged threads in spark plug holes, straight-through holes and blind holes. Both are available as kits which can handle a variety of sizes and thread patterns. Drill the hole, then tap it with the special included tap. Install the Heli-Coil **(see illustration)** and the hole is back to its original diameter and thread pitch.

Regardless of which method you use, be sure to proceed calmly and carefully. A little impatience or carelessness during one of these relatively simple procedures can ruin your whole day's work and cost you a bundle if you wreck an expensive head or block.

Component disassembly

Disassemble components carefully to help ensure that the parts go back together properly. Note the sequence in which parts are removed. Make note of special characteristics or marks on parts that can be installed more than one way, such as a grooved thrust washer on a shaft. It's a good idea to lay the disassembled parts out on a clean surface in the order in which you removed them. It may also be helpful to make sketches or take instant photos of components before removal.

When you remove fasteners from a component, keep track of their locations. Thread a bolt back into a part, or put the washers and nut back on a stud, to prevent mix-ups later. If that isn't practical, put fasteners in a fishing tackle box or a series of small boxes. A cupcake or muffin tin, or an egg crate, is ideal for this purpose - each cavity can hold the bolts and nuts from a particular area (i.e. oil pan bolts, valve cover bolts, engine mount bolts, etc.). A pan of this type is helpful when working on assemblies with very small parts, such as the carburetor or valve train. Mark each cavity with paint or tape to identify the contents.

When you unplug the connector(s) between two wire harnesses, or even two wires, it's a good idea to identify the two halves with numbered pieces of masking tape - or a pair of matching pieces of colored electrical tape - so they can be easily reconnected.

Gasket sealing surfaces

Gaskets seal the mating surfaces between two parts to prevent lubricants, fluids, vacuum or pressure from leaking out between them. Gaskets are often coated with a liquid or paste-type gasket sealing compound before assembly.

Age, heat and pressure can cause the two parts to stick together so tightly that they're difficult to separate. Often, you can loosen the assembly by striking it with a soft-face hammer near the mating surfaces. You can use a regular hammer if you place a block of wood between the hammer and the part, but don't hammer on cast or delicate parts that can be easily damaged. When a part refuses to come off, look for a fastener that you forgot to remove.

Don't use a screwdriver or prybar to pry apart an assembly. It can easily damage the gasket sealing surfaces of the parts, which must be smooth to seal properly. If prying is absolutely necessary, use an old broom handle or a section of hard wood dowel.

Once the parts are separated, carefully scrape off the old gasket and clean the gasket surface. You can also remove some gaskets with a wire brush. If some gasket material refused to come off, soak it with rust penetrant or treat it with a special chemical to soften it, then scrape it off. You can fashion a scraper from a piece of copper tubing by flattening and sharpening one end. Copper is usually softer than the surface being scraped, which reduces the likelihood of gouging the part. The mating surfaces must be clean and smooth when you're done. If the gasket surface is gouged, use a gasket sealer thick enough to fill the scratches when you reassemble the components. For most applications, use a non-drying (or semi-drying) gasket sealer.

Hose removal tips

Warning: *If the vehicle is equipped with air conditioning, do not disconnect any of the A/C hoses without first having the system depressurized by a dealer service department or a service station (see the Haynes Automotive Heating and Air Conditioning Manual).*

The same precautions that apply to gasket removal also apply to hoses. Avoid scratching or gouging the surface against which the hose mates, or the connection may leak. Take, for example, radiator hoses. Because of various chemical reactions, the rubber in radiator hoses can bond itself to the metal spigot over which the hose fits. To remove a hose, first loosen the hose clamps that secure it to the spigot. Then, with slip-joint pliers, grab the hose at the clamp and rotate it around the spigot. Work it back and forth until it is completely free, then pull it off. Silicone or other lubricants will ease removal if they can be applied between the hose and the outside of the spigot. Apply the same lubricant to the inside of the hose and the outside of the spigot to simplify installation. Snap-On and Mac Tools sell hose removal tools - they look like bent ice picks - which can be inserted between the spigot and the radiator hose to break the seal between rubber and metal.

As a last resort - or if you're planning to replace the hose anyway - slit the rubber with a knife and peel the hose from the spigot. Make sure you don't damage the metal connection.

If a hose clamp is broken or damaged, don't reuse it. Wire-type clamps usually weaken with age, so it's a good idea to replace them with screw-type clamps whenever a hose is removed.

Automotive chemicals and lubricants
Cleaners

A wide variety of automotive chemicals and lubricants - ranging from cleaning solvents and degreasers to lubricants and protective sprays for rubber, plastic and vinyl - is available.

Brake system cleaner

Brake system cleaner removes grease and brake fluid from brake parts like disc brake rotors, where a spotless surfaces is essential. It leaves no residue and often eliminates brake squeal caused by brake dust or other contaminants. Because it leaves no residue, brake cleaner is often used for cleaning other parts as well.

Carburetor and choke cleaner

Carburetor and choke cleaner is a strong solvent for gum, varnish and carbon. Most carburetor cleaners leave a dry-type lubricant film which will not harden or gum up. So don't use carb cleaner on electrical components.

Degreasers

Degreasers are heavy-duty solvents used to remove grease from the outside of the engine and from chassis components. They're usually sprayed or brushed on. Depending on the type, they're rinsed off either with water or solvent. **Warning:** *Never wash brake system components with petroleum-based solvents.*

Demoisturants

Demoisturants remove water and moisture from electrical components such as alternators, voltage regulators, electrical connectors and fuse blocks. They are non-conductive, non-corrosive and non-flammable.

Electrical cleaner

Electrical cleaner removes oxidation, corrosion and carbon deposits from electrical contacts, restoring full current flow. It can also be used to clean spark plugs, carburetor jets, voltage regulators and other parts where an oil-free surface is necessary.

Lubricants

Assembly lube

Assembly lube is a special extreme pressure lubricant, usually containing moly, used to lubricate high-load parts (such as main and rod bearings and cam lobes) for initial start-up of a new engine. The assembly lube lubricates the parts without being squeezed out or washed away until the engine oiling system begins to function.

Graphite lubricants

Graphite lubricants are used where oils cannot be used due to contamination problems, such as in locks. The dry graphite will lubricate metal parts while remaining uncontaminated by dirt, water, oil or acids. It is electrically conductive and will not foul electrical contacts in locks such as the ignition switch.

Heat-sink grease

Heat-sink grease is a special electrically non-conductive grease that is used for mounting electronic ignition modules where it is essential that heat is transferred away from the module.

Moly penetrants

Moly penetrants loosen and lubricate frozen, rusted and corroded fasteners and prevent future rusting or freezing.

Motor oil

Motor oil is the lubricant formulated for use in engines. It normally contains a wide variety of additives to prevent corrosion and reduce foaming and wear. Motor oil comes in various weights (viscosity ratings) from 5 to 80. The recommended weight of the oil depends on the season, temperature and the demands on the engine. Light oil is used in cold climates and under light load conditions. Heavy oil is used in hot climates and where high loads are encountered. Multi-viscosity oils are designed to have characteristics of both light and heavy oils and are available in a number of weights from 5W-20 to 20W-50. Some home mechanics use motor oil as an assembly lube, but we don't recommend it, because motor oil has a relatively thin viscosity, which means it will slide off the parts long before the engine is fired up.

Silicone lubricants

Silicone lubricants are used to protect rubber, plastic, vinyl and nylon parts.

Wheel bearing grease

Wheel bearing grease is a heavy grease that can withstand high loads and friction, such as wheel bearings, balljoints, tie-rod ends and universal joints. It's also sticky enough to hold parts like the keepers for the valve spring retainers in place on the valve stem when you're installing the springs.

White grease

White grease is a heavy grease for metal-to-metal applications where water is present. It stays soft under both low and high temperatures (usually from -100 to +190-degrees F), and won't wash off or dilute when exposed to water. Another good "glue" for holding parts in place during assembly.

Sealants

Anaerobic sealant

Anaerobic sealant is much like RTV in that it can be used either to seal gaskets or to form gaskets by itself. It remains flexible, is solvent resistant and fills surface imperfections. The difference between an anaerobic sealant and an RTV-type sealant is in the curing. RTV cures when exposed to air, while an anaerobic sealant cures only in the absence of air. This means that an anaerobic sealant cures only after the assembly of parts, sealing them together.

RTV sealant

RTV sealant is one of the most widely used gasket compounds. Made from silicone, RTV is air curing, it seals, bonds, waterproofs, fills surface irregularities, remains flexible, doesn't shrink, is relatively easy to remove, and is used as a supplementary sealer with almost all low and medium temperature gaskets.

Thread and pipe sealant

Thread and pipe sealant is used for sealing hydraulic and pneumatic fittings and vacuum lines. It is usually made from a teflon compound, and comes in a spray, a paint-on liquid and as a wrap-around tape.

Chemicals

Anaerobic locking compounds

Anaerobic locking compounds are used to keep fasteners from vibrating or working loose and cure only after installation, in the absence of air. Medium strength locking compound is used for small nuts, bolts and screws that may be removed later. High-strength locking compound is for large nuts, bolts and studs which aren't removed on a regular basis.

Anti-seize compound

Anti-seize compound prevents seizing, galling, cold welding, rust and corrosion in fasteners. High-temperature anti-seize, usually made with copper and graphite lubricants, is used for exhaust system and exhaust manifold bolts.

Gas additives

Gas additives perform several functions, depending on their chemical makeup. They usually contain solvents that help dissolve gum and varnish that build up on carburetor, fuel injection and intake parts. They also serve to break down carbon deposits that form on the inside surfaces of the combustion chambers. Some additives contain upper cylinder lubricants for valves and piston rings, and others contain chemicals to remove condensation from the gas tank.

Oil additives

Oil additives range from viscosity index improvers to chemical treatments that claim to reduce internal engine friction. It should be noted that most oil manufacturers caution against using additives with their oils.

Safety first!
Essential Dos and DON'Ts

Regardless of how enthusiastic you may be about getting on with the job at hand, take the time to ensure that your safety is not jeopardized. A moment's lack of attention can result in an accident, as can failure to observe certain simple safety precautions. The possibility of an accident will always exist, and the following points should not be considered a comprehensive list of all dangers. Rather, they are intended to make you aware of the risks and to encourage a safety conscious approach to all work you carry out on your vehicle.

DON'T rely on a jack when working under the vehicle. Always use approved jackstands to support the weight of the vehicle and place them under the recommended lift or support points.

DON'T attempt to loosen extremely tight fasteners (i.e. wheel lug nuts) while the vehicle is on a jack - it may fall.

DON'T start the engine without first making sure that the transmission is in Neutral (or Park where applicable) and the parking brake is set.

DON'T remove the radiator cap from a hot cooling system - let it cool or cover it with a cloth and release the pressure gradually.

DON'T attempt to drain the engine oil until you are sure it has cooled to the point that it will not burn you.

DON'T touch any part of the engine or exhaust system until it has cooled sufficiently to avoid burns.

DON'T siphon toxic liquids such as gasoline, antifreeze and brake fluid by mouth, or allow them to remain on your skin.

DON'T inhale brake lining or clutch disc dust - it is potentially hazardous (see Asbestos below)

DON'T allow spilled oil or grease to remain on the floor-wipe it up before someone slips on it.

DON'T use loose fitting wrenches or other tools which may slip and cause injury.

DON'T push on wrenches when loosening or tightening nuts or bolts. Always try to pull the wrench toward you. If the situation calls for pushing the wrench away, push with an open hand to avoid scraped knuckles if the wrench should slip.

DON'T attempt to lift a heavy component alone - get someone to help you.

DON'T rush or take unsafe shortcuts to finish a job.

DON'T allow children or animals in or around the vehicle while you are working on it.

DO wear eye protection when using power tools such as a drill, sander, bench grinder, etc. and when working under a vehicle.

DO keep loose clothing and long hair well out of the way of moving parts.

DO make sure that any hoist used has a safe working load rating adequate for the job.

DO get someone to check on you periodically when working alone on a vehicle.

DO carry out work in a logical sequence and make sure that everything is correctly assembled and tightened.

DO keep chemicals and fluids tightly capped and out of the reach of children and pets.

DO remember that your vehicle's safety affects that of yourself and others. If in doubt on any point, get professional advice.

Asbestos

Certain friction, insulating, sealing, and other products - such as brake linings, brake bands, clutch linings, torque converters, gaskets, etc. - contain asbestos. Extreme care must be taken to avoid inhalation of dust from such products since it is hazardous to health. If in doubt, assume that they do contain asbestos. Always wear a filtering mask when working on or around disc or drum brakes.

Batteries

Never create a spark or allow a bare light bulb near a battery. They normally give off a certain amount of hydrogen gas, which is highly explosive.

Always disconnect the battery ground (-) cable at the battery before working on the fuel or electrical systems.

If possible, loosen the filler caps or cover when charging the battery from an external source (this does not apply to sealed or maintenance-free batteries). Do not charge at an excessive rate or the battery may burst.

Take care when adding water to a non maintenance-free battery and when carrying a battery. The electrolyte, even when diluted, is very corrosive and should not be allowed to contact clothing or skin.

Always wear eye protection when cleaning the battery to prevent the caustic deposits from entering your eyes.

Fire

We strongly recommend that a fire extinguisher suitable for use on fuel and electrical fires be kept handy in the garage or workshop at all times. Never try to extinguish a fuel or electrical fire with water. Post the phone number for the nearest fire department in a conspicuous location near the phone.

Fumes

Certain fumes are highly toxic and can quickly cause unconsciousness and even death if inhaled to any extent. Gasoline vapor falls into this category, as do the vapors from some cleaning solvents. Any draining or pouring of such volatile fluids should be done in a well ventilated area.

When using cleaning fluids and solvents, read the instructions on the container carefully. Never use materials from unmarked containers.

Never run the engine in an enclosed space, such as a garage. Exhaust fumes contain carbon monoxide, which is extremely poisonous. If you need to run the engine, always do so in the open air, or at least have the rear of the vehicle outside the work area.

Gasoline

Remember at all times that gasoline is highly flammable. Never smoke or have any kind of open flame around when working on a vehicle. But the risk does not end there. A spark caused by an electrical short circuit, by two metal surfaces contacting each other, or even by static electricity built up in your body under certain conditions can ignite gasoline vapors, which in a confined space are highly explosive. Do not, under any circumstances, use gasoline for cleaning parts. Use an approved safety solvent. Also, DO NOT STORE GASOLINE IN A GLASS CONTAINER - use an approved metal or plastic container only!

Always disconnect the battery ground (-) cable at the battery before working on any part of the fuel system or electrical system. Never risk spill in a fuel on a hot engine or exhaust component.

Household current

When using an electric power tool, inspection light, etc., which operates on household current, always make sure that the tool is correctly connected to its plug and that, where necessary, it is properly grounded. Do not use such items in damp conditions and, again, do not create a spark or apply excessive heat in the vicinity of fuel or fuel vapor.

Secondary ignition system voltage

A severe electric shock can result from touching certain parts of the ignition system (such as the spark plug wires) when the engine is running or being cranked, particularly if components are damp or the insulation is defective. In the case of an electronic ignition system, the secondary system voltage is much higher and could prove fatal.

Keep it clean

Get in the habit of taking a regular look around the shop to check for potential dangers. Keep the work area clean and neat. Sweep up all debris and dispose of it as soon as possible. Don't leave tools lying around on the floor.

Be very careful with oily rags. Spontaneous combustion can occur if they're left in a pile, so dispose of them properly in a covered metal container.

Check all equipment and tools for security and safety hazards (like frayed cords). Make necessary repairs as soon as a problem is noticed don't wait for a shelf unit to collapse before fixing it.

Accidents and emergencies

Shop accidents range from minor cuts and skinned knuckles to serious injuries requiring immediate medical attention. The former are inevitable, while the latter are, hopefully, avoidable or at least uncommon. Think about what you would do in the event of an accident. Get some first aid training and have an adequate first aid kit somewhere within easy reach.

Think about what you would do if you were badly hurt and incapacitated. Is there someone nearby who could be summoned quickly? If possible, never work alone just in case something goes wrong.

If you had to cope with someone else's accident, would you know what to do? Dealing with accidents is a large and complex subject, and it's easy to make matters worse if you have no idea how to respond. Rather than attempt to deal with this subject in a superficial manner, buy a good First Aid book and read it carefully. Better yet, take a course in First Aid at a local junior college.

Environmental safety

At the time this manual was being written, several state and federal regulations governing the storage and disposal of oil and other lubricants, gasoline, solvents and antifreeze were pending (contact the appropriate government agency or your local auto parts store for the latest information). Be absolutely certain that all materials are properly stored, handled and disposed of. Never pour used or leftover oil, solvents or antifreeze down the drain or dump them on the ground. Also, don't allow volatile liquids to evaporate - keep them in sealed containers. Air conditioning refrigerant should never be expelled into the atmosphere. Have a properly equipped shop discharge and recharge the system for you.

Notes

3 Troubleshooting

This Chapter provides an easy reference guide to the more common problems which may occur in just about any kind of automotive brake system. In the following table, the bold headings contain the various problems that may be encountered. The left column lists the probable cause(s) of the problem, and the right column reveals the corrective action necessary to restore proper operation. **Note:** *Although the problems included here do apply to the wheel brakes and, to an extent, the hydraulic system of ABS-equipped vehicles, malfunctions that pertain solely to the anti-lock portion of such systems aren't dealt with here. For a description of Anti-lock Brake Systems (ABS) and general troubleshooting procedures, refer to Chapter 10.*

Remember that successful troubleshooting is not a mysterious "black art" practiced only by professional mechanics. It's simply the result of a bit of knowledge combined with an intelligent, systematic approach to the problem. Always work by a process of elimination, starting with the simplest solution and working through to the most complex. If necessary, refer to Chapter 1 and review the basic braking system principles.

If you keep a close eye on the condition of your brakes and perform routine inspections like you're supposed to (as described in Chapter 4), you might not ever need the following troubleshooting information. Brakes work hard, though, probably harder than any other system in your car or truck. They're subjected to lots of abuse. Unexpected problems do occur, and a straightforward, logical analysis of the disorder will save you time and unnecessary expense.

Before assuming that a brake problem exists, make sure the tires are in good condition and properly inflated. Also, the front end must be in proper alignment and the vehicle must not be loaded with weight in an unequal manner.

If, after using this troubleshooting guide, you are still unable to resolve the problem, seek advice from a professional mechanic. If necessary, have the vehicle towed to a repair shop. *Don't drive a vehicle with defective brakes.*

Warning: *The dust created by the brake system may contain asbestos, which is harmful to your health. Never blow it out with compressed air and don't inhale any of it. An approved filtering mask should be worn when working on the brakes. Do not, under any circumstances, use petroleum-based solvents to clean brake parts. Use brake system cleaner only!*

Haynes Automotive Brake Manual

PROBLEM	PROBABLE CAUSE	CORRECTIVE ACTION

No brakes - pedal travels to floor

1 Low fluid level	1 and 2 Low fluid level and air in the system are symptoms of another problem - a leak somewhere in the hydraulic system. Locate and repair the leak.
2 Air in system	
3 Defective seals in master cylinder	3 Rebuild or replace master cylinder
4 Brake lining extremely worn or out of adjustment	4 Inspect brakes, replace parts as necessary. Adjust brakes
5 Fluid overheated and vaporized due to heavy braking	5 Bleed hydraulic system (temporary fix). Replace brake fluid (proper fix).

Brake pedal slowly travels to floor under braking or at a stop

1 Defective seals in master cylinder	1 Rebuild or replace master cylinder
2 Leak in a hose, line, caliper or wheel cylinder	2 Locate and repair leak
3 Air in hydraulic system	3 Bleed the system, inspect system for a leak

Fluid level in master cylinder reservoir low, but no signs of external leakage

Master cylinder leaking fluid into power brake booster or into driver's footwell	Rebuild or replace master cylinder

Brake pedal feels "spongy" when depressed

1 Air in hydraulic system	1 Bleed the system, inspect system for a leak
2 Brake shoes not centered in drum	2 Inspect drum brakes, mount shoes correctly
3 Brake drums machined too thin or excessively worn	3 Inspect drums, replace if diameter exceeds maximum allowable diameter
4 Cracked brake drum	4 Carefully inspect drums, replace if necessary
5 Brake shoes distorted	5 Replace brake shoes
6 Caliper or caliper mount flexing	6 Inspect calipers and mounts for loose fasteners, cracks and other signs of fatigue. Replace as necessary
7 Master cylinder or power booster loose	7 Tighten fasteners
8 Brake fluid overheated (beginning to boil)	8 Bleed the system (temporary fix). Replace the brake fluid (proper fix).
9 Brake fluid contaminated	9 Replace brake fluid
10 Deteriorated brake hoses (ballooning under pressure)	10 Inspect hoses, replace as necessary (it's a good idea to replace all of them if one hose shows signs of deterioration).
11 Soft or swollen caliper seals	11 Replace seals (if seals are swollen due to contamination, flush entire system and replace all rubber components)
12 Defective residual check valve	12 Replace valve, bleed system
13 Broken brake pedal pivot bushing or bracket	13 Replace pivot bushing or repair bracket

PROBLEM	PROBABLE CAUSE	CORRECTIVE ACTION

Brake pedal feels hard when depressed - excessive effort required to stop vehicle

	PROBABLE CAUSE	CORRECTIVE ACTION
	1 Power booster faulty	1 Replace booster
	2 Engine not producing sufficient vacuum, or hose to booster clogged, collapsed or cracked	2 Check vacuum to booster with a vacuum gauge. Replace hose if cracked or clogged, repair engine if vacuum is extremely low
	3 Brake linings contaminated by grease or brake fluid	3 Locate and repair source of contamination, replace brake pads or shoes
	4 Brake linings glazed	4 Replace brake pads or shoes, check discs and drums for glazing, service as necessary
	5 Brake linkage binding	5 Repair linkage, lubricate
	6 Caliper piston(s) or wheel cylinder(s) binding or "frozen"	6 Repair or replace calipers or wheel cylinders
	7 Brakes wet	7 Apply pedal to boil-off water (this should only be a momentary problem)
	8 Kinked, clogged or internally split brake hose or line	8 Inspect lines and hoses, replace as necessary

Excessive brake pedal travel (but will "pump up")

	PROBABLE CAUSE	CORRECTIVE ACTION
	1 Drum brakes out of adjustment	1 Adjust brakes
	2 Air in hydraulic system	2 Bleed system, inspect system for a leak
	3 Wheel bearings out of adjustment	3 Adjust wheel bearings

Excessive brake pedal travel (but will not "pump up")

	PROBABLE CAUSE	CORRECTIVE ACTION
	1 Defective caliper seals	1 Rebuild or replace calipers
	2 Corroded caliper piston	2 Rebuild or replace calipers
	3 Brake linings worn out	3 Inspect brakes, replace pads and/or shoes
	4 Master cylinder pushrod misadjusted	4 Adjust pushrod
	5 Hydraulic system leak	5 Locate and repair leak

Brake pedal pulsates during brake application

	PROBABLE CAUSE	CORRECTIVE ACTION
	1 Brake drums out-of-round	1 Remove drums, have turned by an automotive machine shop
	2 Excessive brake disc runout or disc surfaces out-of-parallel	2 Have discs turned by an automotive machine shop
	3 Cracked disc or drum	3 Inspect and replace, if cracked
	4 Loose wheel bearings	4 Adjust wheel bearings
	5 Loose lug nuts	5 Tighten lug nuts
	6 Bent rear axle	6 Replace rear axle
	7 Caliper not sliding properly due to improper installation or obstruction (floating and sliding calipers only)	7 Install cailper correctly, repair cause of obstruction

Haynes Automotive Brake Manual

PROBLEM	PROBABLE CAUSE	CORRECTIVE ACTION

Brake pedal doesn't return

	PROBABLE CAUSE	CORRECTIVE ACTION
	1 Brake pedal binding	1 Inspect pivot bushing and pushrod, repair or lubricate
	2 Plugged vent holes in master cylinder cap or cover	2 Unplug vent holes

Brakes slow to release

	PROBABLE CAUSE	CORRECTIVE ACTION
	1 Malfunctioning power booster	1 Replace booster
	2 Pedal linkage binding	2 Inspect pedal pivot bushing and pushrod, repair/lubricate
	3 Malfunctioning proportioning valve	3 Replace proportioning valve
	4 Sticking caliper or wheel cylinder	4 Repair or replace calipers or wheel cylinders
	5 Kinked or internally split brake hose	5 Locate and replace faulty brake hose

Brakes "grab" (one or more wheels)

	PROBABLE CAUSE	CORRECTIVE ACTION
	1 Grease or brake fluid on brake lining	1 Locate and repair cause of contamination, replace lining
	2 Brake lining glazed	2 Replace lining, deglaze disc or drum
	3 Loose or defective wheel bearings	3 Adjust or replace wheel bearings
	4 Lining loose on brake shoe	4 Replace lining
	5 Loose brake backing plate	5 Tighten backing plate
	6 Loose caliper mount	6 Tighten caliper mount

Vehicle pulls to one side during braking

	PROBABLE CAUSE	CORRECTIVE ACTION
	1 Grease or brake fluid on brake lining	1 Locate and repair cause of contamination, replace lining
	2 Brake lining glazed	2 Deglaze or replace lining, deglaze disc or drum
	3 Loose or defective wheel bearings	3 Adjust or replace wheel bearings
	4 Lining loose on brake shoe	4 Replace lining
	5 Loose brake backing plate	5 Tighten backing plate
	6 Loose caliper mount	6 Tighten caliper mount
	7 Plugged brake line or hose	7 Unplug line or replace hose
	8 Tire pressures incorrect	8 Adjust tire pressures
	9 Caliper or wheel cylinder sticking	9 Repair or replace calipers or wheel cylinders
	10 Wheels out of alignment	10 Have wheels aligned
	11 Weak suspension spring	11 Replace springs
	12 Weak or broken shock absorber	12 Replace shock absorbers

PROBLEM	PROBABLE CAUSE	CORRECTIVE ACTION

Brakes drag (indicated by sluggish engine performance or wheels being very hot after driving)

1 Brake pedal pushrod incorrectly adjusted	1 Adjust pushrod
2 Master cylinder pushrod (between booster and master cylinder) incorrectly adjusted	2 Adjust pushrod
3 Obstructed compensating port in master cylinder	3 Disassemble master cylinder and clean, reassemble with new rubber parts
4 Master cylinder piston seized in bore	4 Rebuild or replace master cylinder
5 Contaminated fluid causing swollen seals throughout system	5 Flush system, replace all rubber parts of system
6 Clogged brake lines or internally split brake hoses	6 Flush hydraulic system, replace defective brake hoses
7 Sticking calipers or wheel cylinders	7 Overhaul or replace calipers or wheel cylinders
8 Parking brake not releasing	8 Inspect parking brake linkage and parking brake mechanism, repair as required
9 Improper shoe-to-drum clearance	9 Adjust brake shoes
10 Faulty proportioning valve	10 Replace proportioning valve

Brakes fade (due to excessive heat)

1 Brake linings excessively worn or glazed	1 Deglaze or replace brake pads and/or shoes
2 Excessive use of brakes	2 Downshift into a lower gear, maintain a constant slower speed (going down hills)
3 Vehicle overloaded	3 Reduce load
4 Brake drums or discs worn too thin	4 Measure drum diameter and disc thickness, replace drums or discs as required
5 Contaminated brake fluid	5 Flush system, replace fluid
6 Brakes drag	6 Repair cause of dragging brakes
7 Driver resting left foot on brake pedal	7 Don't "ride the brakes"

Brakes noisy (high-pitched squeal)

1 Glazed lining	1 Deglaze or replace lining
2 Contaminated lining (brake fluid, grease, etc.)	2 Repair source of contamination, replace linings
3 Weak or broken brake shoe hold-down or return spring	3 Replace springs
4 Rivets securing lining to shoe or backing plate loose	4 Replace shoes or pads
5 Excessive dust buildup on brake linings	5 Wash brakes off with brake system cleaner
6 Brake drums worn too thin	6 Measure diameter of drums, replace if necessary
7 Wear indicator on disc brake pads contacting disc	7 Replace brake pads
8 Anti-squeal shims missing or installed improperly	8 Install shims correctly

Note: *Other remedies for quieting squealing brakes include the application of an anti-squeal compound to the backing plates of the brake pads, and lightly chamfering the edges of the brake pads with a file. The latter method should only be performed with the brake pads thoroughly wetted with brake system cleaner, so as not to allow any asbestos dust to become airborne.*

Haynes Automotive Brake Manual

PROBLEM	PROBABLE CAUSE	CORRECTIVE ACTION

Brakes noisy (scraping sound)

	1 Brake pads or shoes worn out, rivets, backing plate or brake shoe metal contacting disc or drum	1 Replace linings, have discs and/or drums machined

Brakes chatter

	1 Worn brake lining	1 Inspect brakes, replace shoes or pads as necessary
	2 Glazed or scored discs or drums	2 De-glaze discs or drums with sandpaper (if glazing is severe, machining will be required)
	3 Drums or discs heat checked	3 Check discs and/or drums for hard spots, heat checking, etc. Have discs/drums machined or replace them
	4 Disc runout or drum out-of-round excessive	4 Measure disc runout and/or drum out-of-round, have discs or drums machined or replace them
	5 Loose or worn wheel bearings	5 Adjust or replace wheel bearings
	6 Loose or bent brake backing plate (drum brakes)	6 Tighten or replace backing plate
	7 Grooves worn in discs or drums	7 Have discs or drums machined, if within limits (if not, replace discs or drums)
	8 Brake linings contaminated (brake fluid, grease, etc.)	8 Locate and repair source of contamination, replace pads or shoes
	9 Excessive dust buildup on linings	9 Wash brakes with brake system cleaner
	10 Defective power booster	10 Replace booster
	11 Distorted brake shoes	11 Replace brake shoes
	12 Drum brake shoes out of adjustment	12 Adjust brakes
	13 Surface finish on discs or drums too rough after machining (especially on vehicles with sliding calipers)	13 Have discs or drums properly machined
	14 Brake shoes glazed	14 Deglaze or replace brake shoes

Brake shoes click

	1 Shoe support pads on brake backing plate grooved or excessively worn	1 Replace brake backing plate
	2 Brake pads loose in caliper	2 Tighten pad "ears" against caliper frame
	3 Also see items listed under *Brakes chatter*	

Brakes make groaning noise at end of stop

	1 Brake pads and/or shoes worn out	1 Replace pads and/or shoes
	2 Brake linings contaminated (brake fluid, grease, etc.)	2 Locate and repair cause of contamination, replace brake pads or shoes
	3 Brake linings glazed	3 Deglaze or replace brake pads or shoes
	4 Excessive dust buildup on linings	4 Wash brakes with brake system cleaner
	5 Scored or heat-checked discs or drums	5 Inspect discs/drums, have machined if within limits (if not, replace discs or drums)
	6 Broken or missing brake shoe attaching hardware	6 Inspect drum brakes, replace missing hardware

PROBLEM	PROBABLE CAUSE	CORRECTIVE ACTION

Brakes noisy (scraping sound)

	PROBABLE CAUSE	CORRECTIVE ACTION
	1 Brake pads or shoes worn out, rivets, backing plate or brake shoe metal contacting disc or drum	1 Replace linings, have discs and/or drums machined

Brakes chatter

	PROBABLE CAUSE	CORRECTIVE ACTION
	1 Worn brake lining	1 Inspect brakes, replace shoes or pads as necessary
	2 Glazed or scored discs or drums	2 De-glaze discs or drums with sandpaper (if glazing is severe, machining will be required)
	3 Drums or discs heat checked	3 Check discs and/or drums for hard spots, heat checking, etc. Have discs/drums machined or replace them
	4 Disc runout or drum out-of-round excessive	4 Measure disc runout and/or drum out-of-round, have discs or drums machined or replace them
	5 Loose or worn wheel bearings	5 Adjust or replace wheel bearings
	6 Loose or bent brake backing plate (drum brakes)	6 Tighten or replace backing plate
	7 Grooves worn in discs or drums	7 Have discs or drums machined, if within limits (if not, replace discs or drums)
	8 Brake linings contaminated (brake fluid, grease, etc.)	8 Locate and repair source of contamination, replace pads or shoes
	9 Excessive dust buildup on linings	9 Wash brakes with brake system cleaner
	10 Defective power booster	10 Replace booster
	11 Distorted brake shoes	11 Replace brake shoes
	12 Drum brake shoes out of adjustment	12 Adjust brakes
	13 Surface finish on discs or drums too rough after machining (especially on vehicles with sliding calipers)	13 Have discs or drums properly machined
	14 Brake shoes glazed	14 Deglaze or replace brake shoes

Brake shoes click

	PROBABLE CAUSE	CORRECTIVE ACTION
	1 Shoe support pads on brake backing plate grooved or excessively worn	1 Replace brake backing plate
	2 Brake pads loose in caliper	2 Tighten pad "ears" against caliber frame
	3 Also see items listed under *Brakes chatter*	

Brakes make groaning noise at end of stop

	PROBABLE CAUSE	CORRECTIVE ACTION
	1 Brake pads and/or shoes worn out	1 Replace pads and/or shoes
	2 Brake linings contaminated (brake fluid, grease, etc.)	2 Locate and repair cause of contamination, replace brake pads or shoes
	3 Brake linings glazed	3 Deglaze or replace brake pads or shoes
	4 Excessive dust buildup on linings	4 Wash brakes with brake system cleaner
	5 Scored or heat-checked discs or drums	5 Inspect discs/drums, have machined if within limits (if not, replace discs or drums)
	6 Broken or missing brake shoe attaching hardware	6 Inspect drum brakes, replace missing hardware

Notes

4 Maintenance

Introduction

The key objective of this Chapter is to establish a preventive maintenance program to minimize the chances of brake trouble or system failure. This is done by maintaining a constant awareness of the condition of the entire braking system and by correcting defects or replacing worn-out parts before they become serious problems. The majority of the operations outlined in this Chapter are nothing more than inspections of the various components that make up the brake system.

Secondly, this preventive maintenance program should cause most, if not all, maintenance and repairs to take place by intent, hopefully eliminating all "un-scheduled maintenance." By doing so, all brake system repairs that may become necessary can be performed when and where you choose. Not only will it save you money by catching worn-out brake pads and shoes before they cause serious damage, it'll also save you money by avoiding body shop repair bills, tow truck, ambulance and hospital fees, lawsuits, etc!

The following brake maintenance schedule is very simple, as are the checking procedures that go with it. The maintenance intervals are based on normal operating conditions for a normal passenger vehicle or light truck. If a particular vehicle is subjected to more severe usage, the intervals should be shortened and the inspection procedures performed more often. Vehicles that tow trailers, carry heavy loads, are driven in a lot of stop-and-go traffic or in mountainous regions are going to wear out parts and need service more often than vehicles that enjoy mostly traffic-free highway trips. Mileage isn't always a factor - the important thing is to perform the inspections routinely. If your vehicle is operated under severe conditions, it would be a good idea to at least perform the disc and/or drum brake inspections twice as frequently as outlined below, until you develop a feel for the rate at which your vehicle wears down the brake linings.

The maintenance schedule is basically an outline for the actual inspection and maintenance procedures. The individual procedures are then described in detail. Some of the items on the schedule may be listed at a shorter interval than recommended by some manufacturers, and some of the items listed may not even appear on the maintenance schedules of some manufacturers, but that's OK - it's better to be safe than sorry.

If you're experiencing a specific problem with the brake system, refer to Chapter 3, *Troubleshooting*, for the proper course of action to take.

Maintenance schedule

Every 250 miles or weekly, whichever comes first

Check the tires and tire pressures
Check the lug nut tightness
Check the brake fluid level and condition
Check the brake pedal "feel" and travel (master cylinder check)
Check the power brake booster, if equipped
Check the operation of the brake lights and the brake warning light

Every 6000 miles or 6 months, or whichever comes first

Check the front brakes
Check the brake hoses and lines

Every 12,000 miles or 12 months, whichever comes first

All items listed above, plus:
Check the rear brakes
Check and, if necessary, adjust the parking brake
Check the brake pedal pivot bushing and bracket

Every 30,000 miles or 2 years, whichever comes first

All items listed above, plus:
Replace the brake fluid (this is an item for which many manufacturers don't even specify a maintenance interval, so it can be considered optional. However, replacing the brake fluid on a regular basis will prevent contamination in the hydraulic system and all the problems associated with it)
Check the master cylinder for leakage past the primary piston seals

Every 80,000 miles, or 5 years, whichever comes first

Rebuild or replace the master cylinder, calipers and wheel cylinders (here's another item that most manufacturers don't specifically recommend, but it can prevent a myriad of problems. Master and wheel cylinder cups and caliper pistons and seals don't last forever, so to overhaul or replace them before they pose a problem makes sense. It's up to you).

Inspection and maintenance procedures

Preliminary checks

Tires and tire pressures

The first step in a brake system inspection should be of the tire condition and pressures. If the tires aren't inflated properly, or if they're in bad shape, there's no way the vehicle is going to stop in a straight line.

Check to see how much tread is left on the tires. Tread wear can be monitored with a simple, inexpensive device known as a tread depth indicator **(see illustration)**.

Note any abnormal tread wear. Tread pattern irregularities such as cupping, flat spots and more wear on one side than the other are indications of front end alignment and/or balance problems. If any of these conditions are noted, take the

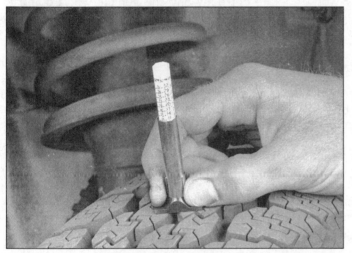

4.1 Use a tire tread depth indicator to monitor tire wear - they are available at auto parts stores and service stations and cost very little

4.2 If a tire loses air on a steady basis, check the valve core first to make sure it's snug (special inexpensive wrenches are commonly available at auto parts stores)

vehicle to a tire shop or service station to correct the problem.

Look closely for cuts, punctures and embedded nails or tacks. Sometimes a tire will hold air pressure for a short time or leak down very slowly after a nail has embedded itself in the tread. If a slow leak persists, check the valve stem core to make sure it's tight **(see illustration)**. Examine the tread for an object that may have embedded itself in the tire or for a "plug" that may have begun to leak (radial tire punctures are repaired with a plug that's installed in a puncture). If a puncture is suspected, it can be easily verified by spraying a solution of soapy water onto the puncture area **(see illustration)**. The soapy solution will bubble if there's a leak. Unless the puncture is unusually large, a tire shop or service station can usually repair the tire.

Carefully inspect the inner sidewall of each tire for evidence of brake fluid leakage. If you see any, inspect the brake hoses, calipers and/or wheel cylinders immediately.

Correct air pressure adds miles to the lifespan of the tires, improves mileage and enhances overall ride quality. Tire pressure cannot be accurately estimated by looking at a tire, especially if it's a radial. A tire pressure gauge is essential. Keep an accurate gauge in the vehicle. The pressure gauges attached to the nozzles of air hoses at gas stations are often inaccurate.

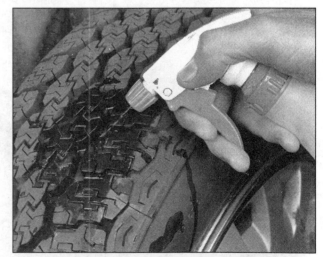

4.3 If the valve core is tight, raise the corner of the vehicle with the low tire and spray a soapy water solution onto the tread as the tire is turned slowly - leaks will cause small bubbles to appear

Always check tire pressure when the tires are cold. Cold, in this case, means the vehicle has not been driven over a mile in the three hours preceding a tire pressure check. A pressure rise of four to eight pounds is not uncommon once the tires are warm.

Unscrew the valve cap protruding from the wheel or hubcap and push the gauge firmly onto the valve stem **(see illustration)**. Note the reading on the gauge and compare the figure to the recommended tire pressure shown on the placard on the driver's side door pillar or in the owner's manual. Be sure to reinstall the valve cap to keep dirt and moisture out of the valve stem mechanism. Check all four tires and, if necessary, add enough air to bring them up to the recommended pressure.

4.4 To extend the life of the tires, check the air pressure at least once a week with an accurate gauge (don't forget the spare!)

Lug nut (or bolt) tightness

Loose lug nuts (or bolts) present a very hazardous situation. Not only is wheel in danger of falling off, the brakes won't operate correctly and the vehicle won't come to a straight stop. It will pull to one side or the other, the pedal will pulsate, and there's a good chance the vehicle will go out of control.

If the vehicle that you're working on is equipped with hubcaps, remove them. Using a lug nut wrench or a breaker bar and socket, check the tightness of all of the lug nuts or bolts. None of them should be able to be tightened more than 1/8-turn. All vehicles have an actual torque specification for their lug nuts or bolts, but the important thing here is that none are loose.

If you find any lug nuts that are loose, be sure to check the wheel studs for damage, and the holes in the wheel for elongation. If any of the studs are damaged, replace them (don't attempt to clean up the the threads with a die or thread file). The same goes for lug bolts. If any of the holes in the wheel are elongated, replace the wheel.

Brake fluid

The brake master cylinder is mounted on the upper left of the engine compartment firewall on most vehicles. On some models it's located on the right (passenger's) side of the engine compartment. On rear-engined vehicles, it's located in the luggage compartment (or inboard of the fenderwell, with a remote reservoir in the luggage compartment).

The fluid inside can be checked after removing the cover or cap **(see illustrations)**. Before removing the cover, be sure to wipe the top of the reservoir with a clean rag to prevent contamination of the brake system.

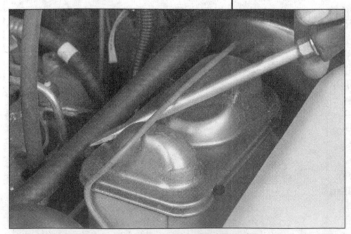

4.5 On vehicles with this type of reservoir cover, pry the clip off the cover with a screwdriver

4.6 This master cylinder has two caps - each one must be removed to check the fluid level of both hydraulic circuits

When adding fluid, pour it carefully into the reservoir to avoid spilling it on surrounding painted surfaces. Vehicles with translucent plastic reservoirs usually have maximum and minimum level marks - it isn't necessary to remove the cover or cap to check the fluid level **(see illustration)**. Be sure to keep the fluid level within this range. On models with integral reservoirs, add fluid until the level is within 1/4-inch from the top **(see illustration)**.

Most vehicles on the road use DOT 3 approved brake fluid, but it would be a good idea to check your owners manual - some of the more expensive foreign automobiles use a type of mineral oil instead of brake fluid. Mixing the brake fluid with the mineral oil in one of these systems would destroy all of the rubber components in the hydraulic system and cause brake failure. **Warning:** *Brake fluid can harm your eyes and damage painted surfaces, so use extreme caution when handling or pouring it. Do not use brake fluid that has been standing open or is more than one year old. Brake fluid absorbs moisture from the air. Excess moisture can cause a dangerous loss of brake performance, because the moisture in the fluid can boil. If you get any fluid in your eyes, immediately flush your eyes with water and seek medical attention.*

At this time, the fluid and master cylinder can be inspected for contamination. The system should be drained and refilled if deposits, dirt particles or water droplets are seen in the fluid.

After filling the reservoir to the proper level, make sure the cover or cap is on tight to prevent fluid leakage. Make sure the rubber diaphragm, if equipped, is in good condition.

The brake fluid level in the master cylinder will drop slightly as the pads at the front wheels wear down during normal operation. If the master cylinder requires repeated additions to keep it at the proper level, it's an indication of leakage in the brake system, which should be corrected immediately. Check all brake lines and connections as described later in this Chapter.

If, upon checking the master cylinder fluid level, you discover one or both reservoirs empty or nearly empty, fill the reservoirs with brake fluid, bleed the system and check the entire hydraulic system for leaks (including behind the master cylinder).

Power brake booster (vacuum-operated)

Begin the power booster check by depressing the brake pedal several times with the engine off to deplete any vacuum remaining in the booster.

Now, depress the pedal and start the engine. If the pedal goes down slightly, operation is normal **(see illustration)**. Release the brake pedal and let the engine run for a couple of minutes.

Turn off the engine and depress the brake pedal several times slowly. If the pedal goes down farther the first time but gradually rises after the second or third depression, the booster is airtight **(see illustration)**.

Start the engine and depress the brake pedal, then stop the engine with the pedal still depressed. If there is no change in the reserve distance (the distance between the pedal and the floor) after holding the pedal for about 30-seconds, the booster is airtight.

If the pedal feels "hard" when the engine is running, the

4.7 The brake fluid level in master cylinders with translucent plastic reservoirs is easily checked - the fluid level should be kept near the MAX mark

4.8 On master cylinders with integral, cast iron reservoirs, the brake fluid level should be kept at approximately 1/4-inch from the top

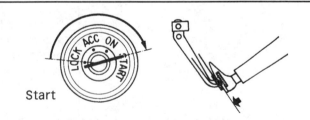

4.9 Push down on the brake pedal then start the engine - the brake pedal should go down slightly, indicating normal booster operation

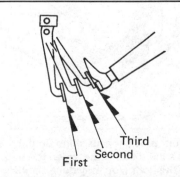

4.10 With the engine turned off, the pedal should build up with each pump if the booster is functioning properly

booster isn't operating properly. Refer to Chapter 9 for more booster checking procedures.

Power brake booster (hydraulically operated)

Check the fluid level in the power steering reservoir. If it's OK, reinstall the cap. If it's low, check the power steering system, the lines leading to the hydraulic booster and the booster itself for leaks. If any leaks are found, repair them.

The hydraulic operates much like the vacuum operated booster. If the pedal is "hard" with the engine running, the booster is faulty. With the engine off, the pedal should lose a little travel with each pump of the pedal (it takes about 20 pumps to bleed down the system). With the engine off, depress the pedal and start the engine. The pedal should go down a little, indicating normal operation. If the booster is defective, don't attempt to rebuild it. Install a new or factory rebuilt unit (see Chapter 9).

Brake pedal freeplay, "feel," and travel (master cylinder check)

Using your hand, push down on the pedal and note how far the pedal travels until resistance is felt **(see illustration)**. This is the pedal freeplay. It's important to have approximately 1/16 to 1/8-inch of freeplay, because this allows the pistons in the master cylinder to return to their at-rest positions when the brake pedal is released, ensuring that the compensating ports are uncovered and can equalize the pressure in the areas in front of the pistons.

Too much freeplay will result in a brake pedal that is too close to the floor under hard braking (it may even hit the floor). Generally, freeplay greater than one inch can be considered too much.

Pedal freeplay on some vehicles is adjusted at the master cylinder pushrod that protrudes from the booster. On other vehicles it's adjusted at the pushrod between the brake pedal and the booster.

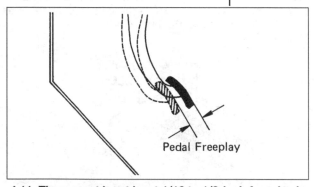

4.11 There must be at least 1/16 to 1/8-inch freeplay in the brake pedal - if not, the master cylinder pistons won't be able to return all the way, which could result in a low pedal (because the pressure chambers in front of the pistons wouldn't fill up completely)

If the vehicle is equipped with power brakes, start the engine. Slowly depress the brake pedal a few times. How does it feel? Is it firm or "spongy." If it feels spongy or springy, chances are there's air in the hydraulic system, in which case it will be necessary to bleed the entire system (see Chapter 9). In addition, the entire system should be inspected for a fluid leak. **Note:** *Air leaks can also develop in hydraulic systems without the presence of a fluid leak. When this happens, defective wheel cylinder cups are usually the culprit, although old caliper piston seals can cause the same problem.*

Now, with the engine still running, push the pedal down with a considerable amount of force (about 100 lbs). The pedal should come to a stop at some point and stay there. If the pedal slowly creeps towards the floor, check the entire hydraulic system for a fluid leak. If the pedal meets no resistance until it's a couple of inches from the floor, there's a strong possibility that there's a problem with one of the hydraulic circuits. If there is no evidence of an external fluid leak, the seals in the master cylinder are bad, in which case it must be rebuilt or replaced. See Chapter 9 for master cylinder removal and installation, as well as the overhaul procedure.

With the pedal firmly depressed, measure the distance between the pedal and the floor - this is the reserve distance **(see illustration)**. This specification varies from vehicle-to-vehicle, but generally there should be at least two inches to spare. Any less than this and you run the risk of the pedal contacting the floor before the brakes are fully utilized.

Inadequate reserve distance can be caused by worn-out brake linings or a problem in the hydraulic circuit.

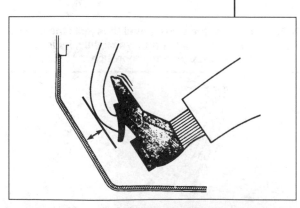

4.12 With your foot firmly pushing on the brake pedal, measure the distance between the bottom of the pedal and the floor (you may need the help of an assistant) - there should be at least two inches of reserve travel left

Brake warning light

The brake warning light on the instrument panel rarely fails, but it is a very important component of the brake system. Although in service it will usually come on only when you're in the midst of a braking problem, it can warn you of some problems before they become catastrophic.

Sit in the driver's seat and release the parking brake. Turn the ignition key to the Start position. As the engine is cranking, the brake warning light on the instrument panel should glow. As soon as you let go of the key, the light should go out, indicating the bulb is functional.

If the light won't go out when you release the key, check the master cylinder. Some vehicles have float-operated switches in their master cylinder reservoirs. If the fluid level drops below a certain level, the switch will complete the circuit and turn the warning light on. If your vehicle is like this, make sure the fluid is at the proper level.

If your master cylinder doesn't have a low fluid level indicator in it but the light refuses to go out, there may be air or a leak in one half of the hydraulic system. Check the entire system for leaks.

If the light doesn't come on, check the fuses. If they're all OK, check for voltage to the bulb as the engine is cranked. If voltage is present, replace the bulb. If there's no voltage, trace the circuit for an open.

Turn the engine off, but leave the ignition key in the On position. Depress the brake pedal with about 100 pounds of force. The warning lamp shouldn't come on. If it does, there's air in the system or a leak in one half of the hydraulic system. Inspect the entire system.

The light should also come on when the parking brake is applied. If it doesn't, check the fuses, the switch at the parking brake pedal, rod or lever, the bulb and if necessary, the circuit.

Brake lights

Have an assistant stand behind the vehicle while you depress the brake pedal. All of the brake lights should come on (on some vehicles the ignition key must be turned to the On position). If any bulbs are burned out, replace them.

If none of the lights come on, refer to Chapter 8 for further diagnosis.

Front brakes

Since your front brakes do most of the work, you should inspect them twice as often as the rear. Before removing the wheels, however, the wheel bearings should be checked for proper adjustment. Loose wheel bearings can cause a host of problems, such as inconsistent pedal travel, brake chatter, pedal pulsation, and the vehicle pulling to one side or the other when the brakes are applied. For this inspection, refer to the sidebar titled *"Wheel bearing, check, repack and adjustment."* Take note, however, that not all wheel bearings are adjustable - some are sealed units which will require replacement if the bearings are sloppy. This includes the front hub bearings on front-wheel drive vehicles.

Drum brakes

If your vehicle is equipped with front drum brakes, refer to the sidebar referenced in the previous paragraph (*"Wheel bearing check, repack and adjustment"*) to remove the hub nut, then slide the drum off. If it won't come off, the shoes have probably begun to wear ridges into the drum and will have to be retracted. For this procedure and the remainder of the drum brake inspection, refer to the *Drum brakes* inspection procedure, under *Rear brakes*, since the procedure is the same as for rear drum brakes. When the inspection is complete, reinstall the drum and wheel bearings and adjust the bearings as described in the aforementioned sidebar. **Note:** *Be sure to clean, inspect and repack the bearings if they appear somewhat dry, dirty, or if it has been more than 20,000 miles since the last time they were packed.* Also, adjust the brake shoes as described in the sidebar entitled *Drum brake adjustment*.

While the wheels are off, be sure to check the brake hoses as described later in this Chapter.

Wheel bearing check, repack and adjustment

Note: *The wheel bearing repack and adjustment portions of this procedure do not apply to vehicles with sealed bearings or to the front axle of front-wheel drive, four-wheel drive vehicles, nor to the rear axle on vehicles with heavy-duty (full-floating) rear axles. If your vehicle falls into any of these categories, consult a shop manual or the appropriate Haynes Automotive Repair Manual.*

1   In most cases the front wheel bearings will not need servicing until the brake pads are changed. However, the bearings should be checked whenever the front of the vehicle is raised for any reason. Several items, including a torque wrench and special grease, are required for this procedure **(see illustration)**.

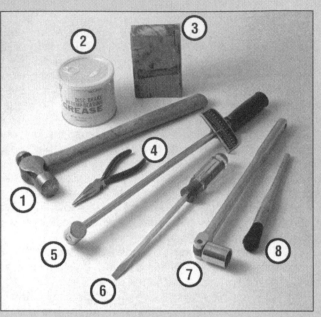

4.13 Tools and materials needed for front wheel bearing maintenance

1   **Hammer** - A common hammer will do just fine
2   **Grease** - High-temperature grease that is formulated specially for front wheel bearings should be used
3   **Wood block** - If you have a scrap piece of 2x4, it can be used to drive the new seal into the hub
4   **Needle-nose pliers** - Used to straighten and remove the cotter pin in the spindle
5   **Torque wrench** - This is very important in this procedure; if the bearing is too tight, the wheel won't turn freely - if it's too loose, the wheel will "wobble" on the spindle. Either way, it could mean extensive damage
6   **Screwdriver** - Used to remove the seal from the hub (a long screwdriver is preferred)
7   **Socket/breaker bar** - Needed to loosen the nut on the spindle if it's extremely tight
8   **Brush** - Together with some clean solvent, this will be used to remove old grease from the hub and spindle

Check

2   Block the wheels at the opposite end of the vehicle being inspected, raise the front of the vehicle and support it securely on jackstands placed under the frame rails, subframe or the seam below the rocker panels where the body is mated with the floorpan (unibody-constructed vehicles only). Spin each wheel and check for noise, rolling resistance and freeplay.

3   Grasp the top of each tire with one hand and the bottom with the other. Move the wheel in-and-out on the spindle. If there's any noticeable movement, the bearings should be checked and then repacked with grease or replaced if necessary. **Note:** *In the case of vehicles with sealed bearing assemblies, the bearings aren't adjustable. Some sealed bearings are integral with the wheel hub, requiring replacement of the entire hub. Others have the sealed bearing assembly pressed into the hub, which will require the services of an automotive machine shop or other repair facility equipped with a hydraulic press and the necessary adapters to remove the old unit and press the new one in.*

Repack

Warning: *The dust created by the brake system may contain asbestos, which is harmful to your health. Never blow it out with compressed air and don't inhale any of it. An approved filtering mask should be worn when working on the brakes. Do not, under any circumstances, use petroleum-based solvents to clean brake parts. Use brake system cleaner only!*

Note: *This procedure applies to adjustable wheel bearing assemblies on the front wheels of rear-wheel drive vehicles and the rear wheels of front-wheel drive vehicles.*

4   Remove the wheel. You'll probably have to lower the vehicle to break the lug nuts (or bolts) loose. After that,

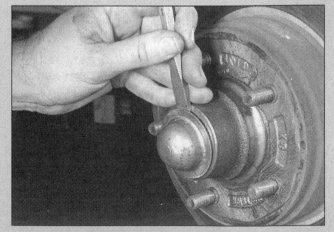

4.14 Dislodge the dust cap by working around the outer circumference with a hammer and chisel

4.15 Remove the cotter pin and discard it - use a new one when the hub nut is reinstalled

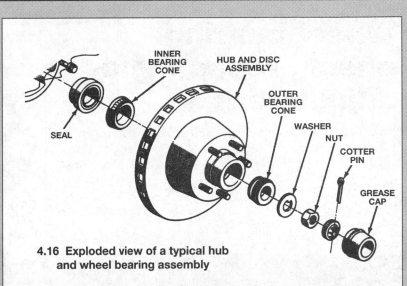

4.16 Exploded view of a typical hub and wheel bearing assembly

SEAL

INNER BEARING CONE

HUB AND DISC ASSEMBLY

OUTER BEARING CONE

WASHER

NUT

COTTER PIN

GREASE CAP

raise the vehicle and support it on jackstands once again. **Note:** *If you're simply adjusting the wheel bearings (not cleaning and repacking them), skip to Step 21.*

5   If the wheel you're working on is equipped with a disc brake, remove the brake caliper (see Chapter 5) and hang it out of the way on a piece of wire. A wood block can be slid between the brake pads to keep them separated, if necessary.

6   Pry the dust cap out of the hub using a screwdriver or hammer and chisel **(see illustration)**.

7   Straighten the bent ends of the cotter pin, then pull the cotter pin out of the nut lock **(see illustration)**. Discard the cotter pin and use a new one during reassembly.

8   Remove the nut lock, nut and washer from the end of the spindle **(see illustration)**.

9   Pull the hub/disc or drum assembly out slightly, then push it back into its original position. This should force the outer bearing off the spindle enough so it can be removed.

10   Pull the hub/disc or drum assembly off the spindle. If you're working on a drum brake and the drum won't pull off, the brake shoes will have to be retracted. Refer to the shoe retraction procedure described in the *Drum brakes* subheading under the *Rear brakes* section

11   Use a screwdriver to pry the seal out of the rear of the hub **(see illustration)**. As this is done, note how the seal is installed.

12   Remove the inner wheel bearing from the hub.

13   Use solvent to remove all traces of the old grease from the bearings, hub and spindle. A small brush may prove helpful; however make sure no bristles from the brush embed themselves inside the bearing rollers. Allow the parts to air dry.

14   Carefully inspect the bearings for cracks, heat discoloration, worn rollers, etc. Check the bearing races inside the hub for wear and damage. If the bearing races are defective, the hubs should be taken to a machine shop with the facilities to remove the old races and press new ones in. Note that the bearings and races come as matched sets and old bearings should never be installed on new races.

4.17 Use a large screwdriver to pry the grease seal out of the rear of the hub

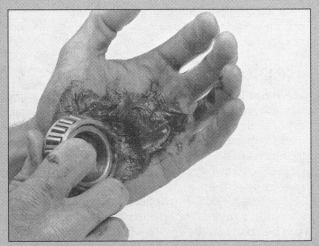

4.18 Work grease into the bearing rollers by pressing it against the palm of your hand

Wheel bearing check, repack and adjustment (continued)

15   Use high-temperature front wheel bearing grease to pack the bearings. Work the grease completely into the bearings, forcing it between the rollers, cone and cage from the back side **(see illustration)**.

16   Apply a thin coat of grease to the spindle at the outer bearing seat, inner bearing seat, shoulder and seal seat.

17   Put a small quantity of grease inboard of each bearing race inside the hub. Using your finger, form a dam at these points to provide extra grease availability and to keep thinned grease from flowing out of the bearing.

18   Place the grease-packed inner bearing into the rear of the hub and put a little more grease outboard of the bearing.

19   Place a new seal over the inner bearing and tap the seal evenly into place with a hammer and blunt punch until it's flush with the hub.

20   Carefully place the hub assembly onto the spindle and push the grease-packed outer bearing into position.

Adjustment

Note: *This procedure applies to adjustable wheel bearing assemblies on the front wheels of rear-wheel drive vehicles and the rear wheels of front-wheel drive vehicles.*

21   Install the washer and spindle nut. Tighten the nut only slightly (no more than 12 ft-lbs of torque).

22   Spin the hub in a forward direction while tightening the spindle nut to approximately 20 ft-lbs to seat the bearings and remove any grease or burrs which could cause excessive bearing play later.

23   Loosen the spindle nut 1/4-turn, then using your hand (not a wrench of any kind), tighten the nut until it's snug. Install the nut lock and a new cotter pin through the hole in the spindle and the slots in the nut lock. If the nut lock slots don't line up, remove the nut lock and turn it slightly until they do. Ideally, when the wheel bearings are properly adjusted there should be no play in the bearings, but no preload either. You should be able to push the washer under the lock nut back and forth with a screwdriver (without prying), but you shouldn't be able to feel any slop in the bearing. **Note:** *If you do feel some play in the bearing, tighten the hub nut a little more than hand-tight, but don't exceed 10 or 15 inch-pounds of preload. If there's still play, the wheel bearings may be excessively worn.*

24   Bend the ends of the cotter pin until they're flat against the nut. Cut off any extra length which could interfere with the dust cap.

25   Install the dust cap, tapping it into place with a hammer.

26   On models with disc brakes, place the brake caliper near the disc and carefully remove the wood spacer. Install the caliper (see Chapter 5).

27   Install the wheel on the hub and tighten the lug nuts.

28   Grasp the top and bottom of the tire and check the bearings in the manner described earlier in Steps 2 and 3.

29   Lower the vehicle.

Disc brakes

Warning: *The dust created by the brake system may contain asbestos, which is harmful to your health. Never blow it out with compressed air and don't inhale any of it. An approved filtering mask should be worn when working on the brakes. Do not, under any circumstances, use petroleum-based solvents to clean brake parts. Use brake system cleaner only!*

Disc brakes are used on the front wheels of most vehicles. Extensive disc damage can occur if the pads are not replaced when needed.

Loosen the wheel lug nuts, raise the front of the vehicle and support it securely on jackstands placed under the frame rails, subframe or the seam below the rocker panels where the body is mated with the floorpan (unibody-constructed vehicles only), if not already done. Block the wheels at the opposite end of the vehicle to prevent it from rolling. The disc brake calipers, which contain the pads, are visible with the wheels removed. There is an outer pad and an inner pad in each caliper. All pads should be inspected.

Before inspecting the brakes, position a drain pan under the brake assembly, clean the entire brake with brake system cleaner and allow it to dry.

Most calipers have a "window" or opening to inspect the pads. Check the thickness of the pad lining by looking into the caliper at each end and down through the inspection window at the top of the housing **(see illustrations)**. Generally, if the pad material has worn to about 3/16-inch or less (or about as thin as the brake pad backing plate), the pads should be replaced. Ideally, the pads should be changed before they become that thin. Some pads are equipped with

4.19 Look through the caliper inspection window to inspect the brake pads (arrow)

4.20 The pad lining which rubs against the disc (arrow) can also be inspected by looking at each end of the caliper

metal wear sensors, consisting of a small, bent piece of metal which is attached to the backing plate of one of the brake pads. When the pad wears down to its specified limit, the metal sensor rubs against the disc and makes a squealing sound **(see illustration)**. If your vehicle is equipped with such sensors, and the tips of the sensors are nearing the disc surface, you might as well go ahead and replace the brake pads.

If you're unsure about the exact thickness of the remaining lining material, remove the pads for further inspection or replacement (see Chapter 5).

Make sure the caliper mounting bolts, guide pins or retaining plates (depending on caliper design) are secure. Also check the tightness of the caliper mounting (or anchor) plate fasteners, if equipped.

Check around the area of the caliper piston bore(s) for wetness, which would indicate a leaking piston seal. On fixed calipers, check around the transfer tube fittings (if equipped) or at the seam where the caliper halves mate. If the caliper is leaking, refer to Chapter 9 and rebuild or replace the calipers.

If the brake pedal has been pulsating, the discs are probably warped. Refer to Chapter 5 and measure the disc runout with a dial indicator.

Before installing the wheels, check for leakage and/or damage (cracks, splitting, etc.) around the brake hoses and connections, as described later in this Chapter.

Check the condition of the disc as described in Chapter 5.

Brake hoses and lines

About every six months, with the vehicle raised and supported securely on jackstands, the rubber hoses which connect the steel brake lines with the front and rear brake assemblies should be inspected for cracks, chafing of the outer cover, leaks, blisters and other damage **(see illustration)**. These are important and vulnerable parts of the brake system and inspection should be complete. A light and mirror will be helpful for a thorough check. If a hose exhibits any of the above conditions, replace it with a new one.

4.21 The brake wear indicator (arrow) will contact the disc and make a squealing noise when the pad is worn out

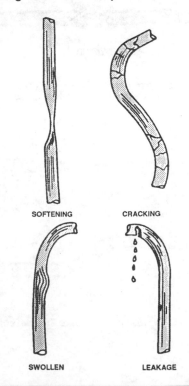

SOFTENING   CRACKING

SWOLLEN   LEAKAGE

4.22 Here are some samples of common brake hose defects

Internally split brake hoses

Some of the trickiest brake problems to diagnose are caused by internally split brake hoses. This malady can restrict pressure from reaching a brake, or it can act like a one-way valve and cause the brake to lock up **(see illustration)**. This is not a common problem, and mostly affects older vehicles with neglected brake systems. It can also be caused by clamping a brake hose shut with locking pliers - something that should never be done.

If you're faced with a brake that won't release, which may be characterized by the vehicle pulling to one side or an overheated brake, open the bleeder valve on that particular caliper or wheel cylinder. If the disc or drum now turns, suspect an internally split hose. Replace the hose and bleed the system (see Chapter 9).

Conversely, if one brake (or both rear brakes on models with a solid rear axle) will not apply but the rest do, there are two possibilities - a "frozen" caliper (or wheel cylinder), or an internally split hose that refuses to pass fluid. Attach a bleeder hose to the bleeder screw on the brake that won't apply. Have an assistant depress the brake pedal while you crack open the bleeder screw. If fluid flows out the bleeder hose, the brake hose is OK - the caliper or wheel cylinder is malfunctioning. If no fluid flows out of the bleeder hose, follow the brake hose to

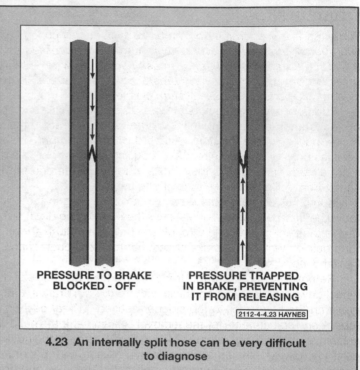

PRESSURE TO BRAKE BLOCKED - OFF **PRESSURE TRAPPED IN BRAKE, PREVENTING IT FROM RELEASING**

2112-4-4.23 HAYNES

4.23 An internally split hose can be very difficult to diagnose

where it meets the metal brake line. Disconnect the line and have your helper push the brake pedal - if fluid squirts from the line, the problem is an internally split brake hose. Replace the hose and bleed the system.

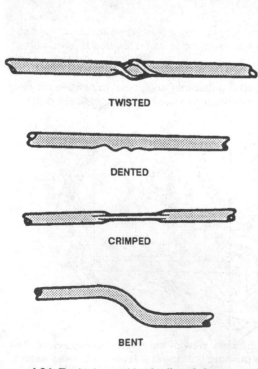

TWISTED

DENTED

CRIMPED

BENT

4.24 Typical metal brake line defects

The metal brake lines should also be inspected. While they aren't subjected to the same punishment that the brake hoses are, they're still fair game to objects thrown up from the tires, and to vibration. Make sure all of the metal clips securing the lines are in place. Check the lines for twists, dents, crimping and kinks **(see illustration)**. Make sure none of the lines travel too close to any hot or moving parts. If you find a damaged section of line, it should be replaced with a genuine factory replacement part, or equivalent, with the lines pre-bent and the fittings already in place. Never attempt to repair a brake line with rubber hose, and never use substandard materials to fabricate a line. Brake lines are made of high-quality seamless, double-thickness steel tubing. They are coated (usually with a copper-lead alloy) to prevent corrosion.

Rear brakes

If you're working on a front-wheel drive vehicle, it's a good idea to check the rear wheel bearing adjustment before proceeding with the brake check. See the sidebar entitled *"Wheel bearing check, repack and adjustment"* for this procedure.

Disc brakes

If your vehicle is fortunate enough to be equipped with rear disc brakes, refer to the disc brake inspection procedure under the *Front brakes* subheading. The procedures are identical.

Drum brakes

Warning: *The dust created by the brake system may contain asbestos, which is harmful to your health. Never blow it out with compressed air and don't inhale any of it. An approved filtering mask should be worn when working on the brakes. Do not, under any circumstances, use petroleum-based solvents to clean brake parts. Use brake system cleaner only!*

Loosen the wheel lug nuts or bolts, raise the vehicle and support it securely on jackstands placed under the frame rails, subframe, the seam below the rocker panels where the body is mated with the floorpan (unibody-constructed vehicles only) or under the rear axle housing (vehicles with solid rear axles only), if not already done. Block the wheels at the opposite end of the vehicle to prevent it from rolling. Release the parking brake.

Some vehicles are equipped with a rubber plug in the brake backing plate (or in the drum on some models) near the edge of the drum. When removed, the brake shoe lining is visible through the hole **(see illustration)**. This is sufficient for a quick check of the brake shoes, but for a thorough check, the drums must be removed.

Remove the brake drum. If you're working on the rear brakes of a front-wheel drive vehicle, or on a vehicle with front drum brakes, and the drum is integral with the hub, refer to the sidebar entitled *"Wheel bearing check, repack and adjustment"* for the removal procedure. If the hubs on your vehicle are sealed, but retained by a lock nut, remove the nut and discard it (a new one should be used upon reassembly). If the drum is not integral with the hub, remove it by pulling it off the axle and brake assembly. **Note:** *Some brake drums are retained to the axle flange by two or three small screws.* If this proves difficult, make sure the parking brake is released, then squirt penetrating oil around the center hub area and around the wheel studs. Allow the oil to soak in and try to pull the drum off again.

If the drum still cannot be pulled off, the brake shoes will have to be retracted. This is done by first removing the small rubber plug from the backing plate or brake drum. On some vehicles there is a lanced cutout in the backing plate or drum instead of a plug. If this is the case, knock the lanced portion out with a hammer and chisel **(see illustration)** (you'll retrieve it after the drum has been removed). **Note:** *Rubber plugs are available from a dealer service department or auto parts store - be sure to install one in the hole when you're done.*

With the plug removed, insert a thin screwdriver and lift or push the adjusting lever off the star wheel, then use an adjusting tool or screwdriver to back off the star wheel several turns **(see illustrations)**. This will move the brake shoes away from the drum. If

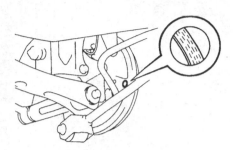

4.25 Some vehicles are equipped with holes in the backing plates (or in the brake drums) that allow you to view the brake lining thickness

4.26 Use a hammer and chisel to remove the plug from the brake backing plate or brake drum

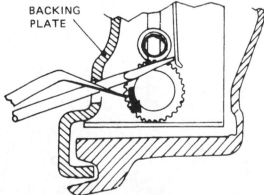

BACKING PLATE

4.28 Here's a cutaway view of the retracting procedure. The screwdriver is pushing the adjuster lever off the star wheel, while the adjusting tool (if you don't have one, another screwdriver can be used) turns the star wheel to reduce the length of the adjuster screw. This allows the shoes to move away from the drum, providing clearance for drum removal.

4.27 Push the self-adjusting lever aside and turn the star wheel

4.29 The lining thickness of a brake shoe with bonded lining is measured from the outer surface of the lining to the metal shoe - be sure to check both ends and in the center

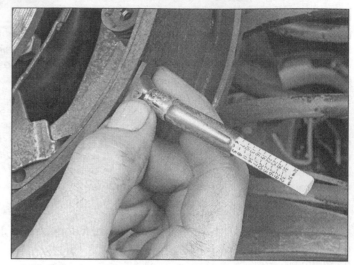

4.30 A tire tread depth gauge works well for measuring the thickness of the remaining material on brake shoes with riveted linings

4.31 Carefully peel the wheel cylinder boot back and check for leaking fluid - any leakage indicates the cylinder must be replaced or rebuilt

the drum still won't pull off, tap around its inner circumference with a soft-face hammer.

With the drum removed, do not touch any brake dust (see the **Warning** at the beginning of this Section). Position a drain pan under the brake assembly, clean the entire brake with brake system cleaner and allow it to dry.

Note the thickness of the lining material on both brake shoes. If the material has worn away to within 1/16-inch of the recessed rivets or metal backing, the shoes should be replaced **(see illustrations)**. The shoes should also be replaced if they're cracked, glazed (shiny surface) or contaminated with brake fluid, grease or gear oil.

Make sure that all the brake assembly springs are connected and in good condition.

Check the brake components for any signs of fluid leakage. If the brake linings are wet, identify the source of the fluid. If it's gear oil, the rear axle seal is leaking and must be replaced. Carefully pry back the rubber dust boots on the wheel cylinders located between the brake shoes **(see illustration)**. Any leakage is an indication that the wheel cylinders should be overhauled immediately (see Chapter 9). Check the brake hoses and connections for signs of leakage.

Clean the inside of the drum with brake system cleaner. Again, be careful not to breathe the dangerous asbestos dust.

Check the inside of the drum for cracks, score marks, deep scratches and hard spots, which will appear as small discolorations. If these imperfections cannot be removed with fine emery cloth, the drum must be taken to a machine shop equipped to resurface the drums.

If, after the inspection process all parts are in good working condition, reinstall the brake drum. If it was necessary to retract the brake shoes to get the drum off, adjust the brake shoes (see the sidebar on drum brake shoe adjustment). If your vehicle has sealed hubs retained by a locknut, tighten the nut securely (preferably you should consult a shop manual or a *Haynes Automotive Repair Manual* that covers your particular vehicle.

Install the wheels and lower the vehicle.

Drum brake adjustment

Most drum brakes are equipped with self-adjusters but some (mainly on older vehicles) are not, and will need periodic adjustment to compensate for wear. Even the brake shoes on self-adjusting drum brakes have to be adjusted when they are installed. This is because the self-adjusters maintain a specified clearance between the brake shoe lining and the drum, but won't compensate for a large gap (not until after many applications of the brake or parking brake, depending on design). Brakes that are out of adjustment will result a low pedal condition and can also cause the vehicle to pull to one side or the other when the brakes are applied.

Loosen the wheel lug nuts or bolts, raise the vehicle and support it securely on jackstands placed under the frame rails, subframe, the seam below the rocker panels where the body is mated with the floorpan (unibody-constructed vehicles only) or under the rear axle housing (vehicles with solid rear axles only), if not already done. Block the wheels at the opposite end of the vehicle to prevent it from rolling. Release the parking brake if you're adjusting rear drum brakes.

Remove the adjustment hole plug(s) in the brake backing plate or drum, if equipped. Rotate the wheel and make sure that the parking brake isn't causing any brake drag. If there is any drag, back off the parking brake adjusting nut (follow the parking brake cable to the equalizer - the point at which the single forward cable connects with the rear cables). Using a screwdriver or brake adjusting tool inserted through the correct hole, turn the adjuster star wheel or the adjuster screw until the drum is locked in place **(see illustrations)**. Some brakes (leading/leading shoe brakes [see Chapter 1]) have two adjusters. If you are working on one of these kinds of brakes, adjust the shoes one at a time.

Now, back-off the star wheel or adjuster screw eight to ten notches, or just to the point where the drum rotates freely without the brake shoes dragging (if the shoes are allowed to drag they may get too hot and expand, causing the drum to bind). To accomplish this it will be necessary to disengage the adjusting lever from the star wheel by inserting a thin screwdriver through the hole in the backing plate or drum, pushing or pulling the lever off the star wheel **(see illustration)**. **Note:** *Leading/leading shoe brakes don't have adjuster levers - simply turn the star wheel in the other direction to back-off the shoes.*

If necessary, adjust the parking brake as described later in this Chapter.

Install the wheels and lower the vehicle.

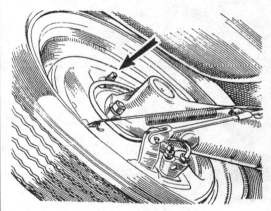

4.32 On some older vehicles, the adjuster is slotted and can be turned with a screwdriver - this turns an eccentric which spreads the shoes apart (or conversely, allows them to retract)

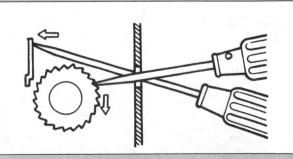

4.33 Other older vehicles also use eccentric adjusters, but the adjuster screw is a square-drive affair protruding from the backing plate

4.34 Here's the most common kind of adjuster setup. Insert a screwdriver through the hole in the backing plate (or brake drum) and turn the adjuster star wheel until the brake shoes drag on the drum ...

4.35 ... then, using another screwdriver or an adjusting tool, push (or pull) the lever off the adjuster wheel and back it off 8 to 10 clicks

Parking brake

The parking brake is pedal, pullrod or lever operated and, on most vehicles, is self-adjusting through the automatic adjusters in the rear brake drums (a very few manufacturers place the parking brake on the front wheels). However, supplementary adjustment may be needed in the event of cable stretch, wear in the linkage or after installation of new components.

Check

Check the entire length of each parking brake cable for wear, fraying, rust, kinks, etc. Apply multi-purpose grease to all friction areas and to the adjuster screw(s)

Many vehicles have a specification as to how many clicks the parking brake lever or pedal is supposed to ratchet, but the most obvious method to check the adjustment of the parking brake is to park the vehicle on a hill or in a steep driveway, apply the parking brake and place the transmission in Neutral (be sure to stay in the car during this check). The lever or pedal should ratchet anywhere between five and eight clicks or so. If the vehicle rolls, the parking brake should be adjusted.

Adjustment

If your parking brake is lever-actuated, look through the slot in the center console or peel back the boot and look for the end(s) of the parking brake cable(s) and the adjusting nut(s). If your vehicle has this setup, apply the parking brake two clicks and tighten the nut until the parking brake lever travel is within reason **(see illustration)**. It may be necessary to remove the center console for access to the nut(s). If there are two cables, tighten each nut equally. If the parking brake is pullrod-actuated, look up under the dash for a turnbuckle-type adjuster **(see illustration)**. On other models, the equalizer (the point where the front cable meets the rear cables) is located under the center console, which will require removal of the console **(see illustration)**. Raise the rear of the vehicle, support it securely on jackstands and release the parking brake. Turn the wheels to confirm that the brakes are not dragging.

On many cars and trucks the adjuster is located under the vehicle. Raise the rear of the vehicle (or the front on vehicles with the parking brake on the front wheels) until the wheels are clear of the ground, support it securely on jackstands and block the front wheels. Release the parking brake pedal or lever.

Locate the parking brake equalizer. Hold the cable from turning with locking pliers, then loosen the locknut, if equipped, and tighten the adjuster

4.36a On some vehicles, the parking brake cable is adjusted at the lever

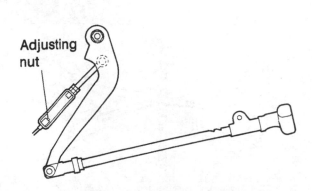

Adjusting nut

4.36b On some vehicles with pullrod-actuated parking brakes, the adjuster is located up under the dash

4.37 The parking brake equalizer on this vehicle is located underneath the center console - once the console has been removed, the parking brake can be adjusted by tightening this nut (arrow) at the equalizer

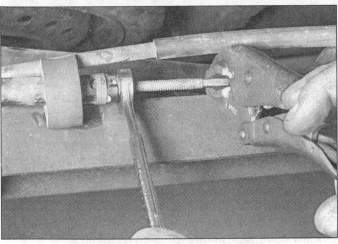

4.38 In this arrangement, the parking brake cable is located at the rear of the vehicle, where the left and right cables meet - to adjust, hold the threaded rod with a pair of pliers and turn the adjusting nut

4.39 On this type of parking brake cable equalizer, simply hold the threaded rod stationary and turn the nut to adjust the cable

nut **(see illustrations)** until a slight drag is felt when the rear wheels are turned. **Note:** *If the threads appear rusty, apply penetrating oil before attempting adjustment.*

Loosen the adjuster nut until there's no longer any drag when the rear wheels are turned, then loosen the cable adjusting nut an additional two turns.

Tighten the locknut, then apply and release the parking brake several times. Confirm that the brakes do not drag. Lower the vehicle to the ground.

Parking brake shoe adjustment (vehicles with rear disc brakes containing a drum style parking brake)

This kind of parking brake setup is almost identical to a full-size service drum brake assembly, only smaller, and its actuation is not hydraulic, but mechanical. It even has an adjuster screw with a star wheel. To adjust the shoes on this kind of parking brake (which will almost never need adjustment, unless you occasionally forget to release the parking brake before driving) refer to the sidebar entitled *"Drum brake adjustment."*

Brake pedal and bracket

Rarely do problems develop in this area, but when they do they have the potential to be very serious. Therefore, it's a good idea to crawl under the dash periodically and inspect the pedal pivot and the support bracket.

Working under the dash, grasp the pedal and wiggle it from side-to-side. If there's any slop in the pedal bushings, the bushings should be replaced **(see illustration)** (see Chapter 8).

On models with hanging brake pedals, inspect the bracketry under the dash that supports the brake pedal. There should be no broken spot welds or distortion in the metal. If there is , take the vehicle to a competent body shop for repair.

Also, be sure to check the pedal for distortion and fatigue.

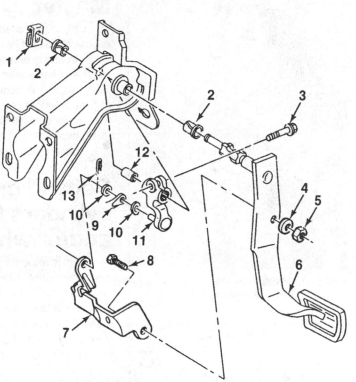

4.40 Typical brake pedal mounting details

1	Retaining clip	8	Bolt
2	Bushing	9	Pushrod
3	Pinch bolt	10	Washer
4	Washer	11	Actuator lever
5	Nut	12	Spacer
6	Brake pedal	13	Retaining clip
7	Actuator		

Brake fluid replacement

As stated in the maintenance schedule, many manufacturers don't actually specify an interval for replacing the brake fluid, but it's a great preventive maintenance step. Dirty, contaminated, moisture-laden fluid will deteriorate brake system components and, besides leaks, can cause dangerous problems like boiling fluid and loss of braking pressure.

To replace the fluid in your hydraulic system, simply bleed the brakes at each wheel until clean, clear brake fluid flows out. If performed at the interval specified, this method is sufficient.

If the recommended interval is exceeded by many thousands of miles, the only sure way to have pure, uncontaminated brake fluid in the system is to remove the master cylinder, calipers and/or wheel cylinders, rebuild or replace them, and flush the hydraulic lines with brake system cleaner. Finally, blow out the hydraulic lines with compressed nitrogen (compressed air, even filtered, has too much moisture in it). Install the new components (including new flexible hoses), fill the system with fresh brake fluid and bleed out the air.

So you see, it's much more cost effective to purge the old brake fluid at the recommended interval. See Chapter 9 for the brake system bleeding procedure.

Master cylinder seal check

Unbolt the master cylinder (without disconnecting the lines), pull it forward off the booster (if equipped) or firewall and check for leakage. You won't find this item on any manufacturer's maintenance schedule, but leaks can develop in the master cylinder and fluid can escape past the secondary cup or O-ring of the primary piston. This can go unnoticed for a long time. The dropping fluid level may at first be mistaken for normal brake pad wear, when in actuality it is leaking into the power booster (which will damage the booster) or onto the floor in the driver's footwell.

Rebuild or replace the master cylinder if you find leakage (see Chapter 9).

Rebuild or replace the hydraulic cylinders (master cylinder, calipers and/or wheel cylinders)

This is another maintenance item that isn't recommended by all manufacturers. You might even think it sounds like a waste of money and time. But, it's a sure way to know that your hydraulic components are in top shape. Brake pads and shoes are easy to check, and you can gauge how they are wearing just by looking at them. You can't see inside of a master cylinder or caliper, though, so you really don't know what kind of shape they're in. Refer to Chapter 9 for the rebuilding or replacement procedures.

5 Disc brakes

Introduction

This Chapter is divided into two sections: *Brake pad replacement* and *Brake disc inspection, removal and installation.*

The first section is subdivided into three parts and depicts the brake pad replacement procedures for fixed calipers, sliding calipers and floating calipers, in that order. If you aren't sure which kind of caliper you're about to work on, refer to Chapter 1 and review the information on caliper design under the *Disc brakes* subheading.

If you're replacing rear brake pads, refer to the appropriate sub-section, depending on caliper design, and follow the procedure outlined there. However, if you're working on a vehicle that has rear calipers with integral parking brake mechanisms, don't attempt to push the piston into its bore with a C-clamp - refer to the sidebar entitled *"Rear brake pad replacement - calipers with integral parking brake mechanisms."*

The second section deals with inspection, removal and installation of the brake disc.

Information on brake caliper overhaul can be found in Chapter 9.

Brake pad replacement

Warning: *Disc brake pads must be replaced on both front or rear wheels at the same time - never replace the pads on only one wheel. Also, the dust created by the brake system may contain asbestos, which is harmful to your health. Never blow it out with compressed air and don't inhale any of it. An approved filtering mask should be worn when working on the brakes. Do not, under any circumstances, use petroleum-based solvents to clean brake parts. Use brake system cleaner only!*

Note 1: *When servicing the brakes, use only high-quality, nationally recognized brand-name pads.*

Note 2: *This procedure also applies to the rear disc brakes on vehicles equipped with a separate parking brake mechanism (not built into the caliper). If you're working on a vehicle that has rear calipers with integral parking brake mechanisms, don't attempt to push the piston into its bore with a C-clamp - refer to the sidebar entitled* "Rear brake pad replacement - calipers with integral parking brake mechanisms."

The brake pad replacement procedures are covered through a sequence of illustrations, laid out in order from removing the old pads to installing the new ones. The captions presented with the illustrations will walk you through the entire procedure, one step at a time. Work only on one brake assembly at a time, using the assembled brake for reference, if necessary.

Regardless of the type of brake you're working on, there are a few preliminary steps to be taken before yanking the old pads out. First, park the vehicle on a level surface, open the hood and locate the master cylinder - it's usually mounted on the firewall or the power brake booster on the driver's side of the firewall (but on some vehicles it's on the other side of the engine compartment). Cover all painted areas around the master cylinder (fender included), remove the reservoir cover or cap(s) and remove about half of the fluid from the reservoir. It is necessary to do this so the reservoir doesn't overflow when the caliper piston(s) is pushed back into its bore to make room for the new pads. This can be accomplished with a suction pump, a siphoning kit or an old turkey baster or hydrometer. **Warning:** *Brake fluid is poisonous - don't start the siphoning action by mouth. If you use a turkey baster, never again use that baster for cooking!* Discard the fluid.

If you haven't already done so, this is a good time to put on your filtering mask, latex gloves and eye protection. Your hands aren't too dirty yet, and brake dust tends to get all over the wheels and inside of the wheel covers, so put them on *before* removing the wheels.

Next, remove the wheel covers (if equipped) from the front wheels and loosen the wheel lug nuts about one-half turn. Raise the vehicle and support it securely on jackstands. Remove the wheels, but remember to only work on one brake at a time.

Before touching or disassembling anything, clean the caliper, pads and disc with brake system cleaner and allow it to dry **(see illustration)**. Don't ever use compressed air or a brush to remove the brake dust.

Now, refer to the appropriate procedure under *Fixed caliper brake pad replacement, Sliding caliper brake pad replacement* or *Floating caliper brake pad replacement.* Be sure to stay in order and read the captions that accompany each illustration. Before installing the new pads, inspect the brake discs as outlined later in this Chapter. If machining is necessary, remove the discs and have them resurfaced by an automotive machine shop (if they are still thick enough to undergo machining). **Note:** *Professionals recommend resurfacing*

5.1 Before removing anything, clean the caliper and pads with brake cleaner and allow it to dry – position a drain pan under the brake to catch the residue – DO NOT USE COMPRESSED AIR TO BLOW THE DUST FROM THE PARTS!

the brake discs whenever replacing the brake pads, as this will produce a smooth, flat surface that will eliminate brake pedal pulsations and other undesirable symptoms related to questionable discs. At the very least, if you elect not to have the discs resurfaced, deglaze them with sandpaper or emery cloth, using a swirling motion to ensure a non-directional finish. Be careful not to get grease on the brake disc or brake pads.

After you have successfully installed the pads on one side of the vehicle, repeat the procedure on the other brake.

There are also a couple of things you must do before you can call the job complete. As soon as you have lowered the vehicle and tightened the lug nuts securely, check the brake fluid level in the master cylinder reservoir. If necessary, add some to bring it to the desired level (see Chapter 4, if necessary). Install the reservoir cover or cap, but don't close the hood yet.

Before starting the engine, depress the brake pedal a few times to bring the pads into contact with the disc. The pedal will go to the floor for the first couple of pumps, but will build up after that. If this is not done, you'll probably crash into something on the other side of the street after you back out of your driveway!

Now that you have a solid brake pedal, recheck the fluid level. It may have gone down some as the piston(s) in the caliper were forced outward, shoving the pads up against the discs.

Lastly, before committing the vehicle to normal service, road test it in an isolated area, if possible. For the first few hundred miles or so, try to avoid hard braking to allow the new pads to bed in.

Fixed caliper brake pad replacement

Warning: *The dust created by the brake system may contain asbestos, which is harmful to your health. Never blow it out with compressed air and don't inhale any of it. An approved filtering mask should be worn when working on the brakes. Do not, under any circumstances, use petroleum-based solvents to clean brake parts. Use brake system cleaner only!*

Tools required (these will suffice for most applications, however, the calipers on your vehicle may differ slightly):

Jack and jackstands
Lug nut (or bolt) wrench
Filtering mask
Safety glasses or goggles
Latex gloves
Hammer
Pin pinch
Needle-nose pliers
Prybar or large screwdriver (two in the case of three- or four-piston calipers)

Materials required

Brake pads
Brake system cleaner
Brake anti-squeal compound
High-temperature brake grease
Brake fluid
Cotter pins (if applicable)

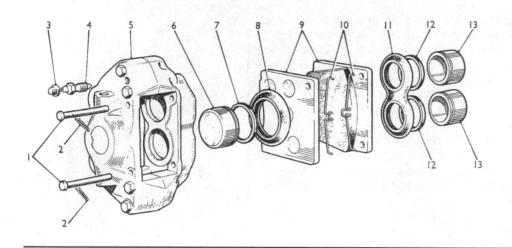

5.2 Exploded view of a typical fixed brake caliper assembly

1   Retaining pins
2   Spring clip or cotter pin
3   Dust cover
4   Bleeder screw
5   Caliper body
6   Piston (inner)
7   Piston seal
8   Dust boot
9   Pad assembly
10   Anti-rattle spring
11   Dust boot
12   Piston seal
13   Piston (outer)

Remove the old brake pads

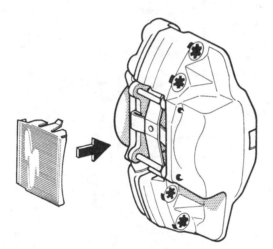

5.3 If equipped, pry off the cover from the caliper, noting how it's installed

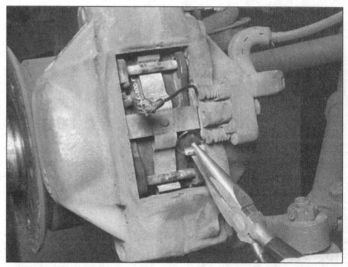

5.4 Some vehicles are equipped with electric wear sensors that plug unto the pads. When the pads wear down to the sensors, the probes on the sensors ground out on the brake disc, completing the circuit and turning on a light on the instrument panel. To remove these sensors, pull them out of their mounting slots with a pair of pliers. Note: *If the pads have worn down to the sensors, be sure to replace the sensors.*

5.5 The brake pads on most fixed calipers are held in the caliper with one or two retaining pins. The retaining pins are usually secured by cotter pins, spring-steel clips or the ends of the anti-rattle springs. If your retaining pin has a cotter pin like this one, remove the cotter pin and discard it (a new cotter pin must be used upon installation). If spring steel clips or the ends of the anti-rattle springs secure the retaining pins, pull them out with a pair of pliers

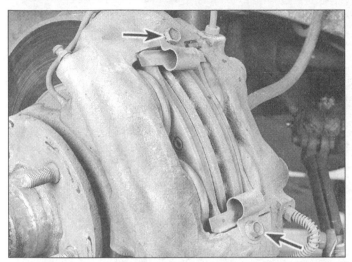

5.6 On some vehicles, the pads aren't secured by retaining pins, but are trapped in the caliper with two hold-down clips. To remove the clips, unscrew the bolts (arrows)

5.7 This caliper has retaining pins secured by spring steel clips. It also is equipped with anti-rattle springs - be careful not to let the springs fly out as you remove the retaining pins.

5.8 Use a hammer and a small punch to knock the retaining pin(s) out of the caliper. Note: *Some retaining pins, like the pins on this Mercedes, don't have retaining clips or cotter pins. Instead, the retaining pins are fitted with spring-steel collars around their heads that fit tightly into the holes in the caliper frame (if they don't fit tightly, replace them with new ones).*

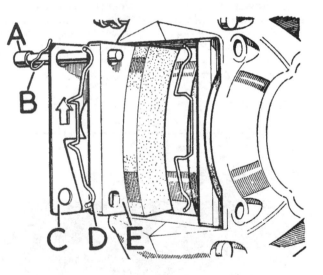

5.9 Typical brake pad arrangement on a fixed caliper

A   Pad retaining pin
B   Spring clip
C   Anti-squeal shim (note the directional arrow)
D   Anti-rattle spring
E   Brake pad

5.10 After removing the pad retaining pin(s), pull out *one* of the old brake pads (not both). You may need to use a pair of pliers to free the pad.

Preparation for installation

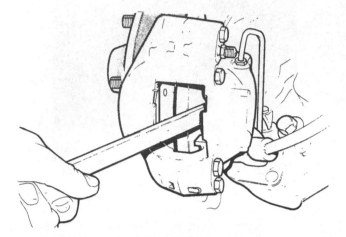

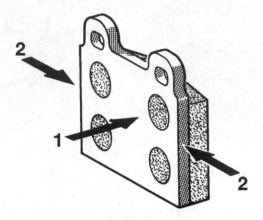

5.11 To make room for the new, thicker pad, push the caliper piston straight back into its bore as far as possible. A prybar or large screwdriver will work. If the caliper half has two pistons in it, you'll need two prybars. Caution: *Be careful not to cock the piston sideways at all - damage to the caliper bore or the piston seal could result.* **Before installing the new pad, check for fluid leakage around the caliper piston. If any is found, rebuild or replace the calipers (see Chapter 9).**

5.12 Prepare the new brake pads for installation by applying an anti-squeal compound to the backing plate (1) and a light film of high-temperature brake grease to the edges of the pads where they ride in the caliper (2). Both of these items can be purchased at your local auto parts store. Allow a few minutes for the compound to "set up" before installing the pads. Note: *Be sure to clean the caliper in the area where the brake pads make contact.* **Warning:** *Do not touch the pad friction material - if you get any grease on the pad, immediately remove it with brake system cleaner.*

Install the new pads

5.13 Insert the new pad into the caliper, along with the anti-squeal shim, if one was present when you took the old pad out. Note: *Many anti-squeal shims are directional - they have an arrow stamped in them that must point in the same direction as the forward rotation of the brake disc.* **Remember, don't get any grease on the pad lining. After installing the pad, repeat Steps 5.10 through 5.13 to remove and install the other pad.**

5.14 Install the retaining pins and anti-rattle springs (if equipped), or the hold-down springs. On this design, the rather stiff anti-rattle spring must be held down as the retaining pin is inserted.

All models

Repeat the entire procedure to replace the pads on the other wheel. Don't forget to pump the pedal a few times to bring the pads into contact with the disc. Be sure to then check the brake fluid level.

Sliding caliper brake pad replacement

Warning: *The dust created by the brake system may contain asbestos, which is harmful to your health. Never blow it out with compressed air and don't inhale any of it. An approved filtering mask should be worn when working on the brakes. Do not, under any circumstances, use petroleum-based solvents to clean brake parts. Use brake system cleaner only!*

Tools required (these will suffice for most applications, however, the calipers on your vehicle may differ slightly):

Jack and jackstands
Lug nut (or bolt) wrench
Filtering mask
Safety glasses or goggles
Latex gloves
Large C-clamp
Needle-nose pliers
Length of wire (a straightened-out coathanger will work)
Wrench, socket, Allen wrench or hammer and punch, depending on caliper design (to remove caliper bolts or pins)

Materials required

Brake system cleaner
Brake pads
Retaining clips, anti-rattle springs, caliper pins or keys (depending on caliper design), if damaged
Brake anti-squeal compound
High-temperature brake grease (slide rail grease)
Thread locking compound (non-hardening) (for calipers that are secured by bolts)
Brake fluid

5.15 Continue to drive the pins fully in. On models that use clips to secure the retaining pins, they can simply be pushed in, but on pins with spring-steel collars like this Mercedes, you'll have to knock them in until they're seated. Install the retaining pin clip(s) or cotter pin(s). If equipped with pad wear sensors, insert the probes of the sensors into the holes in the pad material, then seat the sensor in the cutout in the pad backing plate.

Remove the caliper

5.16 On most calipers, a C-clamp can be used to depress the piston into its bore; this aids removal of the caliper and installation of the new pads. If the design of the caliper doesn't permit this, the piston(s) can be depressed after the caliper has been removed. DO NOT attempt this on rear calipers with integral parking brake mechanisms.

5.17 On calipers held in place with a retainer clip and anti-rattle spring, unscrew the bolt and remove the clip and spring from each end of the caliper. Inspect the clip and anti-rattle spring for cracks or distortion. If any are found, replace all of the anti-rattle springs and clips.

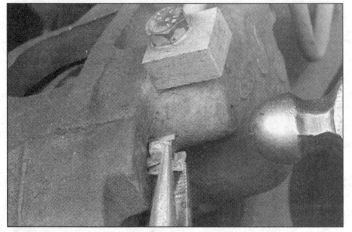

5.18 If your caliper is retained by pins that are driven into the slot between the caliper and the caliper mount (sometimes called the anchor plate, adapter or spindle flank), squeeze the end of the lower caliper pin with a pair of pliers and knock the pin into the groove as far as possible . . .

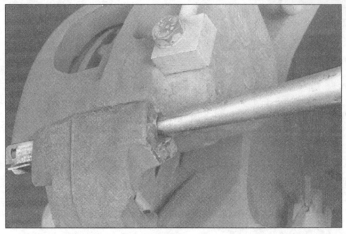

5.19 . . . then use a hammer and punch to drive the pin out completely. Proceed to remove the upper pin in the same manner. Check the pins for rust or other damage. If they were very easy to remove or appear damaged, replace them with new ones.

5.20 If the caliper on your vehicle is secured by a key, remove the key retaining bolt (some use Allen head bolts, others use conventional hex-head bolts) . . .

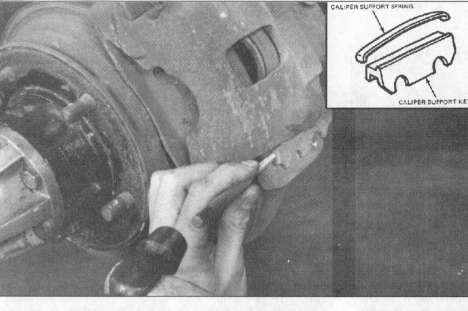

5.21 ... then drive the caliper key out with a hammer and punch (be careful not to lose the caliper support spring. Check the key and support spring for damage and wear, replacing them if necessary.

5.22 Another retention method, although somewhat less common, uses guide plates that wedge the caliper into the adapter (mount). To remove the caliper, pull out the retaining pin from the guide plate ...

5.23 ... then tap out the guide plate with a punch and hammer. Check the retaining pins and guide plates for wear. Replace as necessary.

5.24 After the retention devices have been removed, lift the caliper up and off the brake disc. Take care not to kink or pull on the brake hose

5.25 If the caliper isn't going to be removed for service, suspend it with a length of wire - DON'T let it hang by the brake hose. Now would be a good time to check around the caliper piston for leakage. Carefully pry up the dust boot, where it meets the piston - if any brake fluid or rust is found, rebuild or replace the calipers (see Chapter 9).

Remove the brake pads

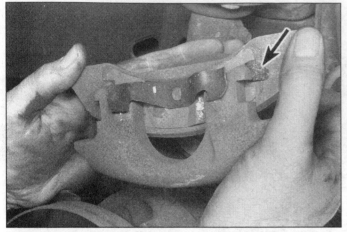

5.26 On some calipers, the outer pad is secured to the caliper by a spring clip. To remove the pad, push it towards the piston to dislodge the torque buttons (arrow), then slide the pad out of the caliper.

5.27 Some outer pads are held in the caliper frame by "ears" on the pad backing plate. Most of the time these can simply be snapped out of the caliper, but if the pad sticks, knock it out of the caliper with a hammer.

5.28 Remove the inner brake pad from the caliper adapter. On some setups the pad will slide right out of the adapter . . .

5.29 . . . but on models with an anti-rattle clip on the pad, it could be a tight fit. Pry the pad out of the adapter with a screwdriver, if necessary.

5.30 On some designs, both pads ride in the caliper mount. Take note of any springs and shims that may be present, remove them, then pry the pads out of the mount

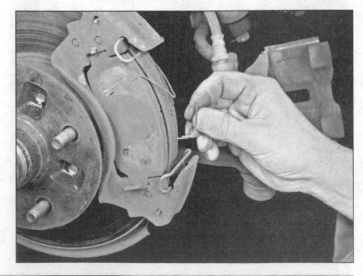

Preparation for installation

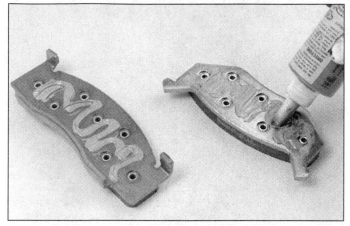

5.31a Apply a coat of anti-squeal compound to the backing plates of the new pads. Allow the compound to dry for a few minutes before installing the pads (follow the directions on the bottle).

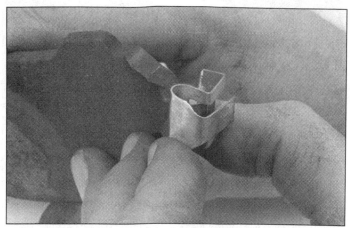

5.31b If applicable, install a new anti-rattle clip (or clips) onto the pad. Most sliding caliper designs use these clips on one end of the inner pad (usually the bottom) only.

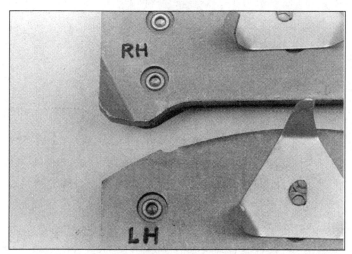

5.32 Some brake pads are marked for left and right sides of the vehicle - be sure they are installed in the correct positions

5.33 Push the caliper piston(s) completely into the caliper bore(s) to ensure plenty of room for the new pads. Use a wood block and a C-clamp, but don't use excessive force. DO NOT attempt this on rear calipers with integral parking brake mechanisms.

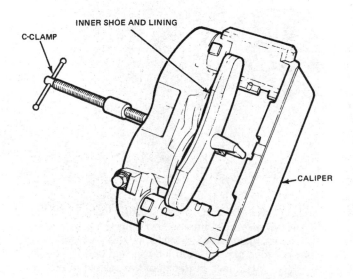

INNER SHOE AND LINING

C-CLAMP

CALIPER

5.34 You can also use the old inner pad if a wood block won't fit, as shown on this twin-piston sliding caliper.

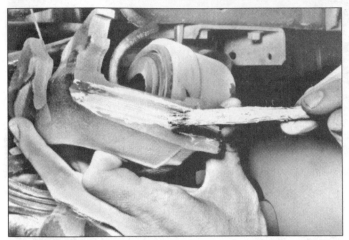

5.35 Clean the grooves on the ends of the caliper where it slides into the adapter (mount). Also clean the grooves or rails on the adapter, then apply a light film of high-temperature brake grease to the sliding surface of the caliper . . .

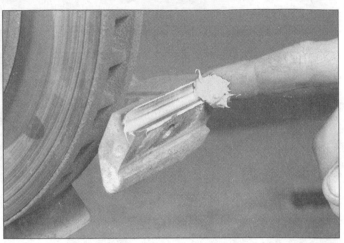

5.36 . . . or to the sliding surfaces of the adapter (these areas are also referred to as "ways" by some manufacturers).

Install the new pads

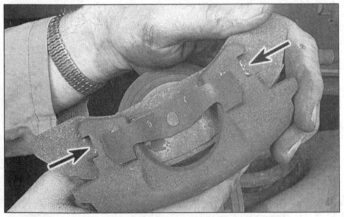

5.37 If the outer pads on your vehicle are secured to the caliper with a spring clip like this, push the pad into position on the caliper frame, making sure the torque buttons seat fully into the cutouts (arrows).

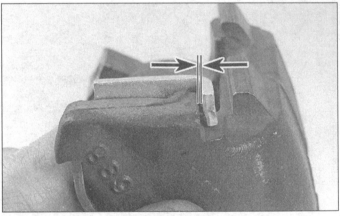

5.38 If the outer pad on your vehicle has bent "ears" like this, set the pad into the caliper frame and check the fit by moving it back and forth, as indicated by the arrows. They should fit tightly - if there's any slop . . .

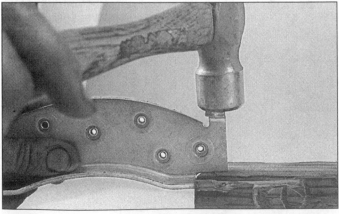

5.39 . . . set the pad against a wood block and tap the ear with a hammer, then re-check the fit (this might take a couple of tries to get right).

5.40 When you install the new inner brake pad on a design like this, make sure the "ears" (arrows) at either end of the pad backing plate are properly seated onto the "ways" of the machined surfaces of the caliper adapter.

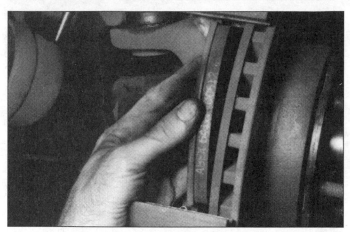

5.41 On models that have an anti-rattle clip (or clips on the pad(s), set the lower end of the pad into the groove or notch in the caliper adapter (mount), compress the clip and slide the upper end of the pad into position. If both pads on your vehicle ride in the caliper mount, see illustration 5.30.

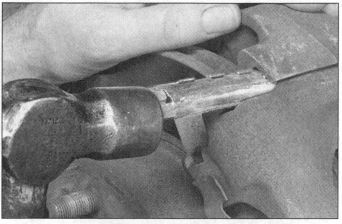

5.42 Install the caliper over the brake pads and disc. On calipers retained by pins, set the pins in their grooves and drive them into place . . .

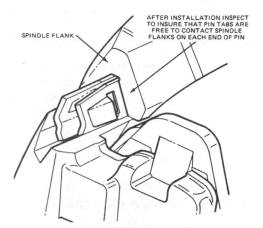

5.43 . . . making sure they are driven in completely. Make sure the pin tabs on each side of the spindle flank (adapter or mount) are exposed.

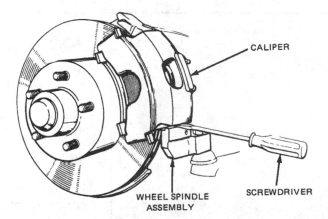

5.44 If the caliper is secured by a caliper key, insert a screwdriver between the spindle flank (adapter or mount) and the caliper to create a gap for the key . . .

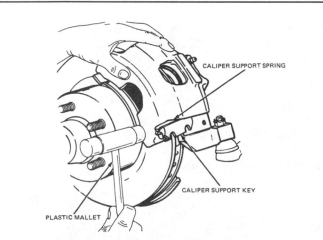

5.45 . . . then insert the caliper support key and spring clip into the slot. Drive the key and spring into place, then install the bolt, tightening it securely.

5.46 If your caliper is retained by an anti-rattle spring and retainer clip, install them (with the anti-rattle spring on top of the retainer clip, like this). Apply a drop of non-hardening thread locking compound and tighten the bolts to the torque listed in the Specifications at the back of the book.

5.47 If your caliper is secured by guide plates, slide the plates into the gaps between the caliper and the caliper mount, then install the retaining pins

All models

Repeat the entire procedure to replace the pads on the other wheel. Don't forget to pump the pedal a few times to bring the pads into contact with the disc. Be sure to then check the brake fluid level.

Floating caliper brake pad replacement

Warning: *The dust created by the brake system may contain asbestos, which is harmful to your health. Never blow it out with compressed air and don't inhale any of it. An approved filtering mask should be worn when working on the brakes. Do not, under any circumstances, use petroleum-based solvents to clean brake parts. Use brake system cleaner only!*

Note: *If the brake pads on your vehicle are held in the caliper by retaining pins, refer to the* Fixed caliper brake pad replacement *procedure.*

Tools required (these will suffice for most applications, however, the calipers on your vehicle may differ slightly):

Jack and jackstands
Lug nut (or bolt) wrench
Filtering mask
Safety glasses or goggles
Latex gloves
Large C-clamp
Needle-nose pliers
Length of wire (a straightened-out coathanger will work)
Wrench, socket, Allen wrench or Torx wrench, depending on caliper design (to remove caliper bolts or pins)

Materials required

Brake system cleaner
Brake pads
Caliper bushings and mounting bolts or guide pins (if worn or damaged)
Brake anti-squeal compound
High-temperature brake grease (slide rail grease)
Brake fluid

Remove the caliper

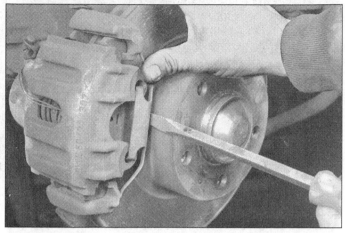

5.48 If equipped, remove the anti-rattle spring - this kind can simply be pried off.

5.49 To remove this kind of anti-rattle spring, depress the ends and pull the spring out.

5.50 Use a large C-clamp to compress the piston(s) into the caliper. DO NOT attempt this on rear calipers with integral parking brake mechanisms.

5.51 On some calipers, all you have to do is remove the lower bolt, or guide pin (arrow) . . .

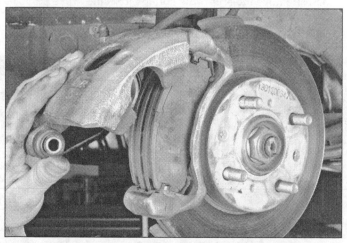

5.52 . . . and swing the caliper up for access to the brake pads . . .

5.53 . . . then insert a bolt or punch into the hole in the caliper mount (also referred to as the adapter or torque plate) to keep the caliper from falling. If there is no provision for supporting the caliper like this, use a piece of wire and attach it between the caliper and a suspension component

Haynes Automotive Brake Manual

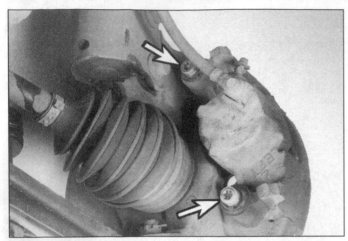

5.54 On some calipers both bolts (guide pins) must be removed (arrows). Some models use Allen or Torx-head bolts - always use the proper tool to avoid damaging the bolts.

5.55 Don't ever allow the caliper to hang by the brake hose - hang it from the coil spring or other suspension member with a piece of wire.

Remove the brake pads - caliper adapter-mounted pads

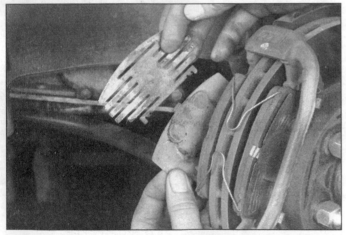

5.56 Remove the anti-squeal shims from the pads, if present. Note how they are positioned - some models have only one shim per pad, but some may have two.

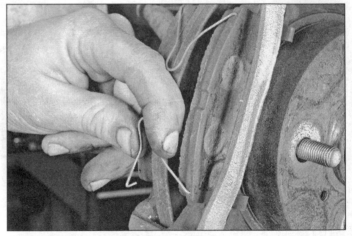

5.57 Remove the anti-rattle springs from the brake pads, if equipped. Replace these springs if they are damaged.

5.58 Remove the outer brake pad from the caliper adapter.

5.59 Remove the inner brake pad

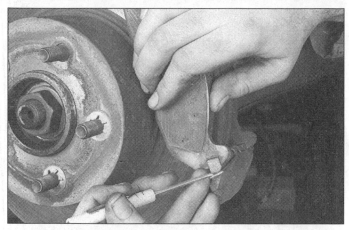

5.60 On some designs, the brake pads are retained by an anti-rattle spring which traverses the width of the caliper adapter. To remove the pads, pry downward on the lower anti-rattle clip and remove the outer brake pad . . .

5.61 . . . then pull the anti-rattle clips away from the pad with your index fingers and force the inner brake pad out with your thumbs

Remove the brake pads - caliper-mounted pads

5.62 Remove the inner pad by snapping it out of the piston

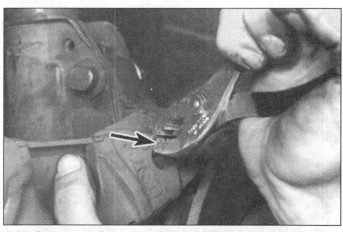

5.63 On some calipers, the outer pad is fastened by tabs that fit through holes in the caliper and are bent over. Remove the pad by bending the tabs straight out with a pair of pliers . . .

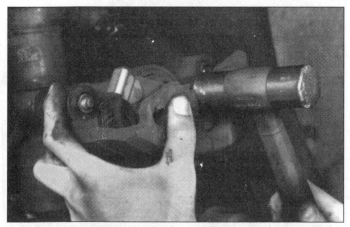

5.64 . . . then dislodge it from the caliper with a hammer

5.65 Most other outer pads are held to the caliper frame with a spring clip. Push the pad away from the caliper frame and slide it off the caliper (it may be necessary to pry it off with a screwdriver.

Haynes Automotive Brake Manual

Preparation for installation

5.66 Carefully peel back the edge of the piston boot and check for corrosion and leaking fluid. If any is present, rebuild or replace the calipers (see Chapter 9).

5.67 Check the anti-rattle clips, if present, for cracks and distortion and install new ones, if necessary.

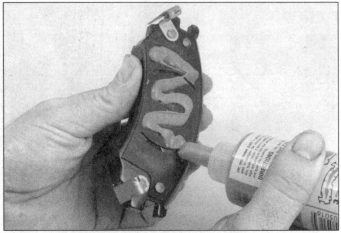

5.68 Apply anti-squeal compound to the back sides of both new brake pads. Allow the compound to "set up" for a few minutes before installing them (follow the instructions on the product).

5.69 If applicable, remove the pad retainer spring from the old inner pad and snap it into the new pad, in the direction shown (arrow).

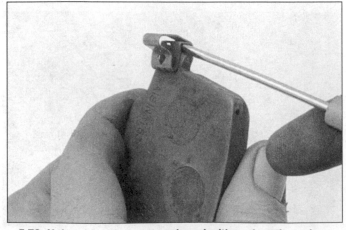

5.70 If the old pads were equipped with anti-rattle springs like this, pry them off and transfer them to the new pads. If they are worn or bent, install new ones.

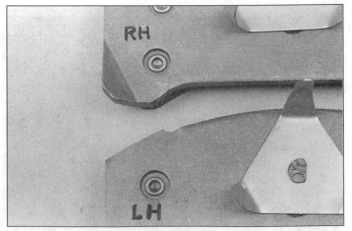

5.71 Some brake pads are marked for left and right sides of the vehicle - be sure they are installed in the correct positions.

5.72 Inspect the caliper bolts and bushings (A) for damage and the contact surfaces (B) for corrosion. Clean the contact surfaces with brake system cleaner and sandpaper if they are dirty.

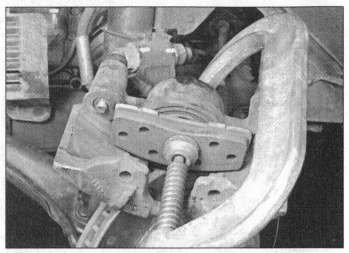

5.73 If the piston isn't completely bottomed in the caliper, use a C-clamp and the old inner pad to bottom the piston in the caliper bore. DO NOT attempt this on rear calipers with integral parking brake mechanisms.

5.74 Apply a light coat of high-temperature brake grease (caliper slide rail grease) to the upper steering knuckle (or caliper adapter)-to-caliper contact surface

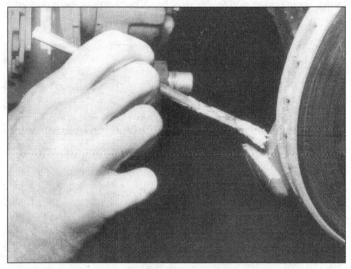

5.75 Also lubricate the lower contact surface with the same grease

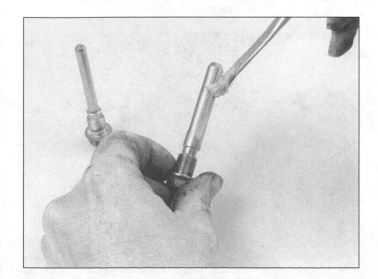

5.76 Wipe off the caliper mounting bolts (guide pins) and apply a thin coat of silicone grease (or high-temperature brake grease) to their sliding surfaces

Install the new brake pads - caliper adapter-mounted pads

5.77 Install the new brake pads in the caliper adapter - make sure their upper and lower notches are properly engaged with the projections on the adapter.

5.78 On pads that have "ears," make sure the ears engage properly with the flats on the caliper adapter

5.79 On models equipped with this type of anti-rattle clip, position the clips on the anchor plate and install the inner brake pad . . .

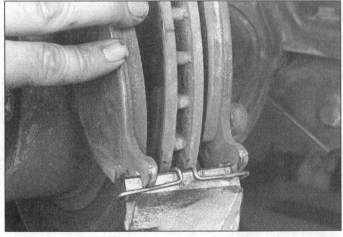

5.80 . . . place the lower end of the outer pad on the anchor plate and push it down against the anti-rattle clip, then lift up on the upper anti-rattle clip and swing the top of the pad into position

5.81 Install the caliper (on this design the notch in the lower edge of the caliper must be engaged with the anchor plate) and tighten the mounting bolts (guide pins) to the torque listed in the Specifications at the back of the manual. Note: *It's a good idea to put a drop of thread-locking compound (the non-hardening type) to the threads of the mounting bolts before installing them.* Install the anti-rattle spring, if equipped. Make sure the brake hose doesn't get twisted when installing the caliper.

Install the new pads - caliper-mounted pads

5.82 Install the inner pad by pushing the clip straight into the piston

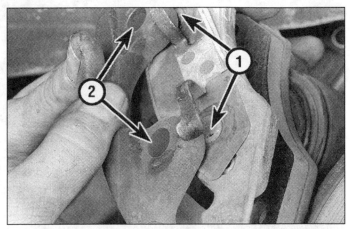

5.83 To install the new outer pad into the caliper, slide the pad into the caliper, making sure the locating lugs on the pad (if equipped) (1) seat into the mounting holes in the caliper frame (2). If your pad has bent "ears" that fit over the top of the caliper frame, see illustrations 5.38 and 5.39.

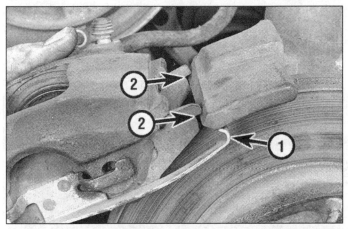

5.84 Install the caliper/brake pad assembly over the disc and onto the caliper adapter (mount). If the outer pad on your vehicle is equipped with an anti-rattle spring (1) like this one, position the anti-rattle spring under the upper arm of the mount with the notches in the upper edge of both pads (2) resting on the upper arm of the mount. Then, rotate the caliper down until the notches in the opposite end of the pads seat against the lower arm of the mount. Make sure the brake hose doesn't get twisted when installing the caliper.

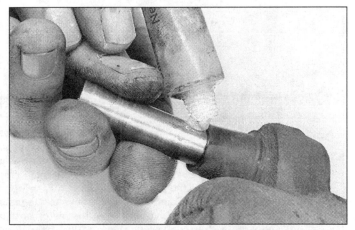

5.85 Apply silicone grease (or high-temperature brake grease) to the caliper pins (mounting bolts) and to the inside the insulators. Insert the pins through the caliper housing and into the caliper mount. Note: *It's a good idea to put a drop of thread locking compound (the non-hardening type) to the threads of the mounting bolts before installing them.* Start the threads by hand, then tighten the pins to the torque listed in the Specifications at the back of the manual.

5.86 On models where the outer pad is positioned by bendable "ears" that protrude through holes in the caliper frame, insert a large screwdriver between the outer pad flange and the brake disc to seat the pad, then bend the tabs over with a hammer.

All models

Repeat the entire procedure to replace the pads on the other wheel. Don't forget to pump the pedal a few times to bring the pads into contact with the disc. Be sure to then check the brake fluid level.

Rear brake pad replacement - calipers with integral parking brake mechanisms.

Warning: *The dust created by the brake system may contain asbestos, which is harmful to your health. Never blow it out with compressed air and don't inhale any of it. An approved filtering mask should be worn when working on the brakes. Do not, under any circumstances, use petroleum-based solvents to clean brake parts. Use brake system cleaner only!*

To replace the rear brake pads on vehicles that have parking brake mechanisms built into the rear calipers, follow the procedure under the appropriate heading (fixed, sliding or floating caliper brake pad replacement) to remove the caliper and extract the brake pads. The only differences you'll encounter when replacing the pads on a caliper with an integral parking brake mechanism are these:

a) Sometimes it's necessary to unclip the brake hose and/or the parking brake cable from a bracket to allow caliper removal. In almost all cases, though, the brake hose can remain attached to the caliper.

b) DO NOT attempt to push the piston into the caliper bore by using a C-clamp, as shown in illustrations 5.16, 5.33, 5.50 and 5.73. To make room for the new pads, the caliper piston will have to be bottomed in its bore, but on most models you can't do it by pushing the piston in. The piston will have to be turned in, like a big screw **(see illustrations)**.

c) On some calipers (notably later model General motors products), the piston can be pushed into the bore, but only after removing the parking brake lever from the back of the caliper **(see illustrations)**.

d) After the job has been completed, operate the parking brake until satisfactory parking brake pedal or lever travel is obtained.

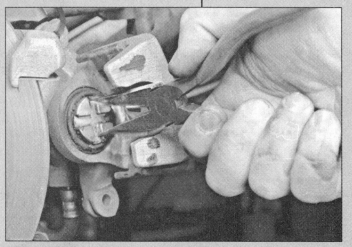

5.87 To make room for the new brake pads on rear calipers with integral parking brake mechanisms, engage the tips of a pair of needle-nose pliers with the slots or cutouts in the piston face, then rotate the piston clockwise until it stops. Be careful not to distort the dust boot - if it sticks to the piston, insert a small, dull screwdriver between the piston and the boot, then gently drag the screwdriver around the piston to break the bond.

5.88 If you need a little more leverage than a pair of needle-nose pliers provides, use a pin spanner. Special tools are also available to turn the piston in, but in most cases the methods shown here work just fine.

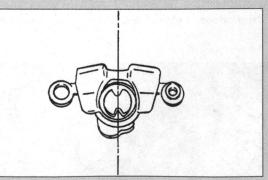

5.89 Before installing the new pads, adjust the position of the piston so the cutouts are aligned like this. Most inner brake pads on these calipers are equipped with a pair of pegs or a D-shaped projection that fits into the cutouts or slot. Make sure the projections on the pads engage properly with the slot or cutouts.

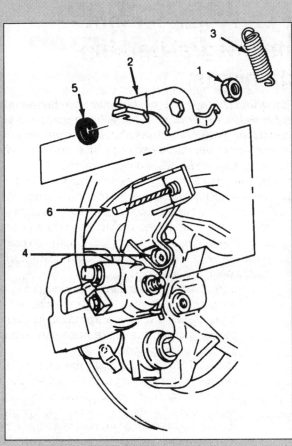

5.90 If the rear caliper on your vehicle doesn't lend itself to rotating the piston (like on this late model GM design), remove the parking brake lever from the back of the caliper . . .

1 Nut
2 Lever
3 Return spring
4 Bracket bolt
5 Lever seal (sealing ridge faces caliper)
6 Parking brake cable

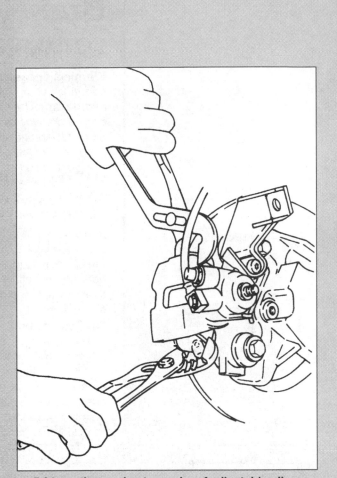

5.91 . . . then, using two pairs of adjustable pliers, squeeze the inner pad and the outer part of the caliper frame together to bottom the piston

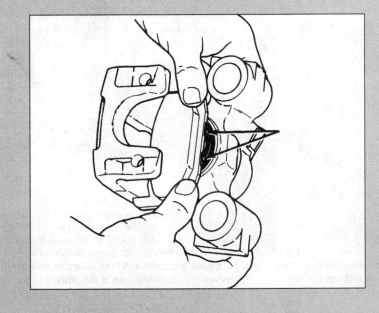

5.92 When installing the new inner pad, position the tangs of the pad retainer as shown, then clip the pad into the retainer - make sure the tab on the retainer fits into the "D" notch in the piston face. After installing the caliper, install the parking brake seal (if removed) and lever.

Brake disc inspection, removal and installation

Warning: *The dust created by the brake system may contain asbestos, which is harmful to your health. Never blow it out with compressed air and don't inhale any of it. An approved filtering mask should be worn when working on the brakes. Do not, under any circumstances, use petroleum-based solvents to clean brake parts. Use brake system cleaner only!*

Inspection

If you haven't already done so, loosen the wheel lug nuts, raise the front of the vehicle and support it securely on jackstands. Apply the parking brake. Remove the wheels.

If the brake disc on your vehicle is not integral with the wheel hub, reinstall the lug nuts and tighten them securely. This clamps the disc tightly against the hub and will allow for an accurate runout check. If you tighten the lug nuts but there's still a gap between the nut and the hub, install washers between the nuts and the hub.

Turn the brake disc - if it is difficult to turn, depress the piston(s) in the caliper with a C-clamp **(see illustration 5.16 or 5.50)** (sliding and floating calipers) or, in the case of fixed calipers, pry the brake pads away from the disc a little.

Visually inspect the disc surface for score marks and other damage **(see illustration)**. Light scratches and shallow grooves are normal after use and won't affect brake operation. Deep grooves - over 0.015-inch (0.38 mm) deep - require disc removal and refinishing by an automotive machine shop. Be sure to check both sides of the disc.

To check disc runout, place a dial indicator at a point about 1/2-inch from the outer edge of the disc **(see illustration)**. Set the indicator to zero and turn the disc. Generally, the indicator reading should not exceed 0.004-inch. If it does, the disc should be refinished by an automotive machine shop. Some manufacturers allow for more runout than 0.004-inch, but runout is not desirable - the less the better. Some manufacturers regard a runout of 0.004-inch to be excessive. **Note:**

5.93 The brake pads on this vehicle were obviously neglected, as they wore down to the rivets and cut deep grooves into the disc - wear this severe means the disc must be replaced

5.94 To check disc runout, mount a dial indicator as shown and rotate the disc

Professionals recommend resurfacing the brake discs regardless of the dial indicator reading (to produce a smooth, flat surface that will eliminate brake pedal pulsations and other undesirable symptoms related to questionable discs). At the very least, if you elect not to have the discs resurfaced, deglaze them with sandpaper or emery cloth **(see illustration)**.

The disc must not be machined to a thickness less than the specified minimum thickness. The minimum (or discard) thickness is cast into the disc **(see illustrations)**. The disc thickness can be checked with a micrometer **(see illustration)**. Measure the thickness in several places around the disc. You shouldn't wind up with readings that are more than 0.0005-inch different. If you do, the disc parallelism is out of the range that most manufacturers specify as acceptable.

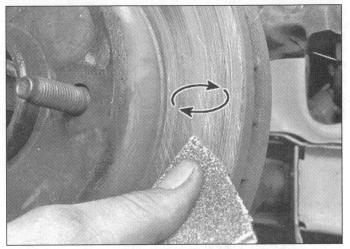

5.95 **Using a swirling motion, remove the glaze from the disc with sandpaper or emery cloth**

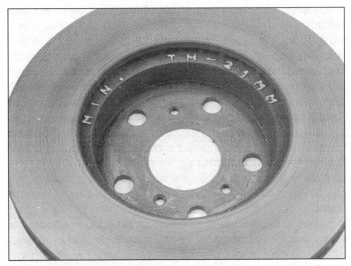

5.96 **The minimum wear dimension is stamped or cast into the center of the disc, or . . .**

5.97 **. . . on the disc "hat" area, or . . .**

5.98 **. . . on the edge of the disc**

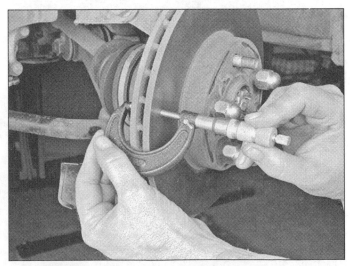

5.99 **Use a micrometer to measure disc thickness**

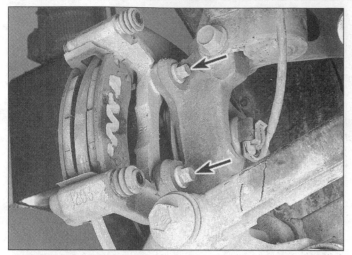

5.100 To remove the caliper adapter (mount), remove these two bolts (arrows)

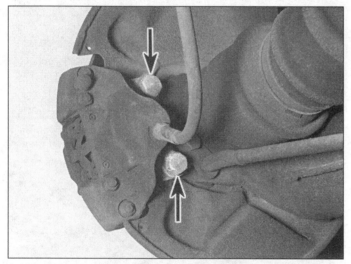

5.101 On vehicles with fixed calipers, remove the two caliper-to-steering knuckle or rear knuckle bolts (arrows)

5.102 Some vehicles have brake discs that are bolted to the hubs

Removal

Remove the brake caliper. On sliding and floating calipers, follow the caliper removal steps described in the *Brake pad replacement* procedure. Then, on models with adapter plates that would interfere with removal of the disc, remove the adapter plate. To do this, remove the two adapter plate-to-steering knuckle (or rear knuckle) bolts **(see illustration)**. On vehicles with fixed calipers, remove the two bolts that retain the caliper to the steering knuckle (or rear knuckle) **(see illustration)**.

On some vehicles, the disc is an integral part of the hub. To remove the disc on these models, refer to the Chapter 4 sidebar entitled *"Wheel bearing check, repack and adjustment"* for the hub/disc removal procedure. **Note:** *If the disc is to be taken to a machine shop for resurfacing, clean out all the old grease from the hub before the machine work is done. Then, when you get it back, clean the hub out again to ensure no metal chips remain.*

Some vehicles have a similar arrangement, where the disc seems to be an integral part of the hub, but is actually detachable. On rear-wheel drive models the disc is removed like any other integral disc/hub assembly, but the disc can be unbolted and separated from the hub **(see illustration)**. Many smaller front-wheel drive and four-wheel drive vehicles use a setup like this, too, but the disc/hub assembly has to be pulled off the front driveaxle and pressed (usually) from the steering knuckle **(see illustration)**.

On other vehicles the brake disc slips over the hub and wheel studs and can simply be pulled off (of course, you'll have to remove the lug nuts if you reinstalled them). An exception to this rule are the vehicles which use screws to retain the disc to the hub. Sometimes these screws become rusted and will have to be soaked with penetrating oil and removed with an impact screwdriver **(see illustration)**. It's also a good idea to apply penetrating oil around the area where the disc meets the hub. **Note:** *Before separating the disc from the hub, mark the relationship of the disc to one of the wheel studs. If the wheels were dynamically balanced on*

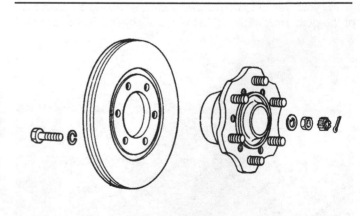

5.103 On the front brakes of many front-wheel drive and four-wheel drive vehicles the discs are also bolted to the hubs, but the entire assembly must be removed because the disc fits against the backside of the hub

the vehicle, returning the disc to the original position will retain the balance.

Even if you use penetrating oil, the disc might still refuse to come off. Check to see if the disc has two or three threaded holes in the area where the wheel seats. If it does, thread two or three bolts of the correct size and thread pitch into the holes and tighten them to force the disc off the hub **(see illustration)**. If you're removing a rear brake disc with a drum-type parking brake inside the center of the disc, and the disc won't come off, remove the plug from the parking brake adjusting access hole. Engage the notches in the adjusting star wheel with a suitable tool (such as a screwdriver) and retract the parking brake shoes. Now pull the disc off the hub.

5.104 If the disc retaining screws are stuck, use an impact screwdriver to loosen them

5.105 If the disc sticks to the hub, thread two bolts (or three, on some models) into the holes in the disc and tighten them to force the disc off the hub

Installation

Installation is the reverse of removal. If you're installing a disc with an integral hub, repack the wheel bearings, install the disc and adjust the bearings as described in the Chapter 4 sidebar *"Wheel bearing check, repack and adjustment."*

Install the caliper adapter (if applicable), tightening the bolts to the torque listed in the Specifications at the rear of this manual. Install the caliper, also tightening the bolts to the proper torque.

If your vehicle is equipped with a drum-type parking brake inside of the hub of the disc, be sure to adjust the parking brake shoes (see Chapter 7).

Notes

6 Drum brakes

Introduction

This Chapter is divided into two sections: *Brake drum removal, inspection and installation* and *Brake shoe replacement.*

The first section deals with removal of the brake drum for access to the brake shoes, adjuster(s), parking brake mechanism and wheel cylinder, and inspection and installation of the brake drum.

The second section is subdivided into two parts and depicts the brake shoe replacement procedures for the two most typical drum brake designs: Duo-servo and leading/leading-type drum brakes. If your vehicle is equipped with uni-servo drum brakes, refer to the duo-servo section. If your vehicle is equipped with leading/leading-type drum brakes, refer to the leading/trailing sequence. If you aren't sure which kind of brake you're about to work on, refer to Chapter 1 and review the information under the *Drum brakes* subheading.

Information on wheel cylinder overhaul or replacement can be found in Chapter 9.

Brake drum removal, inspection and installation

Warning: *The dust created by the brake system may contain asbestos, which is harmful to your health. Never blow it out with compressed air and don't inhale any of it. An approved filtering mask should be worn when working on the brakes. Do not, under any circumstances, use petroleum-based solvents to clean brake parts. Use brake system cleaner only!*

Removal

Loosen the wheel lug nuts, raise the vehicle and support it securely on jackstands. Block the wheels at the opposite end of the vehicle so it can't roll. If a rear brake drum is to be removed, release the parking brake. Remove the wheel(s).

If you're removing a drum that is integral with the hub assembly (one that can't be separated from the hub), refer to the Chapter 4 sidebar entitled *"Wheel bearing check, repack and adjustment"* for the drum removal procedure.

If you're removing a brake drum that simply slips off of an axle or hub flange, apply penetrating oil around the area where the drum meets the hub or axle flange, and also around the wheel studs. Allow the penetrant a few minutes to seep in.

Check the drum for the presence of screws that secure the drum to the hub or axle flange. If present, remove them. It may be necessary to use an impact screwdriver to break the screws loose **(see illustration 5.104** in Chapter 5**)**.

Mark the relationship of the brake drum to one of the wheel studs or to the hub **(see illustration)**. This will preserve the balance if the wheels have been dynamically balanced on the vehicle. Try to pull the drum off. If it sticks, the brake shoes will have to be retracted. Refer to Chapter 4 and follow the procedure described under the *Drum brakes* portion of the *Rear brakes* section.

If the drum is stuck to the hub or axle flange, look for two or three threaded holes in the drum that look like they have no apparent purpose. If your drum is equipped with such holes, install two (or three) bolts of the correct size and thread pitch into the threaded holes in the drum. Tighten the bolts, a little at a time, until the drum is forced off **(see illustration)**.

6.1 Mark the relationship of the drum to the hub (or one of the wheel studs) so the balance will be retained

6.2 If the brake drum refuses to slide off the hub, screw two bolts of the proper size and thread pitch into the threaded holes and tighten them evenly, a little at a time. The drum will be pushed off by the bolts

Inspection

Check the drum for cracks, score marks, deep scratches and hard spots, which will appear as small discolored areas. If the hard spots cannot be removed with sandpaper or emery cloth, or if any of the other conditions listed above exist, the drum must be taken to an automotive machine shop to have it turned. **Note:** *Professionals recommend resurfacing the drums whenever a brake job is done. Resurfacing will eliminate the possibility of out-of-round or tapered drums. If the drums are worn so much that they can't be resurfaced without exceeding the maximum allowable diameter (stamped into the drum)* **(see illustration)**, *then new ones will be required. At the very least, if you elect not to have the drums resurfaced, remove the glazing from the surface with emery cloth or sandpaper using a swirling motion* **(see illustration)**.

If you have a brake drum micrometer or a vernier caliper that's large enough, measure the inside diameter of the drum at four different locations around the drum (45-degree intervals) **(see illustration)**. Write down each measurement. Generally, a drum that is 0.010-inch out-of-round needs to be machined. Some manufacturers would consider 0.010-inch out-of-round way too much.

Also, check the diameter of the brake drum at the innermost portion of the friction surface and at the outermost portion of the friction surface. Subtract the two measurements to calculate drum taper. If the drum exhibits a taper of 0.005-inch or more, it should be resurfaced.

Note: *If the drum is to be taken to a machine shop for resurfacing, clean out all of the old grease from the hub (if applicable) before the machine work is done. Then, when you get it back, clean the hub out again to ensure no metal chips remain.*

6.3 The maximum drum diameter is cast or stamped into the brake drums. You'll usually find this dimension on the inside of the drum, but some manufacturers cast it into the outside of the drum.

6.4 Remove the glaze from the drum surface with sandpaper or emery cloth.

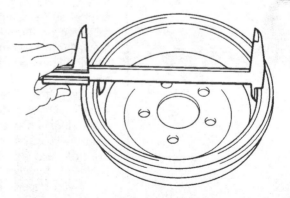

6.5 Measuring the diameter of the brake drum with a large vernier caliper.

Installation

Before installing the brake drum, make sure the friction surface is perfectly clean. If it isn't, or if in doubt, wash the drum thoroughly with brake system cleaner. Also confirm that the brake shoes are clean and all of the components of the drum brake assembly are in order.

If you're installing a drum that slips over the wheel studs, do just that. Make sure the match mark you applied to the drum and wheel stud are aligned. If you're installing a brake drum that's integral with the wheel hub, clean and repack the wheel bearings as described in the Chapter 4 sidebar entitled *"Wheel bearing check, repack and adjustment."* Install the hub/drum assembly and adjust the wheel bearings as described in Chapter 4.

Regardless of the type of drum you've just installed, adjust the drum brakes. This is also covered the Chapter 4 sidebar entitled *"Drum brake adjustment."*

Brake shoe replacement

Warning: *Drum brake shoes must be replaced on both wheels at the same time - never replace the shoes on only one wheel. Also, the dust created by the brake system may contain asbestos, which is harmful to your health. Never blow it out with compressed air and don't inhale any of it. An approved filtering mask should be worn when working on the brakes. Do not, under any circumstances, use petroleum-based solvents to clean brake parts. Use brake system cleaner only!*

Caution: *Whenever the brake shoes are replaced, the retracting and hold-down springs should also be replaced. Due to the continuous heating/cooling cycle the springs are subjected to, they lose their tension over a period of time and may allow the shoes to drag on the drum and wear at a much faster rate than normal. When replacing the rear brake shoes, use only high-quality, nationally-recognized brand-name parts.*

Note: *Professionals recommend resurfacing the drums whenever a brake job is done. Resurfacing will eliminate the possibility of out-of-round or tapered drums. If the drums are worn so much that they can't be resurfaced without exceeding the maximum allowable diameter (stamped into the drum)* **(see illustration 6.3)**, *new ones will be required. At the very least, if you elect not to have the drums resurfaced, remove the glazing from the surface with sandpaper or emery cloth using a swirling motion.*

Regardless of the type of brake you're working on, there are a few preliminary steps to be taken before disassembling the brake. First, park the vehicle on a level surface, open the hood and locate the master cylinder - it's usually mounted on the firewall or the power brake booster on the driver's side of the firewall (but on some vehicles it's on the other side of the engine compartment). Cover all painted areas around the master cylinder (fender included), remove the reservoir cover or cap(s) and remove about one-quarter of the fluid from the reservoir. It is necessary to do this so the reservoir doesn't overflow when the wheel cylinder pistons are pushed back into their bores to make room for the new linings. This can be accomplished with a suction pump, a siphoning kit or an old turkey baster or hydrometer. **Warning:** *Brake fluid is poisonous - don't start the siphoning action by mouth. If you use a turkey baster, never again use that baster for cooking!* Discard the fluid.

If you haven't already done so, this is a good time to put on your filtering mask, latex gloves and eye protection. Your hands aren't too dirty yet, and brake dust tends to get all over the wheels and inside of the wheel covers, so put them on *before* removing the wheels.

Next, remove the wheel covers (if equipped) from the wheels and loosen the wheel lug nuts about one-half turn. Raise the vehicle and support it securely on jackstands. Remove the wheels, but remember to only work on one brake at a time. You can use the other side as a model if you aren't sure about something. Another good idea is to make sketches or take instant photos of the areas you think you might have trouble with. The parking brake actuator and the adjuster mechanism are good examples of areas to concentrate on. Always note the orientation of the adjuster screw star wheel and all of the various springs.

Remove the brake drum by following the procedure under the heading *Brake drum removal, inspection and installation.*

Before touching or disassembling anything, clean the brake assembly with brake system cleaner and allow it to dry **(see illustration)**. Don't ever use compressed air or a brush to remove the brake dust.

Now, refer to the appropriate procedure under *Duo-servo brake shoe replacement* or *Leading/trailing brake shoe replacement*. **Note:** *If the vehicle you are working on is equipped with uni-servo brakes, follow the duo-servo brake shoe replacement procedure. If it is equipped with leading/leading-type brakes, follow the pertinent steps in the leading/trailing shoe replacement sequence.* Be sure to stay in order and read the captions that accompany each illustration. Before installing the brake drum, inspect it as outlined previously in this Chapter. If machining is necessary, remove the drums and have them resurfaced by an automotive machine shop (if they are still thick enough to undergo machining). **Note:** *Professionals recommend resurfacing of brake drums whenever replacing the brake shoes, as this will produce a smooth, flat surface that will eliminate brake pedal pulsations and other undesireable symptoms related to questionable drums. At the very least, if you elect not to have the drums resurfaced, de-glaze them with sandpaper or emery cloth, using a swirling motion to ensure a non-directional finish. Be careful not to get grease on the brake drum or brake shoes.*

After you have successfully installed the shoes on one side of the vehicle, repeat the procedure on the other brake. Then refer to Chapter 4 and adjust the shoes, referring to the sidebar entitled *"Drum brake adjustment."*

As soon as you have lowered the vehicle and tightened the lug nuts securely, check the brake fluid level in the master cylinder reservoir. If necessary, add some to bring it to the desired level (see Chapter 4, if necessary). Pump the brake pedal a few times, then recheck the fluid level.

Adjust the parking brake (see Chapter 4).

Lastly, before committing the vehicle to normal service, road test it in an isolated area, if possible. For the first few hundred miles or so, try to avoid hard braking to allow the new shoes to bed in.

6.6 **Before beginning work, wash away all traces of brake dust with brake system cleaner - DO NOT use compressed air or a brush.**

Duo-servo brake shoe replacement

Warning: *The dust created by the brake system may contain asbestos, which is harmful to your health. Never blow it out with compressed air and don't inhale any of it. An approved filtering mask should be worn when working on the brakes. Do not, under any circumstances, use petroleum-based solvents to clean brake parts. Use brake system cleaner only!*

Note: *The following procedure depicts the brake shoe replacement sequence for drum brakes with lever-actuated self-adjusters and cable-actuated self adjusters. Follow the photo sequence and read the accompanying caption, but simply ignore the steps which do not apply to the type of brake you are working on. The same goes for front drum brakes - just ignore all of the references to parking brake components.*

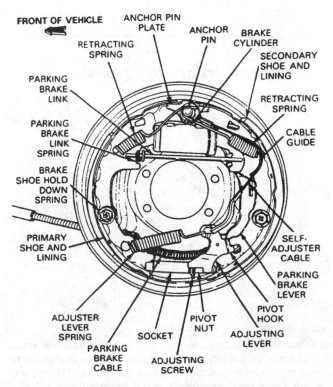

6.7 **Rear drum brake components (with cable-actuated self-adjuster mechanism) - left (driver's) side**

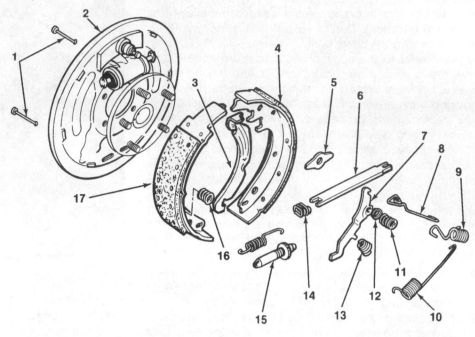

Tools required:

Jack and jackstands
Lug nut (or bolt) wrench
Filtering mask
Safety glasses or goggles
Latex gloves
Brake spring pliers
Brake hold-down spring tool
Brake shoe adjusting tool
   (or a suitable screwdriver)

Materials required:

Brake shoes
Brake hardware kit
Brake system cleaner
High-temperature brake grease
Brake fluid

6.8 Exploded view of a typical rear drum brake (with lever-actuated self-adjuster mechanism) - left (driver's) side

1	Hold-down pins	10	Return (or retracting) spring
2	Backing plate	11	Hold-down spring
3	Parking brake lever	12	Lever pivot
4	Secondary brake shoe	13	Lever return spring
5	Shoe guide (also called "anchor pin plate")	14	Strut spring
6	Parking brake strut	15	Adjusting screw assembly (with "star wheel")
7	Actuator lever	16	Adjusting screw spring
8	Actuator link	17	Primary brake shoe
9	Return (or retracting) spring		

Remove the old brake shoes

6.9 Pull back on the adjuster lever and turn the star wheel to retract the brake shoes (this will make removal of the return springs easier and, on models with cable-actuated parking brakes, will ease removal of the adjuster cable).

6.10 Remove the shoe return springs - the tool shown here is available at most auto parts stores and makes this job much easier and safer.

6.11 On models with cable-actuated adjusters, remove the return spring and cable guide from the secondary shoe . . .

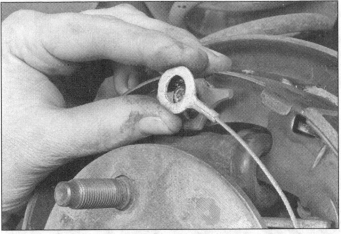

6.12 . . . then remove the self adjuster cable and anchor pin plate.

6.13 On models with lever-actuated adjusters, pull the bottom of the actuator lever toward the secondary brake shoe, compressing the lever return spring. The actuator link can now be removed from the top of the lever . . .

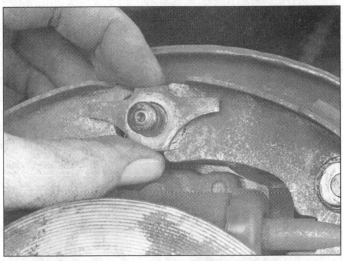

6.14 . . . followed by the anchor pin plate.

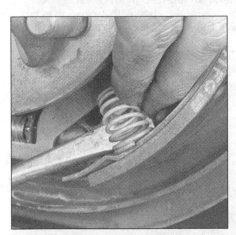

6.15 Also on models with lever-actuated adjusters, pry the actuator lever spring out with a large screwdriver.

6.16 Slide the parking brake strut out from between the axle flange and the primary shoe.

6.17 Remove the hold-down springs and pins. The hold-down spring tool shown here is available at most auto parts stores and greatly simplifies this step. To remove the springs, push down on the retainer, turn it 1/4-turn to align the slot in the retainer with the blade on the pin, then release.

6.18 On models with lever-actuated adjusters, remove the actuator lever and pivot - be careful not to let the pivot fall out of the lever.

6.19 Spread the top of the shoes apart and slide the assembly around the axle.

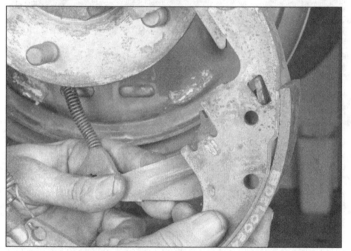

6.20 Unhook the parking brake lever from the secondary shoe, if the design of the brake permits.

6.21 Some parking brake levers simply hook onto the cable and are easily detached. On this design, the lever is usually fastened to the secondary brake shoe with a clip and is not easily separated (it's easier to do it on the workbench).

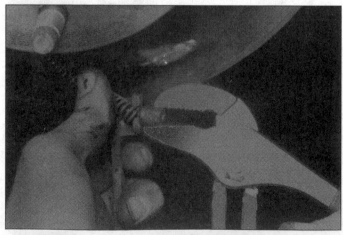

6.22 Other parking brake levers are fastened to the secondary shoe with a clip, but don't readily separate from the cable, either. In this case, pull on the cable end with a pair of pliers and swing the cable out of the slot.

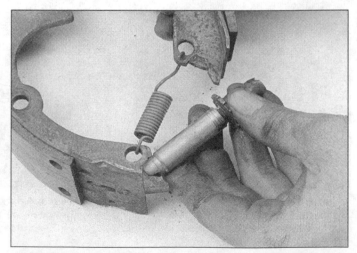

6.23 Spread the bottom of the shoes apart and remove the adjusting screw assembly.

Preparation for installation

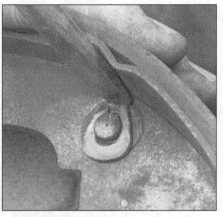

6.24 If the parking brake lever is still fastened to the secondary shoe, remove it by prying off the C-clip. Take note of any washers that may be present.

6.25 Here's another kind of clip that is commonly used to retain the parking brake lever. Spread the ends of the clip and pry it off the pin.

6.26 Clean the adjusting screw with brake system cleaner, dry it off and lubricate the threads and end with high-temperature brake grease. The screw portion of the adjuster will need to be threaded in further than before to allow the drum to fit over the new shoes.

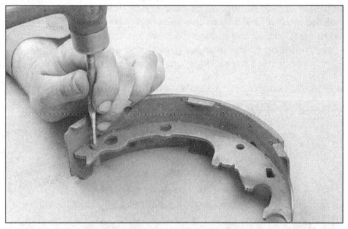

6.27 Feel the brake shoe contact points on the backing plate. If they are rough or a ledge has worn into them, file them smooth. Lubricate these contact points with high-temperature grease.

6.28 If the parking brake lever was mounted to a pin on the old secondary shoe, install a new adjuster lever pin in the new shoe . . .

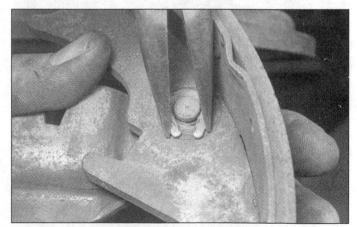

6.29 . . . then install the lever and a new C-clip

6.30 If your brake uses this type of clip, install a new clip over the pin and squeeze the ends together (note the spring washer under the clip)

Install the new shoes

6.31 Note: *On models with cable-actuated adjusters, you may find it easier to install the shoes on the backing plate, **then** install the adjuster screw, adjuster spring and lever.* **Connect the adjuster spring between the new shoes (with the long portion of the spring pointing to the secondary [rear] shoe on most designs), install the adjuster screw assembly (with the long part pointing to the primary [front] shoe on most designs), then connect the parking brake cable (if disconnected from the lever).**

6.33 Insert the parking brake lever into the opening in the secondary brake shoe. If you separated the parking brake cable from the end of the lever, connect it now if you haven't already done so.

6.32 Spread the shoes apart and slide them into position on the backing plate.

6.34 Install the hold-down pin through the backing plate in the primary shoe, then install the spring and retainer. If you're working on a brake with a cable-actuated self-adjuster, install the hold-down pin, spring and retainer through the secondary shoe also.

6.35 On models with lever-actuated adjusters, insert the lever pivot into the actuator lever, place the lever over the secondary shoe hold-down pin and install the hold-down spring and retainer.

6.36 Guide the parking brake strut behind the axle flange and engage the rear end of it in the slot on the parking brake lever. Spread the shoes enough to allow the other end of the strut to seat against the primary shoe.

6.37 Place the anchor pin plate over the anchor pin

6.38 On models with lever-actuated adjusters, hook the lower end of the actuator link to the actuator lever, then loop the top end over the anchor pin . . .

6.39 . . . and install the lever return spring over the tab on the actuator lever. Push the bottom of the spring up onto the brake shoe.

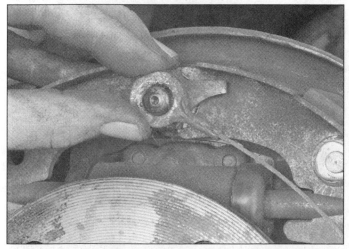

6.40 If you're working an a model with cable-actuated adjusters, place the end of the cable over the anchor pin . . .

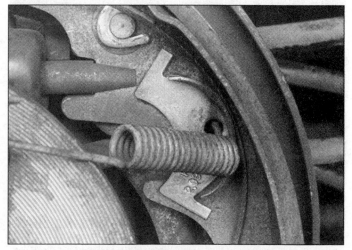

6.41 . . . then hook the end of the secondary shoe return spring through the cable guide and into the hole in the shoe. Make sure the cable sits in the guide.

6.42 Install the shoe retractor springs. The tool shown here is actually the handle of the brake spring pliers. This tool makes this step much easier and safer.

6.43 Bend the ends of the return springs around the anchor pin, so the end of the spring is parallel with the long part of the spring.

6.44 If you're working on a model with cable-actuated adjusters, hook the adjuster lever spring into the proper hole at the bottom of the primary shoe . . .

6.45 . . . then hook the adjuster lever spring and cable into the adjuster lever. Pull the cable down and to the rear, inserting the hook on the lever into the hole in the secondary shoe.

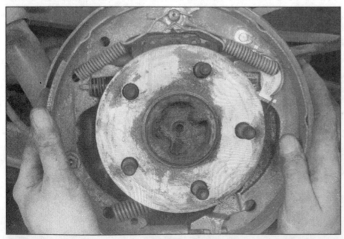

6.46 Wiggle the brake assembly to ensure that the shoes are centered on the backing plate. Make sure the parking brake strut and the wheel cylinder pushrods (if applicable) engage properly with the brake shoes.

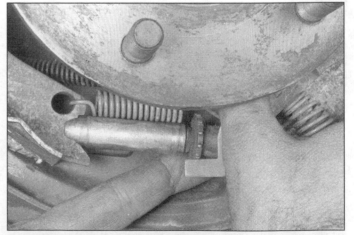

6.47 Turn the adjusting screw star wheel to adjust the shoes in or out as necessary. The brake drum should slide over the shoes and turn with a very slight amount of drag.

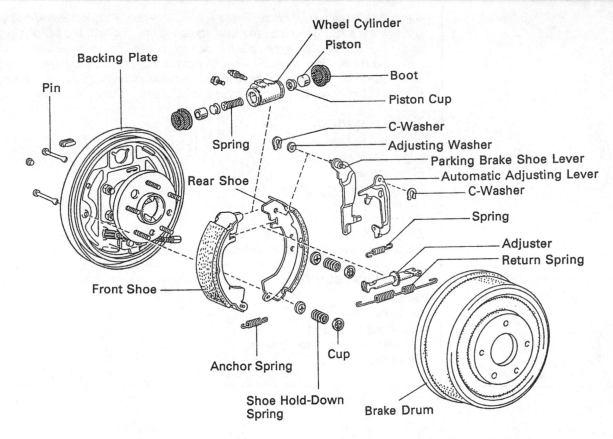

Pin

Backing Plate

Wheel Cylinder

Piston

Boot

Piston Cup

Spring

C-Washer

Adjusting Washer

Parking Brake Shoe Lever

Automatic Adjusting Lever

C-Washer

Rear Shoe

Spring

Adjuster

Return Spring

Front Shoe

Anchor Spring

Cup

Shoe Hold-Down Spring

Brake Drum

6.48 Exploded view of a typical leading/trailing drum brake assembly (Toyota)

Leading/trailing brake shoe replacement

Warning: *The dust created by the brake system may contain asbestos, which is harmful to your health. Never blow it out with compressed air and don't inhale any of it. An approved filtering mask should be worn when working on the brakes. Do not, under any circumstances, use petroleum-based solvents to clean brake parts. Use brake system cleaner only!*

Note 1: *The following procedure depicts a **rear** brake shoe replacement sequence for leading/trailing type drum brakes. Follow the photo sequence and read the caption below each illustration, but simply ignore the steps which do not apply to the type of brake you are working on. The same goes for front drum brakes - just ignore all of the references to parking brake components.*

Note 2: *Most leading/trailing drum brakes are similar in design, regardless of manufacturer. The differences lie primarily in the parking brake actuating mechanism. As there are many different parking brake actuator setups, it is not possible to cover every variation in a manual of this size. Before disassembling any components, be sure to pay close attention to the way all of the components are situated. Note the direction the retractor springs are installed - is there a long, straight portion of the spring passing over the*

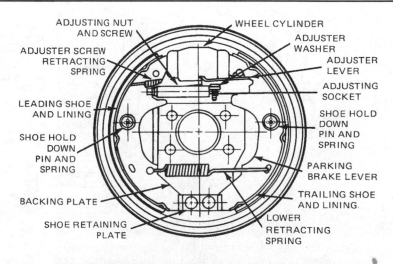

ADJUSTING NUT AND SCREW

WHEEL CYLINDER

ADJUSTER SCREW RETRACTING SPRING

ADJUSTER WASHER

ADJUSTER LEVER

ADJUSTING SOCKET

LEADING SHOE AND LINING

SHOE HOLD DOWN PIN AND SPRING

SHOE HOLD DOWN PIN AND SPRING

PARKING BRAKE LEVER

BACKING PLATE

TRAILING SHOE AND LINING

SHOE RETAINING PLATE

LOWER RETRACTING SPRING

6.49 Assembled view of the components of a typical leading/trailing drum brake (Ford compact)

self-adjuster star wheel? If so, make a note. Also note the mounting holes of all springs. It's a real good idea to make sketches or even take instant photographs of the brake assembly, especially the parking brake components, before actually removing them. Remember to work only on one brake at a time, referring to the other brake for reference, if necessary. At the beginning of the sequence are a few exploded views of typical leading/trailing drum brake assemblies.

Tools required:

Jack and jackstands
Lug nut (or bolt) wrench
Filtering mask
Safety glasses or goggles
Latex gloves
Brake spring pliers
Brake hold-down spring tool
Brake shoe adjusting tool (or a suitable screwdriver)

Materials required:

Brake shoes
Brake hardware kit
Brake system cleaner
High-temperature brake grease
Brake fluid

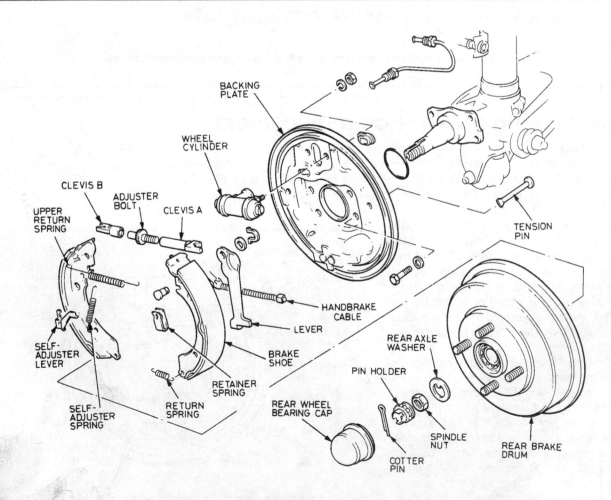

6.50 Exploded view of a typical leading/trailing drum brake assembly (Honda)

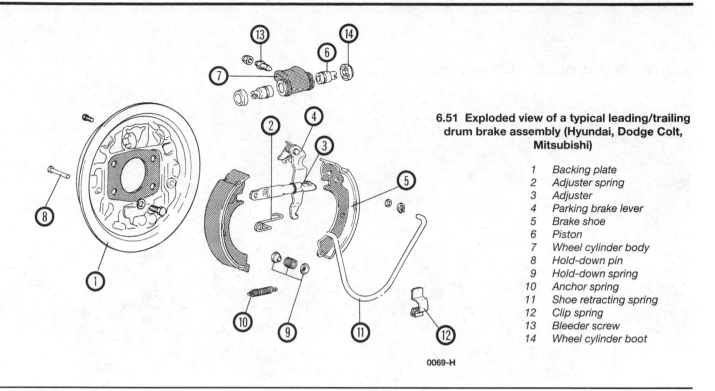

6.51 Exploded view of a typical leading/trailing drum brake assembly (Hyundai, Dodge Colt, Mitsubishi)

1   Backing plate
2   Adjuster spring
3   Adjuster
4   Parking brake lever
5   Brake shoe
6   Piston
7   Wheel cylinder body
8   Hold-down pin
9   Hold-down spring
10 Anchor spring
11 Shoe retracting spring
12 Clip spring
13 Bleeder screw
14 Wheel cylinder boot

0069-H

6.52 Exploded view of a typical leading/trailing drum brake assembly (General Motors)

1 Actuator spring	13 Adjuster nut	25 Wheel cylinder
2 Upper return spring	14 Adjuster screw	26 Bleeder screw cap
3 Spring connecting link	15 Retaining clip	27 Backing plate
4 Adjuster actuator	16 Pin	28 Access hole plug
5 Spring washer	17 Spring washer	
6 Lower return spring	18 Parking brake lever	
7 Hold-down spring assembly	19 Screw and lock washer	
8 Hold-down pin	20 Boot	
9 Leading brake shoe	21 Piston	
10 Trailing brake shoe	22 Piston cup	
11 Adjuster socket	23 Spring assembly	
12 Spring clip	24 Bleeder screw	

Remove the old brake shoes

6.53 Unhook the return spring (also called retractor spring) from the leading brake shoe. A pair of locking pliers can be used to stretch the spring and pull the end out of the hole in the shoe.

6.54 Remove the shoe hold-down springs. On this type, depress the hold-down spring and turn the retainer 90-degrees, then release it. This special tool makes removal much easier - they're available at most auto parts stores and aren't very expensive.

6.55 If your brake is equipped with a spring steel type of hold-down clip, push down on the clip and turn the hold-down pin with a pair of pliers. A special tool is also available, which makes this step even quicker and easier (see illustration 2.78 in Chapter 2).

6.56 Some brakes aren't equipped with hold-down springs, but have a "horseshoe" type spring which serves as a hold-down spring and a shoe return spring. If your brake is like this, wedge a flat-bladed screwdriver under the spring and pry it out of the leading brake shoe, then remove the shoe.

6.57 Remove the leading shoe from the backing plate and unhook the anchor spring from the bottom end of the shoe. If the brake you are working on has a lower shoe-to-shoe return spring instead of a spring that passes under the anchor plate, remove it using the technique shown in illustration 6.53, if necessary.

6.58 Remove the hold-down spring from the trailing shoe.

6.59 If the brake you're working on is equipped with a horseshoe spring, pry it out of the trailing shoe.

6.60 Pull the trailing shoe (and adjuster assembly, if applicable) away from the backing plate . . .

6.61 . . . then hold the end of the parking brake cable with a pair of pliers and pull it out of the parking brake lever.

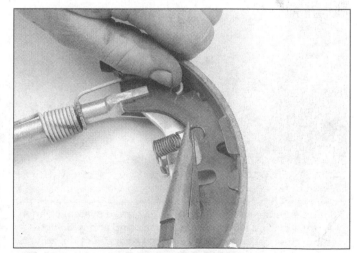

6.62 If your brake has a parking brake setup like this, remove the adjusting lever spring.

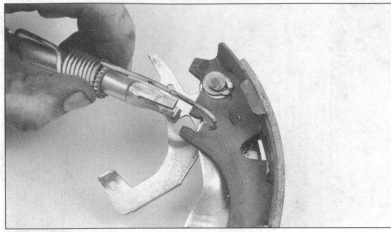

6.63 If the brake is equipped with a captive adjuster screw (retained by the return spring), unhook the spring from the shoe and slide the adjuster and spring off.

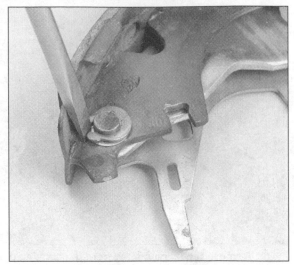

6.64 Pry the C-clip apart and remove it to separate the parking brake lever and adjuster lever from the rear shoe. Take note of any washers or shims that may be present.

6.65 Assemble the parking brake lever and adjuster lever to the new rear shoe and crimp the C-clip closed with a pair of pliers. Always use a new clip and be sure to install any shims or washers, if any were present upon separation.

Preparation for installation

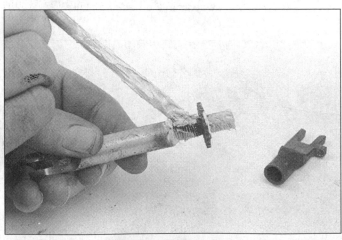

6.66 Clean the adjusting screw with brake system cleaner, dry it off and lubricate the threads and end with high-temperature brake grease. The screw portion of the adjuster will need to be threaded in further than before to allow the drum to fit over the new shoes.

6.67 Feel the brake shoe contact points on the backing plate. If they are rough or a ledge has worn into them, file them smooth. Lubricate these contact points with high-temperature grease.

Install the new shoes

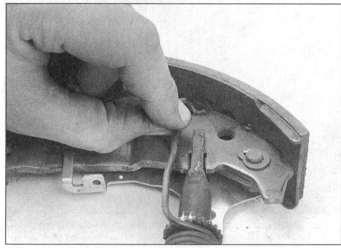

6.68 If the adjuster screw is a captive type, install it onto the rear shoe. Make sure the end fits properly into the slot in the shoe and hook the spring into the opening in the shoe.

6.69 Install the adjuster lever spring, if applicable.

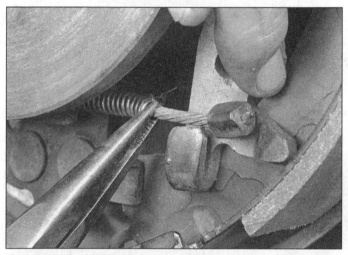

6.70 Hold the trailing shoe assembly up to the backing plate, pull the parking brake cable spring back and hold it there with a pair of pliers, then place the cable into the hooked end of the parking brake lever.

6.71 Place the trailing shoe against the backing plate, being careful not to get any grease on the friction linings. Pass the hold-down pin through its hole in the brake shoe, then install the hold down spring and retainer.

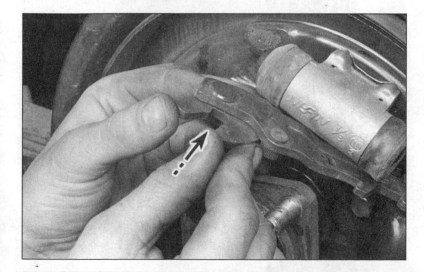

6.72 If your brake has a quadrant-type adjuster mechanism, install it now and push the lever in about half-way (arrow).

6.73 Connect the anchor spring (if applicable) to the bottom of each shoe and mount the leading shoe to the backing plate (again, be careful not to get any grease on the lining material). Install the hold-down spring and retainer.

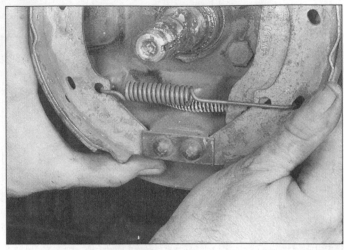

6.74 If the brake has a shoe-to-shoe return spring, install it between the two shoes, making sure the straight portion of the spring is connected to the trailing shoe (it's designed like this so as not to interfere with the parking brake lever).

6.75 If the adjuster assembly on your brake is non-captive (not retained by the return spring), install it now. Spread the shoes apart and insert each end of the adjuster in the proper slots in the shoes. If the slot in the parking brake lever end of the adjuster is stepped, like this one, make sure the longer part of the slot is positioned over the parking brake lever. Note: *Some adjuster screws are marked to indicate which side of the vehicle they go on. If this is the case, be sure to install the adjuster on the proper side. Also, this mark usually faces up.*

6.76 Install the adjuster lever (if it is a detachable one like this) on the parking brake lever pivot pin . . .

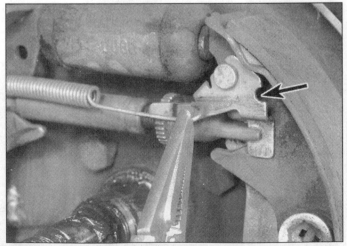

6.77 . . . then install the leading shoe hold-down pin, spring and retainer - stretch the return spring, with the straight part of the spring over the adjuster lever, and hook it in the notch on the adjuster lever (arrow).

6.78 Make sure the shoes are seated properly in the slots in the wheel cylinders and are engaged with the adjuster screw assembly.

6.79 Using a screwdriver, stretch the return spring into its hole in the front shoe.

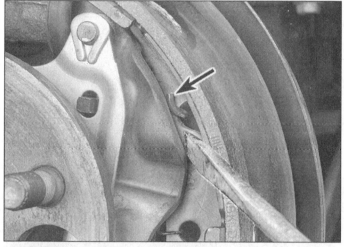

6.80 Pry the parking brake lever forward and check to see that the return spring didn't come unhooked from the trailing shoe.

6.81 Wiggle the assembly to make sure it is seated properly against the backing plate. Install the brake drum, then adjust the brakes as described in Chapter 4.

6.82 If your brake has a quadrant-style adjuster mechanism and the brake drum won't go back on after replacing the shoes, it will be necessary to retract the shoes. Put a screwdriver between the two parts of the ratchet and push in - spring pressure will retract the shoes.

Notes

7 Parking brakes

Introduction

The parking brake, as the name implies, is used for preventing the vehicle from moving when it is parked. It is sometimes referred to as the *emergency brake*, but on most vehicles it would be quite useless in an emergency situation where the regular service brakes have failed, unless there is plenty of straight, unoccupied road ahead.

Almost all auto manufacturers place the parking brake at the rear wheels. The two most popular exceptions to this rule are Subaru and Saab. They place the parking brakes on the front wheels. Some older vehicles had the parking brake mounted on the transmission extension housing, with a small brake drum on the driveshaft.

There are four basic types of parking brakes in use today, but regardless of design they share two things in common. One, they are all actuated by a lever, a pedal or a pullrod in the driver's compartment. Two, a cable or set of cables (sometimes one, sometimes two or more) connects the pedal, lever or pullrod with the parking brakes at the wheels.

On models with drum brakes at the rear wheels, the parking brake is simply incorporated into the drum brake assembly. A lever is connected to the top of the secondary, or trailing brake shoe (depending on brake design) and attached to the parking brake cable at the bottom. A strut (duo-servo type) or strut/adjuster screw (leading trailing type) is placed between the two brake shoes, near the top. When the lever attached to the brake shoe is pulled by the parking brake cable it pushes (via the parking brake strut) the primary, or leading shoe against the brake drum. This causes the pivot of the lever to bear against the secondary, or

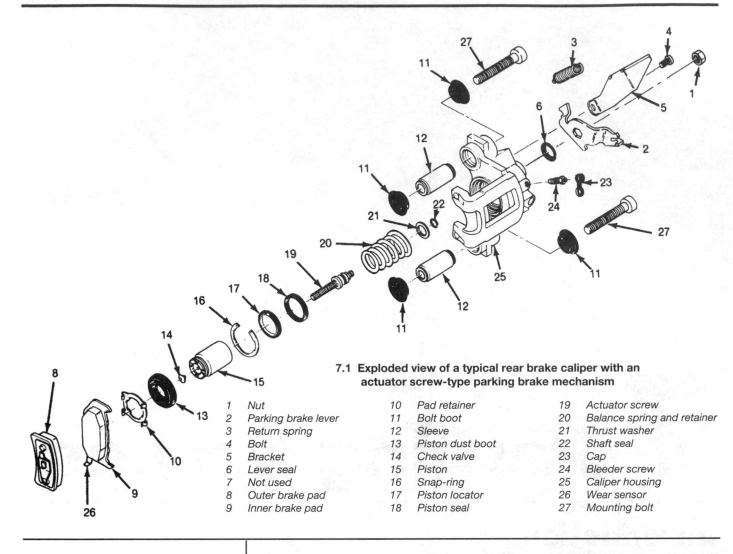

7.1 Exploded view of a typical rear brake caliper with an actuator screw-type parking brake mechanism

1	Nut	10	Pad retainer	19	Actuator screw
2	Parking brake lever	11	Bolt boot	20	Balance spring and retainer
3	Return spring	12	Sleeve	21	Thrust washer
4	Bolt	13	Piston dust boot	22	Shaft seal
5	Bracket	14	Check valve	23	Cap
6	Lever seal	15	Piston	24	Bleeder screw
7	Not used	16	Snap-ring	25	Caliper housing
8	Outer brake pad	17	Piston locator	26	Wear sensor
9	Inner brake pad	18	Piston seal	27	Mounting bolt

trailing brake shoe, forcing it against the brake drum, thereby locking the wheel. See Chapter 6 for more information on the drum brakes.

There are three different kinds of parking brake setups on vehicles with disc brakes on the rear wheels (actually, if the parking brakes are situated on the front wheel brakes, only one of these designs is used). The most common type is fitted to floating or sliding calipers and operates the caliper through a large actuator screw that winds into the back of the caliper piston **(see illustration)**. When the parking brake is set, the cable pulls on a lever attached to the actuator screw, causing the screw to turn. The helical thread on the screw tries to turn the piston, but the piston is prevented from turning because it is engaged with the brake pad by one or two lugs that fit into the backing plate of the inner brake pad. This causes the piston to move out, applying the brake pads the same way it does when hydraulic pressure pushes it out. This system works well, but with age the actuator screws tend to freeze up in the caliper piston, and the seal around the actuator screw is a potential source of brake fluid leakage.

A similar type of parking brake arrangement that is integral with the caliper uses a cam-driven adjuster spindle (or pushrod) to force the piston out from the caliper **(see illustration)**.

Another type of parking brake on vehicles with disc brakes at the rear wheels is like a little drum brake mounted inside the hub, or "hat" portion of the brake disc **(see illustration)**. The brake shoes are anchored to a small backing plate and operate just like the parking brake mechanism on a vehicle with drum brakes. The nice thing about this design is that it is virtually trouble-free, unless you forget to release the parking brake from time to time.

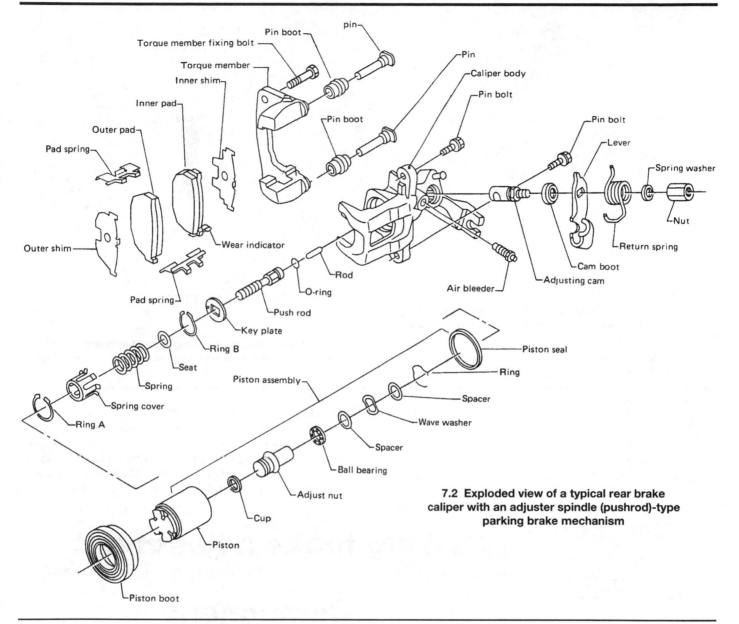

Torque member fixing bolt
Torque member
Inner shim
Inner pad
Outer pad
Pad spring
Outer shim
Pad spring
Wear indicator
Pin boot
Pin boot
Pin boot
pin
Pin
Caliper body
Pin bolt
Pin bolt
Pin bolt
Lever
Spring washer
Nut
Return spring
Cam boot
Adjusting cam
Air bleeder
Rod
O-ring
Push rod
Key plate
Ring B
Seat
Spring
Spring cover
Ring A
Piston assembly
Piston seal
Ring
Spacer
Wave washer
Spacer
Ball bearing
Adjust nut
Cup
Piston
Piston boot

7.2 Exploded view of a typical rear brake caliper with an adjuster spindle (pushrod)-type parking brake mechanism

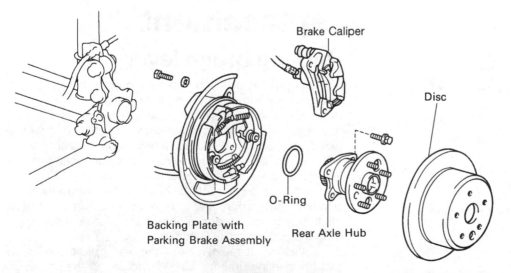

Brake Caliper
Disc
O-Ring
Backing Plate with Parking Brake Assembly
Rear Axle Hub

7.3 Typical drum-in-disc type parking brake mechanism

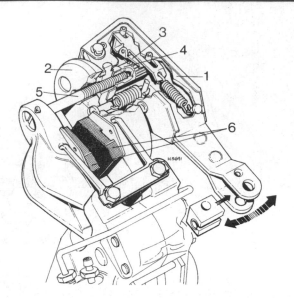

7.4 Cutaway view of a parking brake mechanism that is attached to a rear caliper and uses separate pads

1	Operating lever	4	Adjuster nut
2	Pad carrier assembly	5	Adjuster bolt
3	Pawl assembly	6	Parking brake pad

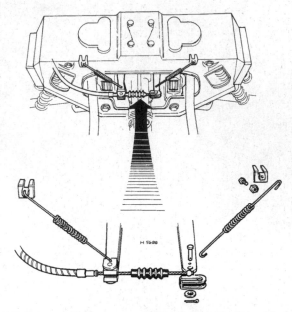

7.5 Some vehicles use a single parking brake cable to operate both parking brake mechanisms, like on this Jaguar

A less common kind of parking brake setup employs an extra set of small brake pads mounted in a fixture attached to each rear caliper **(see illustration)**. They're kind of like small, mechanically actuated calipers mated to the hydraulic calipers. Automatic adjusters in the parking brake pad carriers compensate for wear. On some vehicles that use this system, a single parking brake cable is connected to an operating lever on each mechanism **(see illustration)**, or to a linkage that operates both mechanisms.

Parking brake adjustment

Refer to Chapter 4, *Maintenance*, for the parking brake adjustment procedure.

Component replacement

Parking brake lever

On most vehicles you'll have to remove the center console for access to the parking brake lever bolts. Once this has been done, unscrew the upper parking brake cable nut (the adjusting nut) and pull the cable from the lever **(see illustration)**. Some vehicles have both cables attached to the lever. It may be necessary to remove a trim piece from the lever for access to the adjuster nut.

Remove the nuts or bolts holding the lever to the floorpan **(see illustration)** and detach the lever. On some models it may be necessary to unplug the electrical connector from the parking brake light switch.

Installation is the reverse of removal, but be sure to adjust the parking brake cable(s) by following the procedure outlined in Chapter 4.

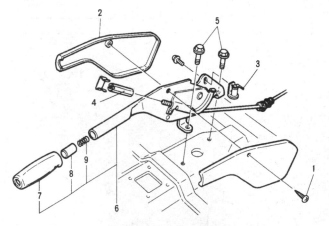

7.6 Typical parking brake lever details

1	Screw	5	Bolts
2	Covers	6	Lever assembly
3	Parking brake switch connector	7	Grip
4	Adjusting nut	8	Release button
		9	Return spring

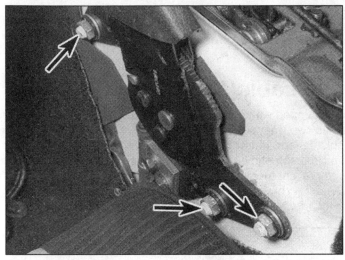

7.7 On some vehicles, the parking brake lever is mounted to the side of the center hump

Parking brake pedal or pullrod

If necessary for access to the parking brake pedal, remove the under-dash panel and/or the kick panel trim. Unplug the electrical connector from the parking brake light switch.

If you're removing a parking brake pedal assembly, disconnect the release rod or cable from the pawl lever on the parking brake mechanism **(see illustration)**. If equipped with a vacuum-operated release, disconnect the vacuum hose or, if the vacuum release diaphragm is not mounted on the pedal assembly, disconnect the link rod from the assembly **(see illustration)**.

If you're removing a parking brake pullrod, detach the cable from the end of the pullrod **(see illustration)** or re-

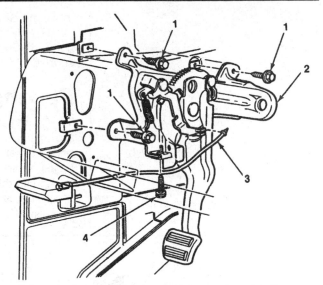

7.8 typical parking brake pedal mounting details

1	Bolt	3	Release rod
2	Pedal assembly	4	Bolt

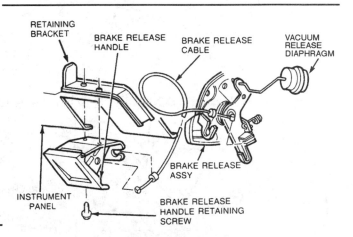

7.9 Typical parking brake pedal assembly with a vacuum release diaphragm

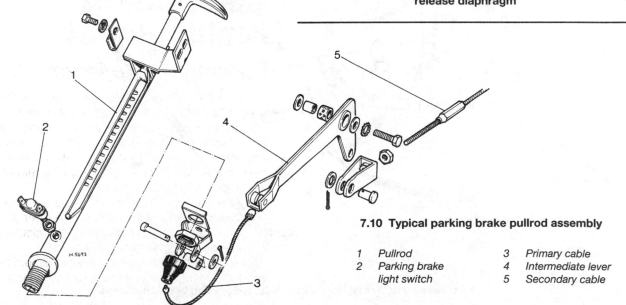

7.10 Typical parking brake pullrod assembly

1	Pullrod	3	Primary cable
2	Parking brake light switch	4	Intermediate lever
		5	Secondary cable

move the clevis pin and detach the rod from the lever **(see illustration)**.

Remove the bolts or nuts and detach the pedal or pullrod assembly. Installation is the reverse of removal, but be sure to adjust the parking brake as described in Chapter 4.

Parking brake cable(s)

Some vehicles are equipped with only one parking brake cable, but chances are that your vehicle has two, maybe even three or four cables to actuate the parking brake mechanisms **(see illustrations)**. These cables are retained by a series of clips or clamps and sometimes also by guide wires and springs, to keep tension on the cables to prevent them from interfering with other parts.

Before replacing a parking brake cable, loosen the adjusting nut(s) to provide some slack. See Chapter 4 for a sampling of parking brake cable adjusters.

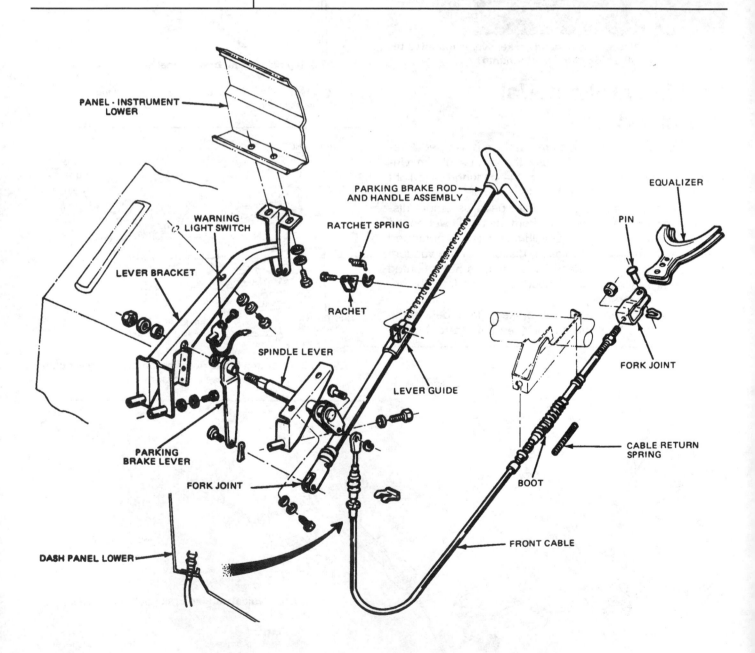

7.11 Here's another typical pullrod-actuated parking brake mechanism

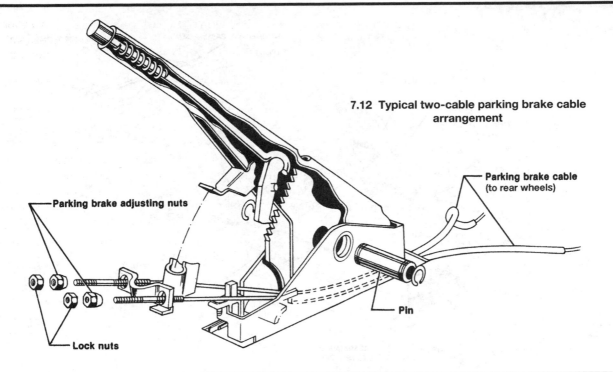

7.12 Typical two-cable parking brake cable arrangement

Parking brake adjusting nuts

Parking brake cable
(to rear wheels)

Lock nuts

Pin

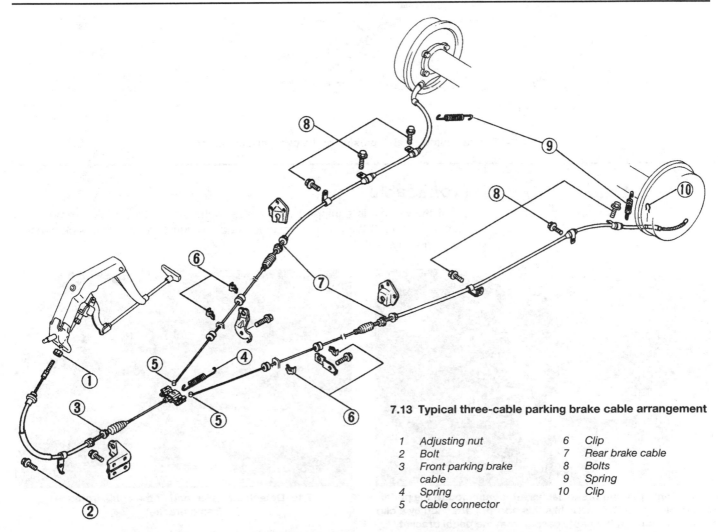

7.13 Typical three-cable parking brake cable arrangement

1	Adjusting nut	6	Clip
2	Bolt	7	Rear brake cable
3	Front parking brake cable	8	Bolts
4	Spring	9	Spring
5	Cable connector	10	Clip

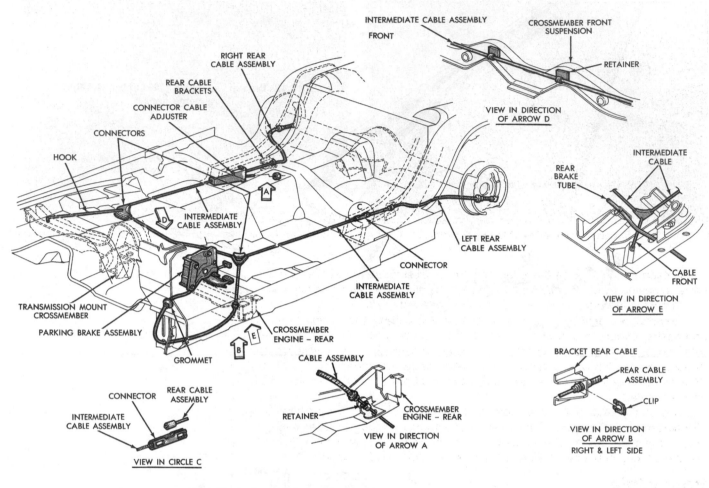

7.14 Typical multiple-cable parking brake cable arrangement

Front cable

If the vehicle is equipped with a lever-actuated parking brake, remove the center console. If the parking brake is pedal-actuated, remove the kick panel, if it's in the way.

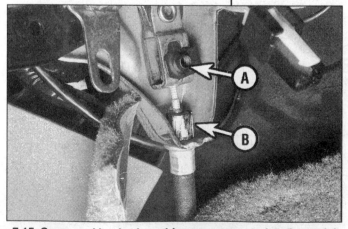

7.15 Some parking brake cables are connected to the pedal assembly by a nut. Others, like this one, are attached by a clip (A). To detach the cable casing from the pedal bracket, depress the tangs (B) and slide the casing from the bracket

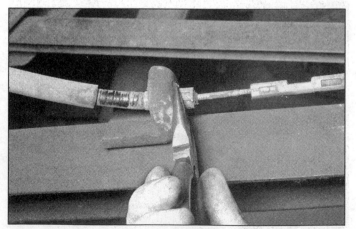

7.16 Detach the other end of the cable from the frame bracket

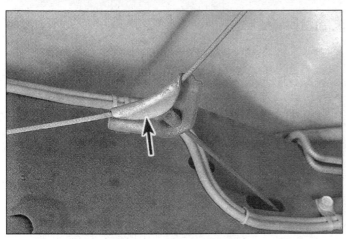

7.17 Separate the cables at the equalizer . . .

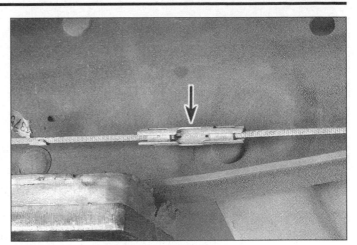

7.18 . . . or at the cable connector (push the cable toward the center opening and pull it out)

Detach the cable from the lever, pedal or pullrod. In most cases this will involve removing a nut or a clip from the end of the cable **(see illustration)**.

Raise the vehicle and support it securely on jackstands.

Remove any bolts or clips retaining the cable casing, then push the cable and casing through the grommet and out from the floorpan. Some casings are retained by locking tangs which must be depressed to allow them to pass through the bracket **(see illustration)**. If the cable casing does not continue out through the floorpan, however, pull the cable and casing into the interior of the vehicle instead.

Detach the other end of the cable from the equalizer or cable connector **(see illustrations)**. Installation is the reverse of removal. Make sure the grommet around the cable casing where it passes through the floor is properly seated. Adjust the parking brake by following the procedure described in Chapter 4.

Rear cable

Loosen the rear wheel lug nuts, raise the vehicle and support it securely on jackstands. Remove the wheel.

Detach the cable from the connector, equalizer or adjuster **(see illustration)**.

Disconnect the cable casing from any mounting brackets or clips.

If the vehicle has rear drum brakes, remove the brake drum (see Chapter 5)

7.19 On this type of rear cable setup, prevent the rod from turning by holding it with a pair of locking pliers, then remove the adjusting nut. The rear cables can now be separated at the adjuster

7.20 Unclip the parking brake cable end from the parking brake lever on the trailing shoe, then remove the shoe and lever assembly

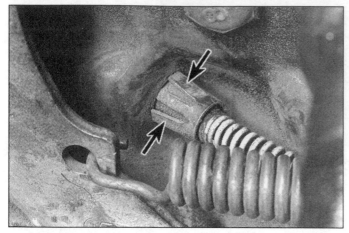

7.21 Depress the retention tangs (arrows) to free the cable and casing from the backing plate. A pair of pliers can be used for this. Alternative methods include a close fitting box-end wrench slipped over the retainer, which will compress the tangs, or a small hose clamp tightened over the tangs.

7.22 On some vehicles the parking brake cable casing is attached to the backing plate by a bolt (arrow)

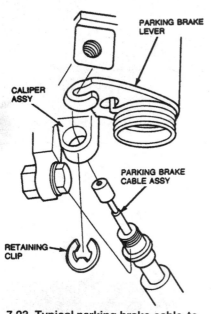

7.23 Typical parking brake cable-to-caliper attachment details

and detach the end of the cable from the parking brake lever **(see illustration)**. In some cases this will require removal of the brake shoes. Depress the tangs on the cable casing retainer and push the cable and casing out through the backing plate **(see illustration)**, or unscrew the bolt retaining the casing to the backing plate **(see illustration)**.

If the vehicle has disc brakes, detach the cable from the parking brake lever on the caliper (or from the actuating lever on vehicles with drum-in-disc parking brake assemblies), separate the cable casing from its bracket (it's usually retained by a clip) and remove the cable **(see illustration)**.

Installation is the reverse of removal. Be sure to adjust the parking brake as outlined in Chapter 4.

Parking brake shoes (drum-in-disc type parking brake)

Warning: *The dust created by the brake system may contain asbestos, which is harmful to your health. Never blow it out with compressed air and don't inhale any of it. An approved filtering mask should be worn when working on the brakes. Do not, under any circumstances, use petroleum-based solvents to clean brake parts. Use brake system cleaner only!*

Note: *All four parking brake shoes should be replaced at the same time, but only work on one brake assembly at a time, using the other side for reference, if necessary.*

Loosen the rear wheel lug nuts, raise the vehicle and support it securely on jackstands. Remove the wheels.

Remove the brake disc (see Chapter 5). Before removing anything, clean the parking brake assembly with brake system cleaner. Position a drain pan under the brake to catch the fluid and residue.

Remove the parking brake shoe return springs from the anchor pin **(see illustration)**. Some designs don't have an anchor pin, but have two shoe-to-shoe springs instead.

Spread the tops of the shoes apart and remove the shoe strut and spring assembly, if equipped **(see illustration)**. Take note as to how the strut is positioned.

Remove the hold-down springs from the shoes **(see illustration)**. Pull the front shoe from the backing plate and remove the adjuster **(see illustration)**, then detach the shoe-to-shoe spring **(see illustration)**.

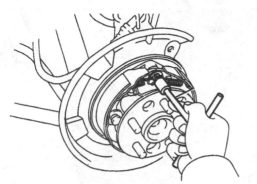

7.24 Unhook the parking brake shoe return springs - a special brake spring removal tool makes this much easier and is available at most auto parts stores (some designs use a shoe-to-shoe spring at the top and the bottom of the shoes, instead of the arrangement shown here).

7.25 Turn and lift out the shoe strut and spring, if equipped.

7.26 Remove the shoe hold-down springs. This particular type requires an Allen wrench. Other styles are like the more conventional hold-down spring retainers as found in a drum brake assembly.

7.27 Pull the front shoe out and remove the adjuster screw

If the parking brake is actuated by a lever attached to the rear shoe, detach the cable from the lever **(see illustration)**, then remove the shoe. Using a screwdriver, spread the C-clip on the parking brake lever pivot pin then remove the lever, shim (if equipped) and pin. Transfer the parts to the new rear shoe and

7.28 Twist the shoe and remove the shoe-to-shoe spring

7.29 Pull the cable out and up to free it from the lever on the rear shoe (some designs don't use a lever attached to the shoe - the shoes are spread apart by an actuator on the backing plate)

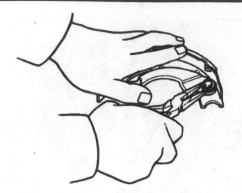

7.30 Use a pair of pliers to crimp the C-clip to the pivot pin

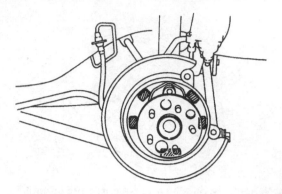

7.31 Apply a light coat of high-temperature grease to the parking brake shoe contact areas (shaded areas) on the backing plate

crimp the C-clip to the pin using a pair of pliers **(see illustration)**.

Thoroughly wash the backing plate with brake system cleaner. Feel the shoe contact areas on the backing plate - it they're rough, smooth them out with a file or sandpaper. Apply a thin coat of high-temperature grease to the contact areas **(see illustration)**.

Clean the adjuster screw threads, then apply a film of high-temperature grease to the moving parts of the adjuster **(see illustration)**.

If it was necessary to detach the cable from the parking brake lever on the rear shoe, reattach it now. Mount the rear shoe to the backing plate and install the hold-down spring.

Connect the shoe-to-shoe spring between the bottom of each shoe, then install the adjuster **(see illustration)**. Position the front shoe on the backing plate and install the hold-down spring. Install the parking brake strut. Make sure you install it with the spring end facing the proper direction. Install the shoe return springs.

Install the brake disc **(see illustration)**. If the disc isn't retained by screws, temporarily install a couple of lug nuts to hold the disc in place. **Note:** *If the axle flange has a notch machined into it that corresponds with the adjuster hole in the disc, be sure to install the disc with the adjuster hole over the notch.*

Adjust the parking brake shoes by referring to the Chapter 4 sidebar entitled *"Drum brake adjustment."*

Install the brake caliper and wheel. Lower the vehicle and tighten the lug nuts securely.

7.32 Clean the adjuster screw and apply high-temperature grease to the indicated areas (arrows)

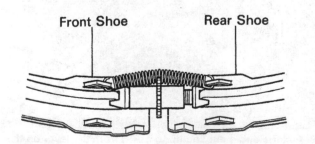

Front Shoe   Rear Shoe

7.33 When assembled, the tension spring and adjuster screw should be arranged like this

7.34 Be sure to align the access hole with the adjuster cutout (arrows) when installing the disc over the parking brake assembly

8 Brake pedals and brake light switches

Brake pedal assemblies

There are two basic designs of brake pedal assemblies - ones that are mounted on the floor, and the hanging type. Most operate the master cylinder (via the power brake booster, if equipped) through a pushrod, but some are connected to a pullrod and lever assembly, which in turn actuates the master cylinder. Some vehicles, especially ones that are marketed in countries that have right-hand drive vehicles as well as in countries that have left-hand drive vehicles, have brake pedals attached to a shaft that traverses the firewall to actuate the master cylinder, which is mounted on the side opposite the driver.

On some vehicles equipped with a manual transmission, the brake pedal pivots on the same shaft as the clutch pedal.

Brake pedals seldom require attention, but, as stated in Chapter 4, it's a good idea to crawl under the dash occasionally to inspect the pedal bushings and mounting hardware. Refer to the accompanying illustrations for examples of typical brake pedal assemblies **(see illustrations)**.

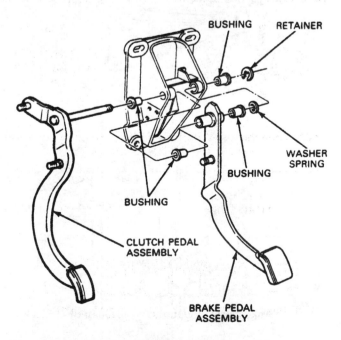

8.1 Brake pedal mounting details (Ford Aerostar with manual transmission)

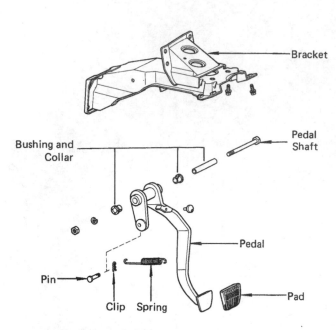

8.2 Brake pedal mounting details (Toyota pick-up)

Bracket

Pedal Shaft

Bushing and Collar

Pedal

Pin

Clip   Spring

Pad

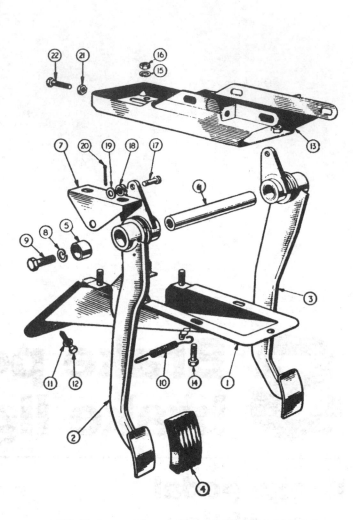

8.3 Brake pedal mounting details (Triumph)

1	Pedal shaft cover	13	Master cylinder support bracket
2	Clutch pedal		
3	Brake pedal	14	Bolt
4	Rubber pad	15	Lockwasher
5	Pivot bushing	16	Nut
6	Pedal shaft	17	Clevis pin
7	Bracket	18	Spring washer
8	Lockwasher	19	Plain washer
9	Bolt	20	Cotter pin
10	Return spring	21	Locknut
11	Lockwasher	22	Pedal limit stop
12	Bolt		

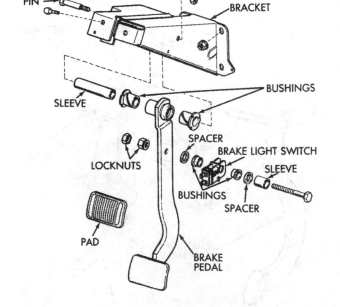

PIVOT PIN

BRACKET

SLEEVE

BUSHINGS

SPACER

BRAKE LIGHT SWITCH

SLEEVE

LOCKNUTS

BUSHINGS

SPACER

PAD

BRAKE PEDAL

8.4 Brake pedal mounting details (Jeep Cherokee)

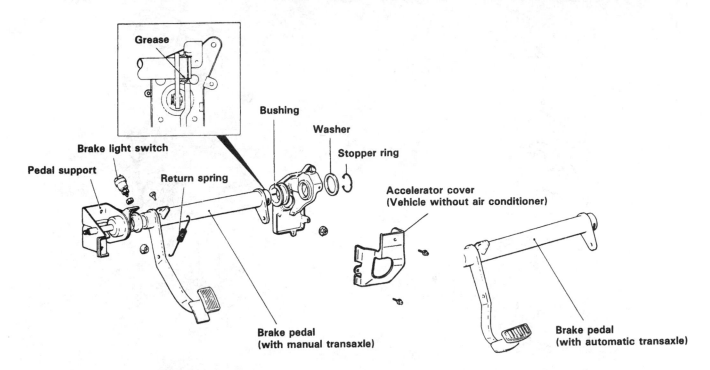

8.5 Brake pedal mounting details (Hyundai Excel, Dodge Colt, Mitsubishi Precis) - this design uses a cross-shaft to actuate the master cylinder (because the master cylinder is mounted on the right (passenger's) side of the firewall)

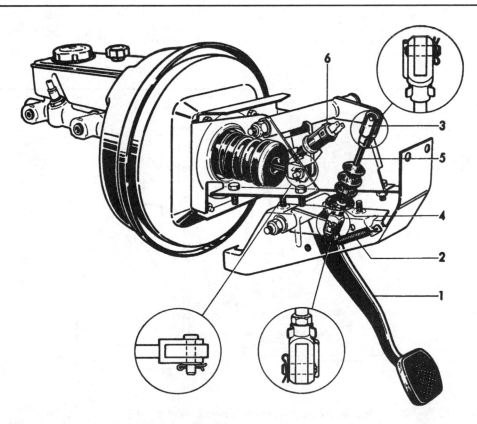

8.6 Brake pedal mounting details (Saab) - this design uses a pullrod to operate a lever mechanism, which depresses the booster pushrod

1	Pedal	3	Clevis	5	Pullrod
2	Return spring	4	Locknut	6	Brake light switch

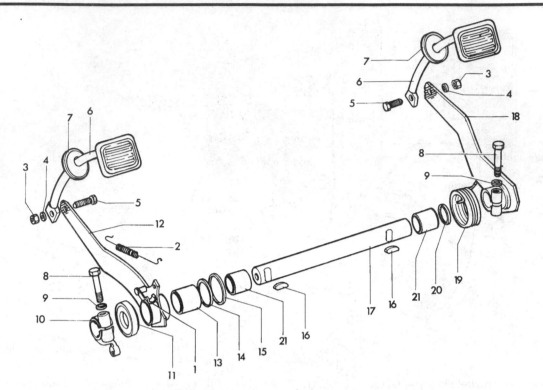

8.7 Brake pedal mounting details (VW Transporter, right-hand drive)

1	Pin and clip	7	Seal	13	Bushing	19	Return spring
2	Return spring	8	Clamp screw	14	Seal	20	Seal
3	Nut	9	Lockwasher	15	Thrust washer	21	Bushing
4	Lockwasher	10	Brake pedal lever	16	Woodruff key		
5	Bolt	11	Seal	17	Shaft		
6	Pedal	12	Clutch pedal lever	18	Brake pedal lever		

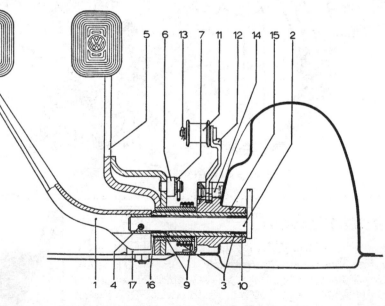

8.8 Brake pedal mounting details (VW Beetle)

1	Clutch pedal	6	Master cylinder pushrod	11	Accelerator pedal roller	14	Accelerator pedal lever pin
2	Pedal shaft	7	Pushrod lock plate	12	Accelerator connecting lever	15	Mounting bracket
3	Bushing	8	Not used	13	Clip	16	Snap-ring
4	Locating pin	9	Bushing			17	Stop plate
5	Brake pedal	10	Mounting tube				

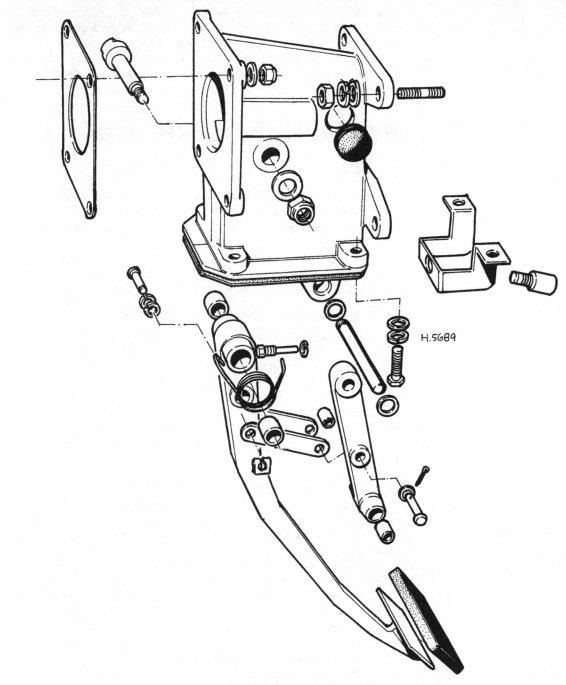

H.5689

8.9 Brake pedal mounting details (Jaguar)

Removal and installation

Removal and installation of the brake pedal usually involves nothing more than detaching the return spring, disconnecting the master cylinder (or booster) pushrod from the pedal, removing a nut or clip and withdrawing the pivot shaft. On some vehicles it will be necessary to remove the clutch pedal, if the brake pedal hangs from the clutch pedal shaft.

If it becomes necessary to remove the brake pedal on your vehicle, carefully study the arrangement of the brake pedal and bushings, the mounting of the brake light switch, how the brake pedal is attached, and, if your vehicle is equipped with a manual transmission, how the clutch pedal is mounted. Make a sketch of all the components, both installed and as you remove them. Lay all the parts out in order after they're removed. Inspect the pivot bushings and shaft for wear, replacing parts as necessary. If any cotter pins or spring clips were removed, be sure to replace them with new ones.

Before installing the pedal assembly, lubricate the pivot bushings and shaft with multi-purpose grease. Adjust the pedal freeplay and check the pedal reserve distance as outlined in Chapter 4.

Brake light switches

Check

Have an assistant stand behind the vehicle while you depress the brake pedal. All of the brake lights should come on (on some vehicles the ignition switch must be turned to the On position). If any bulbs are burned out, replace them.

If none of the lights come on, check the fuses. If the fuses are OK, check the brake light switch. There are two kinds of brake light switches - mechanically operated and hydraulically operated.

Hydraulically operated switches

Hydraulically operated switches are usually screwed into the end of the master cylinder. When pressure is built up in the system, the switch completes the circuit and sends power to the brake lights. With the ignition key On, use a test light to check for power on one side of the switch. If there's no power on either terminal, trace the circuit for a broken or unplugged wire. If there is power, have an assistant depress the brake pedal while you probe the other side of the switch with the test light. If the test light comes on, the switch is good. The problem lies somewhere between the switch and the brake lights. If power is not available when the pedal is depressed, the switch is bad.

Mechanically operated switches

Mechanically operated switches are mounted on the brake pedal arm or on a bracket near the top of the brake pedal. When the pedal is at rest, the switch plunger is depressed by the brake pedal arm, which raises one of the contacts inside the switch and opens the circuit. When the brake pedal is depressed, the spring-loaded plunger extends from the switch body, the contacts inside the switch connect, and the circuit is completed. To test the switch, probe the terminals of the switch with a test light - one of the terminals should be "hot" (on some vehicles it may be necessary to turn the ignition key to the On position). Now, probe the other terminal with the test light and depress the brake pedal - the test light should light up. If it doesn't, the switch is defective. If it does, but the brake lights don't come on, the problem lies somewhere between the switch and the brake lights. **Note:** *On some vehicles that have combination brake lights/turn signal lights, the electrical circuit for the brake lights runs through the turn signal switch.*

Replacement

Hydraulically operated switch

Unplug the wires from the switch. Unscrew the switch and install the new one, tightening it securely. Bleed the circuit of the brake system from which the switch was removed (see Chapter 9). This can usually be done right at the master cylinder, but if after doing so the brake pedal feels spongy, bleed the circuit at the wheels.

Mechanically operated switch

Detach the wires from the switch.

a) If the switch is retained by a locknut, unscrew the nut and detach the switch from its bracket **(see illustration)**. If the switch has an adjusting nut on it, measure the distance from the end of the switch to the adjusting nut, then install the adjusting nut on the new switch in the same location. Install the switch in its bracket, then install the locknut. If necessary, loosen the locknut and adjusting nut and adjust the switch in or out to provide proper brake light operation. Make sure the switch allows the pedal to fully return to its at-rest position. **Note:** *Some brake light switches regulate the brake pedal height (at-rest position). Each vehicle with this type of setup has its own pedal height specification, but a general setting can be made while still ensuring that the switch isn't prohibiting the pedal from returning far enough, which could prevent the master cylinder pistons from returning fully. Detach the master cylinder (or power booster) pushrod from the brake pedal. Adjust the switch so the pushrod mounting hole or pin in the pedal arm is in alignment with the hole in the pushrod clevis (or mounting) hole. Reattach the pushrod to the pedal arm. (This adjustment probably won't be necessary if the dimensions of the new switch are the same as the old one, and the adjusting nut has been installed in the same position on the new switch as it was on the old one.)*

b) Some switches look like they're threaded, but upon closer inspection you notice that the "threads" are serrations. This type of switch is held in its bracket by a plastic bushing or a pressed-metal clip **(see illustration)**. To remove it, simply pull it out of the bracket (if it's stubborn, you may have to pry it out). Install the new switch by pushing it into its mounting hole. Depress the brake pedal, then push the switch in a little more. Release the brake pedal - the pedal will push the switch back into its bracket, adjusting it automatically.

8.10 This brake light switch (arrow) is attached to a mounting bracket and is retained by a locknut and an adjusting nut. Some switches like this are threaded into the bracket and only have a locknut (Toyota)

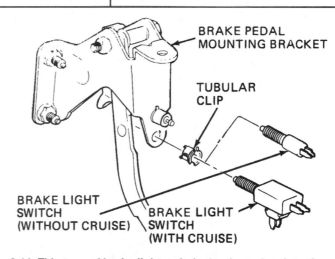

8.11 This type of brake light switch simply pushes into the clip on the mounting bracket. The brake pedal automatically adjusts the position of the switch (General Motors)

c) If the switch is attached to the brake pedal where the master cylinder (or booster) pushrod is attached, remove the nuts and bolt (or clip) from the pedal and detach the pushrod and switch, carefully noting how the switch and pushrod fit together **(see illustration 8.4)**. Also note the locations of any spacers or bushings. Install the switch by reversing the removal procedure. Tighten the fasteners securely.

d) Another type of switch simply clips to the brake pedal and to the steering column **(see illustration)** (on one variation of this design, the switch clips to the pedal and the arm of the switch is retained to the steering column with a spring clip or cotter pin) **(see illustration)**. Detach the electrical connector, remove the clip or cotter pin from the switch arm (if applicable), then unclip the switch from the pedal and steering column. Install the switch by reversing the removal procedure. Some of these switches have a "set lever" for adjusting the switch. To adjust, depress the pedal and, using a hooked tool, pull on the set lever while listening for a click. Release the pedal.

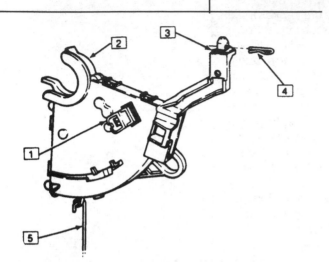

8.12 This kind of brake light switch clips onto the brake pedal. The actuating arm is attached to the steering column by a retaining clip (General Motors)

1	Cruise control connector	4	Retaining clip
2	Brake light switch	5	Wire hook (for adjusting
3	Wave washer		the switch)

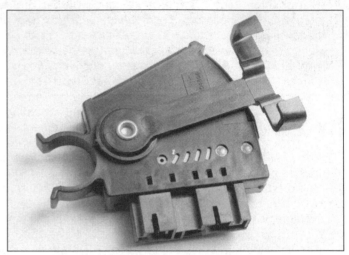

8.13 This brake light switch snaps onto the steering column and the brake pedal and uses no bolts or clips (General Motors)

9 Hydraulic systems and power boosters

Introduction

This Chapter deals with the removal, installation, bleeding and, where applicable, the overhaul procedures of the components that make up the brake hydraulic system. Because of their close relationship with the master cylinder, power boosters are also included in this Chapter.

There are a few important things to remember when working on the hydraulic system:

a) Cleanliness cannot be over-emphasized. Before opening any part of the hydraulic system, clean the fittings and surrounding area with brake system cleaner. This will prevent dirt from contaminating the fluid in the lines.

b) Never re-use old brake fluid. It contains moisture which could boil as heat is conducted through the brake linings to the hydraulic components. When brake fluid boils, gas bubbles are produced. Since gas is compressible, a spongy brake pedal or a complete loss of braking pressure can result. Also, old brake fluid might be tainted with contaminants, which can cause problems. Always use new brake fluid from a small, sealed brake fluid container. Since brake fluid is hygroscopic (meaning it is able to absorb moisture), fluid stored in an open container, or even in a sealed large container (if it's been sitting on the shelf for a long time) may contain too much moisture.

c) **Warning:** *Brake fluid can harm your eyes and damage painted surfaces, so use extreme caution when handling or pouring it. If you get any fluid in your eyes, immediately flush your eyes with water and seek medical attention.*

d) Before attempting to overhaul a hydraulic component, check on the availability and cost of parts, as well as the availability and cost of a new or rebuilt unit. Sometimes rebuilt parts are cheaper than an overhaul kit! Also, not all master cylinders, calipers and wheel cylinders are rebuildable.

e) When overhauling a hydraulic component, work in a spotlessly clean environment. A single grain of dirt can potentially ruin the entire job.

f) Never use petroleum-based solvents for cleaning brake system components. Use brake system cleaner - it leaves no residue when it dries. Brake fluid will also clean hydraulic components, but if you're cleaning parts that are very dirty, it takes quite a bit of fluid to remove it all. Denatured alcohol can also be used, but it contains water - if this is used, you'll have to wait a little longer to let the parts dry. Brake system cleaner is the best way to go.

g) Don't use compressed air to dry off brake parts. Even filtered compressed air may contain too much moisture. It might even contain traces of oil.

h) Never hone an aluminum (or any non-cast iron) hydraulic cylinder.

i) Whenever any part of the hydraulic circuit is opened, that part of the system must be bled of air (refer to the *Brake system bleeding* procedure in this Chapter.

Master cylinder

Warning: *This procedure does not apply to master cylinders that are directly attached to, or integral with, an Anti-lock Brake System (ABS) hydraulic control unit. Many ABS units require a special tool to cycle the valves in the ABS modulator or hydraulic control unit to ensure complete system bleeding. Vehicles equipped with such ABS systems should be taken to a dealer service department or other properly equipped repair shop for any master cylinder work.*

The following procedure depicts the removal, overhaul and installation procedures for the brake master cylinder. Follow the sequence of illustrations, being careful to read the caption under each one. Also be sure to read any accompanying text.

Remember, not all master cylinders are rebuildable. Before beginning work, check with auto parts stores and dealer parts departments for parts availability and prices. Rebuilt master cylinders are available on an exchange basis, which makes this job quite easy, and much less time consuming.

Removal

9.1 If equipped, unplug the electrical connector from the fluid level sensor

9.2 Place rags under the brake line fittings and prepare caps or plastic bags to cover the ends of the lines once they're disconnected. Caution: *Brake fluid will damage paint. Cover all painted surfaces and avoid spilling fluid during this procedure.* **Loosen the tube nuts at the ends of the brake lines where they enter the master cylinder. To prevent rounding off the flats on these nuts, a flare-nut wrench, which wraps around the nut, should be used. Pull the brake lines away from the master cylinder slightly and plug the ends to prevent contamination. Also plug the openings in the master cylinder to prevent fluid spillage.**

9.3 Remove the master cylinder mounting nuts - some vehicles have two nuts, some have three and others have four. Remove the master cylinder from the vehicle, being careful not to spill any fluid on the vehicle's paint. Remove the reservoir caps, or cover and gasket, then discard any fluid remaining in the reservoir.

Overhaul

Warning: *It's essential that you obtain the correct rebuild kit for the master cylinder you're servicing. Be sure to state the year, make and model of the vehicle when buying a kit. If you're still unsure, take the master cylinder with you when buying parts. DO NOT try to use the rebuild kit for a cast iron master cylinder in an aluminum unit, or vice versa. And don't open a new parts kit until you've compared the contents to the old parts you're removing from the master cylinder you intend to rebuild. Always verify parts availability before beginning a master cylinder rebuild.*

Note: *When disassembling the master cylinder, lay out all of the components in order. If the overhaul kit contains piston seals rather than entire piston assemblies, replace the seals on one piston at a time. Before disassembling a piston, make a sketch, noting the location of all seals, washers, springs, retainers, etc. Most importantly, pay attention to the direction the seal lips face. Refer to the accompanying illustrations to familiarize yourself with the master cylinder components* **(see illustrations).**

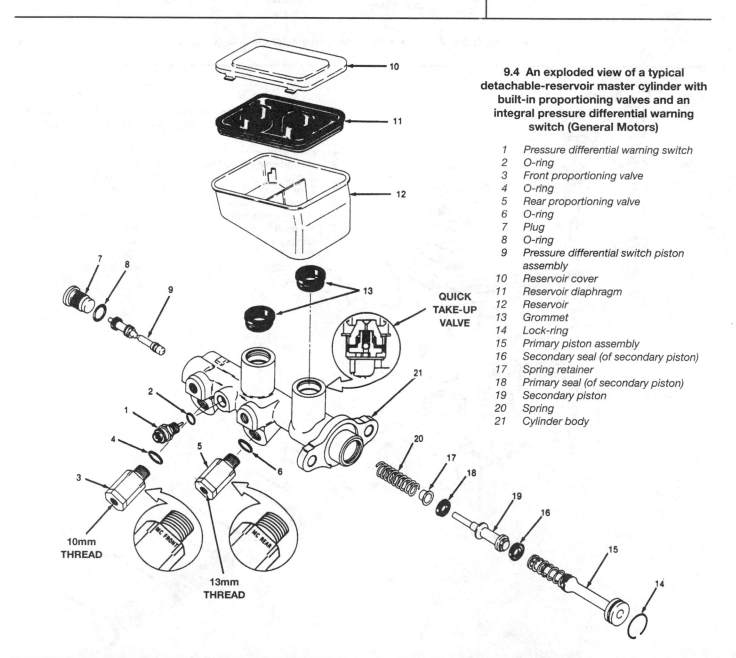

9.4 An exploded view of a typical detachable-reservoir master cylinder with built-in proportioning valves and an integral pressure differential warning switch (General Motors)

1	Pressure differential warning switch
2	O-ring
3	Front proportioning valve
4	O-ring
5	Rear proportioning valve
6	O-ring
7	Plug
8	O-ring
9	Pressure differential switch piston assembly
10	Reservoir cover
11	Reservoir diaphragm
12	Reservoir
13	Grommet
14	Lock-ring
15	Primary piston assembly
16	Secondary seal (of secondary piston)
17	Spring retainer
18	Primary seal (of secondary piston)
19	Secondary piston
20	Spring
21	Cylinder body

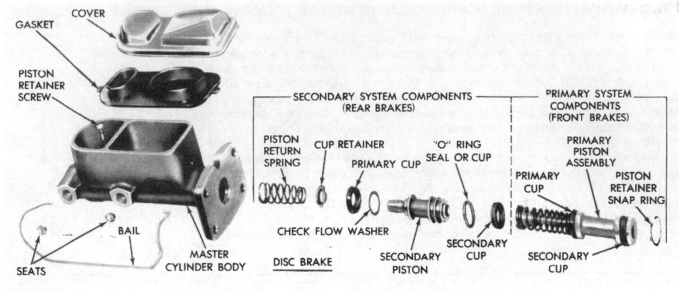

9.5 An exploded view of a typical cast iron master cylinder (Dodge)

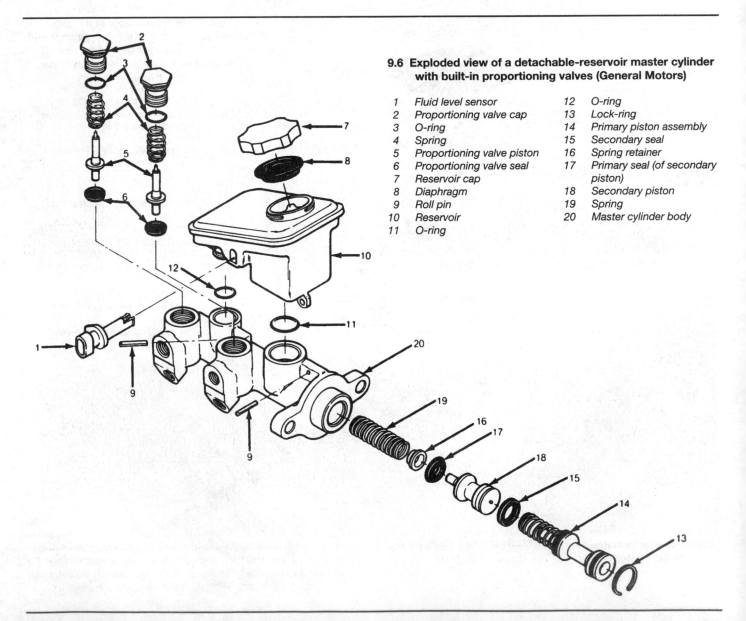

9.6 Exploded view of a detachable-reservoir master cylinder with built-in proportioning valves (General Motors)

1	Fluid level sensor	12	O-ring
2	Proportioning valve cap	13	Lock-ring
3	O-ring	14	Primary piston assembly
4	Spring	15	Secondary seal
5	Proportioning valve piston	16	Spring retainer
6	Proportioning valve seal	17	Primary seal (of secondary piston)
7	Reservoir cap		
8	Diaphragm	18	Secondary piston
9	Roll pin	19	Spring
10	Reservoir	20	Master cylinder body
11	O-ring		

Remove the reservoir (detachable reservoir models only)

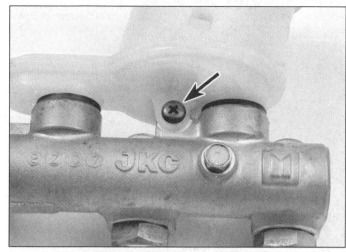

9.7 Some reservoirs are secured by a retaining screw (arrow). Remove the screw and carefully pull the reservoir off the master cylinder body.

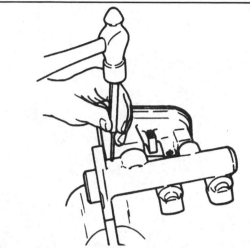

9.8 On some master cylinders, the reservoir is attached with a pair of roll pins which can be removed with a hammer and a small punch. Once the pins are removed, pull the reservoir off.

9.9 If the reservoir sticks in the grommets, carefully pry it off.

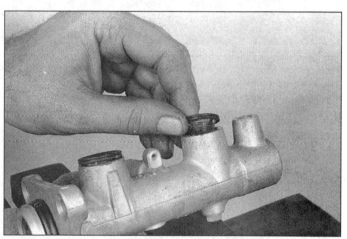

9.10 Remove the grommets from the master cylinder body.

9.11 Some detachable reservoirs are retained by a large bolt in the center of the reservoir. Remove the bolt . . .

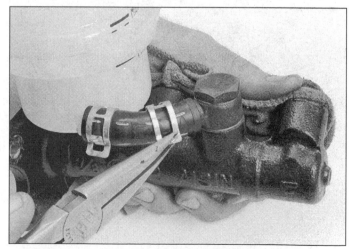

9.12 . . . then slide the hose clamp back on the hose and detach the hose from the fitting. If the sealing washers on the fitting bolt have been leaking, leave the hose attached and remove the fitting bolt and sealing washers, instead.

Remove the master cylinder pistons

Note: *Mount the master cylinder in a vise, with the jaws of the vise clamping on the cylinder mounting flange. Use padded jaws or a rag to prevent damage to the flange mounting surface.*

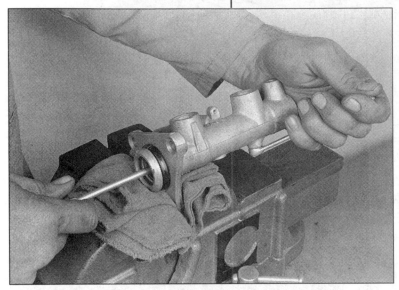

9.13 Using a Phillips screwdriver, depress the pistons, then remove the piston stop bolt or screw (be sure to use a new sealing washer on the stop bolt when you're reassembling the master cylinder). **Note:** *Not all master cylinders have a stop bolt or screw.*

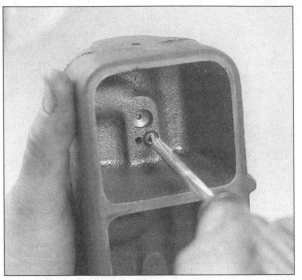

9.14 On some master cylinders, the piston stop is a screw or pin which is accessed from inside of the reservoir.

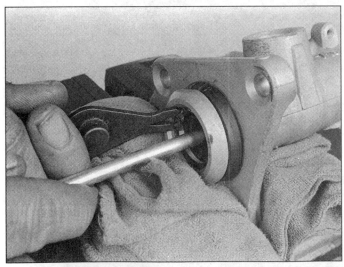

9.15 Depress the pistons again and remove the snap-ring or retainer

9.16 Some master cylinders use a stopper cap like this instead of a snap-ring. Depress the pistons and use a hooked tool to bend the tang up, then remove the stopper cap.

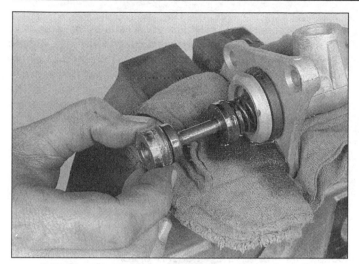

9.17 Remove the primary piston assembly from the master cylinder . . .

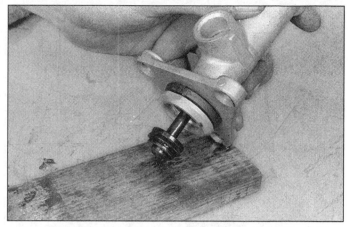

9.18 . . . followed by the secondary piston assembly. If the piston is stuck inside the bore, remove the cylinder from the vise and tap it against a wood block. Pull the piston straight out - the bore could be damaged if the piston becomes cocked. **Note:** *Some (although very few) manufacturers call this piston the primary piston, and the one removed in the previous step the secondary piston.*

Inspect and prepare the master cylinder

Clean the cylinder bore with brake system cleaner and inspect the bore for scoring and pitting. **Warning:** *DO NOT, under any circumstances, use petroleum-based solvents to clean brake parts.* Light scratches and corrosion on the cylinder bore walls can be usually be removed with crocus cloth or with a hone. However, deep scratches or score marks mean the cylinder must be replaced with a new unit. **Warning:** *Never attempt to hone an aluminum master cylinder.* If the pistons or bore are severely corroded, replace the master cylinder. Always use new piston cups and seals (or piston assemblies) when overhauling a master cylinder.

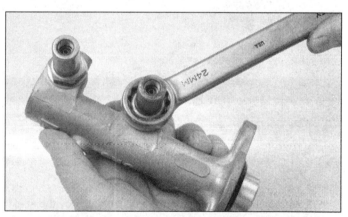

9.19 If equipped, unscrew the proportioning valves from the master cylinder

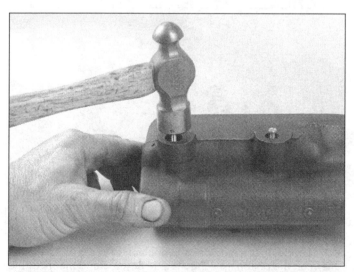

9.20 If you have a cast iron unit and you received new tube seats in the overhaul kit, tap a sheetmetal screw into each tube seat . . .

9.21 . . . then use a prybar as a fulcrum and remove the tube seats by prying them out with a pair of needle-nose pliers. The best way to drive the new seats into place is with a spare section of brake line (a flared end) with a tube fitting. Place the flared portion of the line over the seat, slide the tube nut into place and thread it into the outlet. Tighten the fitting to push the seat into place.

Overhaul the pistons

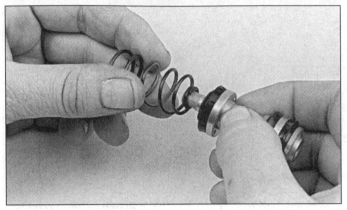

9.22 Begin overhaul of the secondary piston assembly by removing the spring, which can be pulled straight off.

Note 1: *Some master cylinder rebuild kits come with all of the seals for each piston. Others come with the seals for one of the pistons (usually the secondary piston) and an entire new piston assembly. Some kits simply furnish two new piston assemblies. Use all of the parts in the overhaul kit. The following portion of this sequence depicts typical seal replacement procedures for a variety of master cylinders - it may differ slightly from the pistons in your master cylinder. As you disassemble your pistons, BE SURE to note how the components are arranged (make a sketch if necessary). Ignore any steps which don't apply to your pistons.*

Note 2: *Some (although very few) manufacturers have named the pistons exactly opposite that of most manufacturers - they call the piston nearest the front of the vehicle the primary piston, and the piston nearest the power booster (or firewall) the secondary piston.*

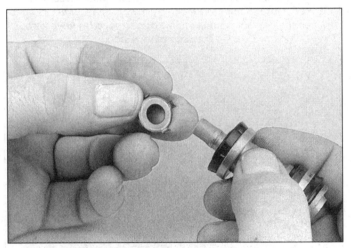

9.23 Remove the piston seal retainer, followed by the seal.

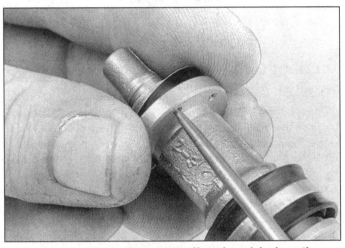

9.24 Remove the seal protector, if equipped, by inserting a thin instrument through one of the holes in the piston.

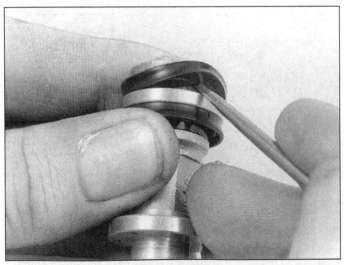

9.25 Pry the remaining piston seals off the piston

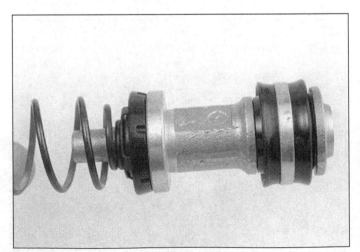

9.26 Lubricate the new seals with clean brake fluid and install them on the piston, making sure all of the components of the piston are situated in the right place, and the seal lips are facing the proper direction (the two outer seals must face the ends of the pistons. If there are three seals, the one in the center faces the spring end). When installing the seals, use your fingers only, or you may damage them.

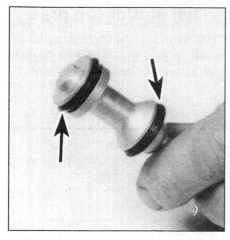

9.27 Some secondary pistons only have two seals - the seal lips must face the ends of the pistons.

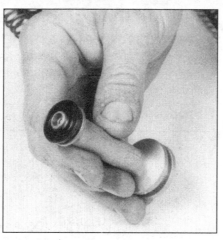

9.28 The seals on the primary piston are removed in the same manner as the seals on the secondary piston were (see illustrations 9.22 through 9.25). When installing the primary seal on the end of the primary piston, the lip must face away from the center of the piston. Remember, lubricate the seals with clean brake fluid before installing them.

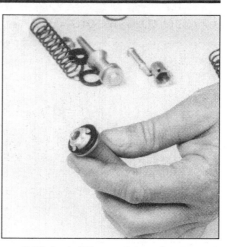

9.29 Install the seal guard over the seal . . .

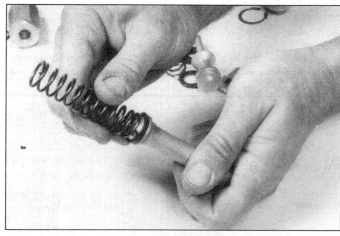

9.30 . . . then push the spring onto the end of the piston.

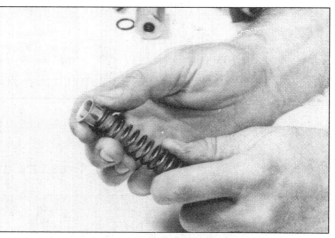

9.31 Insert the spring retainer into the spring

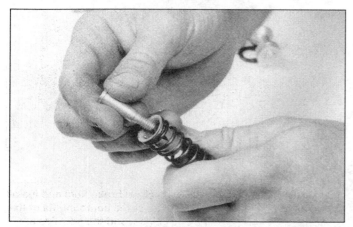

9.32 Insert the spring retaining bolt through the retainer and spring and thread it into the piston. Tighten it securely.

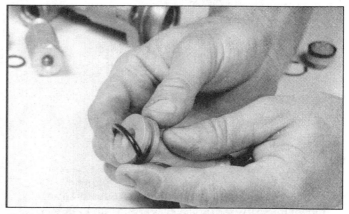

9.33 Lubricate the O-ring with clean brake fluid and install it on the piston. Some primary pistons use a secondary seal instead of an O-ring (some use both) - the seal lip should face the spring end of the piston (the same way the lips of the primary seal face).

Haynes Brake Repair Manual

Overhaul the proportioning valve and/or the pressure differential warning switch

Note: *This portion of the overhaul procedure only applies to master cylinders with an integral pressure differential warning switch and/or integral proportioning valves.*

Proportioning valve

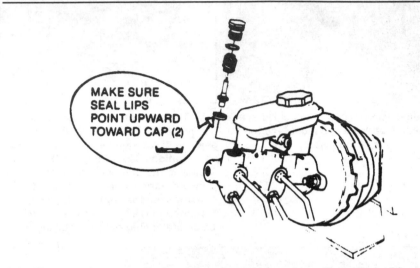

9.34 Unscrew the proportioning valves from the master cylinder body. Disassemble the valves (one at a time), then lubricate the seals with clean brake fluid and install the new components, making sure the seal lips are pointing in the right direction. Install the valves and tighten the caps securely.

Pressure differential warning switch

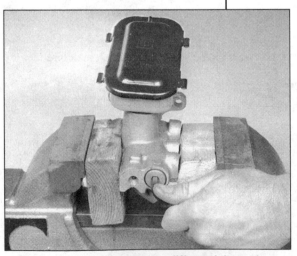

9.35 Remove the pressure differential warning switch from the side of the master cylinder, then mount the cylinder in a vise (use wood blocks as shown). Unscrew the switch piston plug.

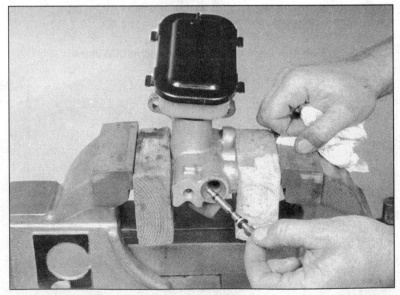

9.36 Pull out the switch piston assembly. Remove the components from the piston.

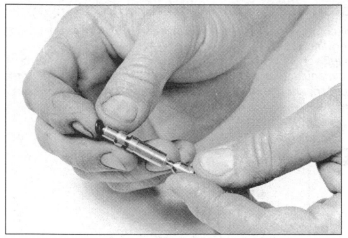

9.37 Lubricate the small O-ring with clean brake fluid and install it on the switch piston

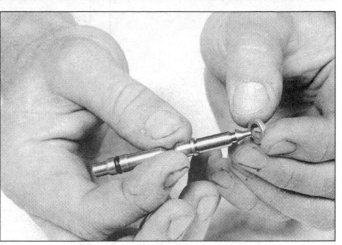

9.38 Install the metal retainer on the switch piston

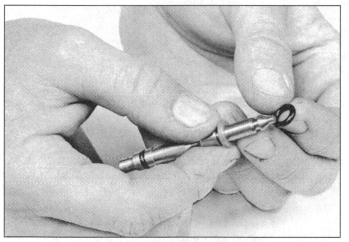

9.39 Lubricate the large O-ring with clean brake fluid and install it on the switch piston

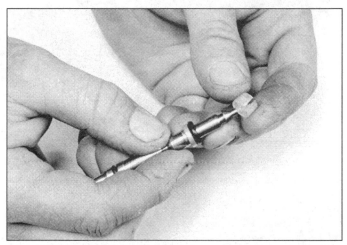

9.40 Install the plastic retainer on the switch piston

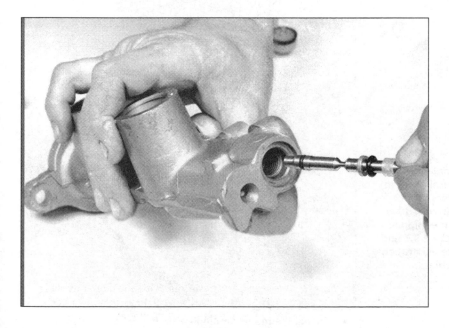

9.41 Lubricate the switch piston assembly and install it into its bore in the master cylinder body. Remove the master cylinder from the vise, then push the piston in until the small-diameter portion of the piston is visible through the pressure differential warning switch hole. Install the pressure differential warning switch (using a new O-ring) and tighten it securely. Install a new O-ring on the switch piston plug (be sure to lubricate it with clean brake fluid) and install the plug, tightening it securely.

Install the piston assemblies and reservoir

Note 1: *Lubricate all components with lots of clean brake fluid before installing them into the master cylinder.*

Note 2: *Some (although very few) manufacturers have named the pistons exactly opposite that of most manufacturers - they call the piston nearest the front of the vehicle the primary piston, and the piston nearest the power booster (or firewall) the secondary piston.*

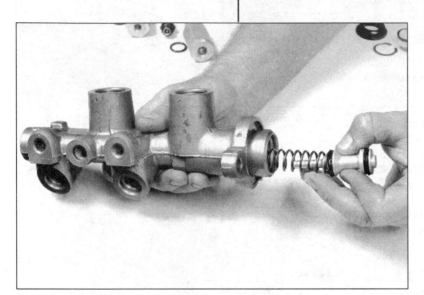

9.42 Install the secondary piston . . .

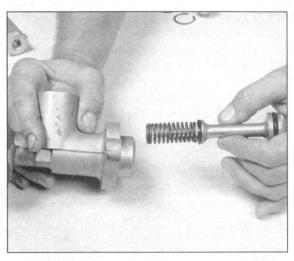

9.43 . . . followed by the primary piston. Depress the pistons and install the snap-ring or retainer as shown in illustration 9.15, then depress the pistons again and install the piston stop bolt, screw or pin (see illustrations 9.13 and 9.14). Be sure to use a new sealing washer on the stop bolt or screw.

9.44 On models with detachable reservoirs, lubricate the new reservoir grommets with clean brake fluid and install them in the openings in the master cylinder. Lay the reservoir face down on a bench and push the master cylinder straight down over the reservoir fittings with a rocking motion. If equipped, install the retaining screw or roll pins (see illustrations 9.7 and 9.8, If your reservoir is retained by a large bolt (see illustration 9.11), use new sealing washers, if applicable.

9.45 Install the reservoir diaphragm in the cover, if applicable.

Bench bleed and install the master cylinder

Whenever the master cylinder is removed, the entire hydraulic system must be bled. The time required to bleed the system can be reduced if the master cylinder is filled with fluid and bench bled before it's installed on the vehicle. Since you'll have to apply pressure to the master cylinder piston, the master cylinder should be mounted in a vise, with the jaws of the vise clamping on the mounting flange.

Bleeder tube method

If available, attach a pair of master cylinder bleeder tubes to the outlet ports of the master cylinder **(see illustration)**. These can be purchased at most auto parts stores and aren't very expensive. Fill the reservoir with fluid, then slowly push the pistons into the master cylinder (a large Phillips screwdriver can be used for this) - air will be expelled from the pressure chambers and into the reservoir. Because the tubes are submerged in fluid, air won't be drawn back into the master cylinder when you release the pistons. Repeat the procedure until no more air bubbles are present.

Remove the bleed tubes, one at a time, and install plugs in the open ports to prevent fluid leakage and air from entering. Install the reservoir cover or cap.

Alternative methods

An alternative to the bleed tube method is to insert threaded plugs into the brake line outlet holes and snug them down so that air won't leak past them - but not so tight that they can't be easily loosened. Fill the reservoir with brake fluid of the recommended type.

Remove one plug and push the pistons into the bore to expel the air from the master cylinder. A large Phillips screwdriver can be used to push on the piston assembly.

To prevent air from being drawn back into the master cylinder, the plug must be replaced and snugged down before releasing the pressure on the piston.

Repeat the procedure until only brake fluid is expelled from the brake line outlet hole. When only brake fluid is expelled, repeat the procedure at the other outlet hole and plug. Be sure to keep the master cylinder reservoir filled with brake fluid to prevent the introduction of air into the system.

Since high pressure isn't involved in the bench bleeding procedure, an alternative to the removal and replacement of the plugs with each stroke of the piston assembly is available. Before pushing in on the piston assembly, remove the one of the plugs completely. Before releasing the piston, however, instead of replacing the plug, simply put your finger tightly over the hole to keep air from being drawn back into the master cylinder. Wait several seconds for brake fluid to be drawn from the reservoir into the bore, then depress the piston again, removing your finger as brake fluid is expelled. Be sure to put your finger back over the hole each time before releasing the piston, and when the bleeding procedure is complete for that outlet, replace the plug and tighten it before going on to the other port.

Master cylinder installation

Install the master cylinder by reversing the removal procedure. Don't tighten the mounting nuts until after the fittings on the hydraulic lines have been threaded into the ports by hand. This will allow you to wiggle the master cylinder back and forth, if necessary, to connect the lines without cross-threading the fittings.

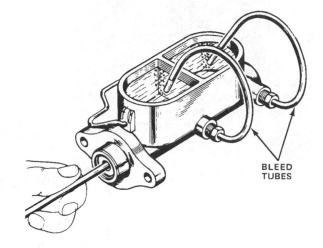

9.46 The best way to bleed air from the master cylinder before installing it on the vehicle is with a pair of bleed tubes

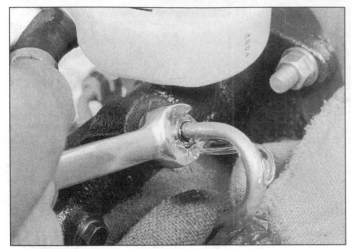

9.47 After the master cylinder is installed, it can be bled by loosening the fittings (one at a time) and applying the brakes

If air entered the master cylinder after the bench bleeding procedure (or if you didn't bench bleed it), bleed the cylinder on the vehicle. To do this, have an assistant push the brake pedal to the floor and hold it there. Loosen the fitting nut (start with the one closest to the booster or firewall) to allow air and fluid to escape, then tighten the nut. Repeat this procedure on all fittings until the fluid is clear of bubbles **(see illustration)**. **Note:** *Some master cylinders have bleeder valves, making it unnecessary to loosen the brake line fittings.*

Bleed the entire brake system as described later in this Chapter.

Power boosters

Vacuum-operated boosters

Check

Perform the power booster checking procedures described in Chapter 4. If a defective booster is not diagnosed from those checks, inspect the check valve. To do this, disconnect the vacuum hose where it connects to the metal pipe or the intake manifold (don't disconnect it at the booster). Apply pressure and suction to the end of the hose, making sure air only flows away from the booster. If it flows in both directions or if there is no airflow at all, replace the check valve. On some vehicles the check valve is located inside the hose, requiring replacement of the hose.

If the check valve is functioning properly and the hose doesn't have a leak, the engine may be the culprit. If the engine is old and tired, if the ignition timing is way off or if there's a large vacuum leak, it may not be producing enough vacuum to enable the booster to give satisfactory power assist. A restricted exhaust system could also be the cause of low vacuum. **Note:** *Some engines are fitted with vacuum pumps which help generate vacuum to power the brake booster and other accessories. A defective vacuum pump will most often make a loud rapping sound that rises and falls with engine speed. You can also test the operation of the vacuum pump as outlined in the engine vacuum check in the next paragraph.*

Connect a vacuum gauge to the booster hose. Place the shifter in park (if the transmission is an automatic), block the wheels and start the engine. Allow the engine to reach normal operating temperature, then look at the gauge - there should be at least 15 in-Hg indicated at idle. If not, diagnose and repair the cause of low vacuum before condemning the power booster. **Note:** *Engines with high-performance camshafts will have a lower and somewhat erratic reading at idle. Emission controlled vehicles from the mid 1970s tend to have slightly lower vacuum due to modified valve timing and retarded ignition timing.*

Removal

Note: *Power booster units should not be disassembled. They require special tools not normally found in most service stations or repair shops. They are fairly complex and because of their critical relationship to brake performance it is best to replace a defective booster unit with a new or remanufactured one.*

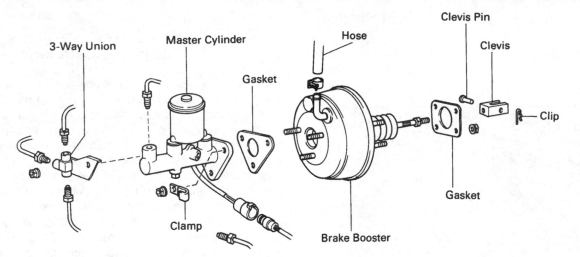

9.48 Installation details of a typical vacuum-operated power brake booster (not all vehicles have a gasket between the master cylinder and the booster).

Unbolt the master cylinder from the booster and pull it forward off the mounting studs, being careful not to kink the brake lines. This will usually provide enough room for booster removal. If you can't do this on your vehicle, remove the master cylinder completely.

Disconnect the vacuum hose where it attaches to the brake booster.

Working under the dash, disconnect the brake pedal return spring, then remove the clip (and clevis pin on most vehicles) to disconnect the pushrod from the brake pedal **(see illustrations)**. On some vehicles it may be necessary to remove a panel from under the dash, and maybe even an air conditioning or heater duct, for access to the pedal.

Remove the mounting nuts and withdraw the booster from the engine compartment **(see illustration)**. The mounting nuts on most vehicles are located under the dash. On some vehicles, however, the mounting nuts are removed from the engine compartment, directly behind the booster. Also, it may be necessary on some vehicles to remove one or more support braces.

9.49 To disconnect the brake booster pushrod from the brake pedal, remove the retaining clip (right arrow) and pull out the clevis pin (left arrow)

9.50 Most vehicles have four booster mounting nuts, accessible from under the dash

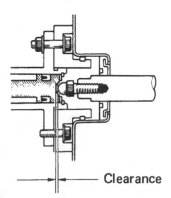

9.51 Ideally, there should be zero clearance between the booster pushrod and the master cylinder pushrod, but there shouldn't be any interference either - if there is any interference, the brakes may drag; if there's much clearance, brake pedal travel will be excessive.

Installation

Installation procedures are basically the reverse of those for removal. Tighten the booster mounting nuts securely. If a cotter pin was used to secure the clevis pin in the pushrod clevis, be sure to use a new one.

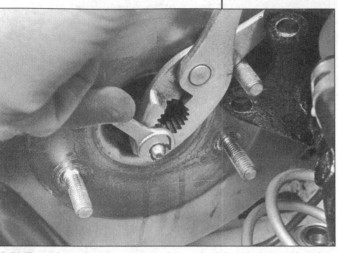

9.52 Turn the adjusting nut on the end of the brake pushrod to change the clearance - hold the pushrod with pliers to keep it from turning

If the power booster is being replaced, the clearance between the master cylinder and the pushrod in the booster must be measured and, if necessary, adjusted (this applies only to boosters with adjustable pushrods) **(see illustration)**. Using a depth micrometer or vernier calipers, measure the distance from the piston seat (recessed area) in the master cylinder to the master cylinder mounting flange. Next, using a hand-held vacuum pump, apply 20 in-Hg of vacuum and measure the distance from the end of the vacuum booster pushrod to the mounting face of the booster (including the gasket, if any) where the master cylinder mounting flange seats. The measurements should be the same (indicating zero clearance). If not, subtract one measurement from the other - a slight amount of clearance is acceptable (approximately 0.004 to 0.020-inch). An interference fit is *not* acceptable because this means the booster pushrod would be depressing the master cylinder pistons slightly, even at rest.

If it is necessary to adjust the length of the pushrod, turn the adjusting screw on the end of the pushrod until the clearance is within the desired range **(see illustration)**.

Check the operation of the brakes before returning the vehicle to normal service.

Hydraulically operated boosters

Check

Refer to Chapter 4 for the booster checking procedures. Checks that are any more involved than the ones described there require special testing equipment. Take the vehicle to a dealer service department or other repair shop for further diagnosis, if necessary.

Removal and installation

Power steering fluid-operated booster

With the engine off, depress and release the brake pedal several times (at least 20) to discharge all pressure from the accumulator. Unbolt the master cylinder from the booster and pull it gently forward without straining the brake lines.

Disconnect the hydraulic lines from the booster unit. Plug all lines and openings to prevent excessive fluid loss and contamination.

Working inside the vehicle, above the brake pedal, disconnect the booster pushrod from the pivot lever, then remove the nuts securing the booster to the firewall **(see illustration)**.

Installation is the reverse of removal. Be sure to tighten the line fittings securely and bleed the power steering system.

To bleed the power steering system, begin by checking the fluid level in the power steering fluid reservoir. Add fluid if necessary, until the level is at the Cold mark on the dipstick.

Start the engine and allow it to run at fast idle. With the wheels in the straight ahead position recheck the fluid level, adding more, if necessary, to reach the Cold mark.

Bleed the system by turning the steering wheel from side-to-side without hitting the stops. This will work the air out of the system. Keep the reservoir full of fluid as this is done. You'll know when the system is fully bled, as the steering wheel action will feel smooth and all strange noises coming from the power

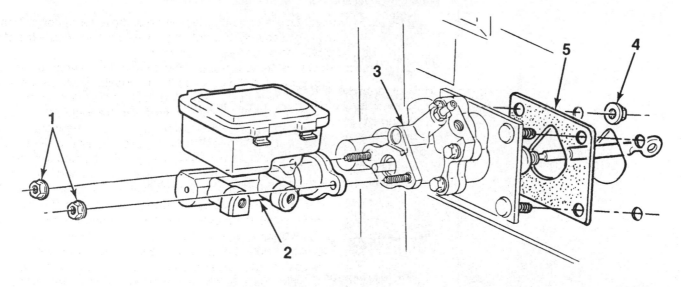

9.53 Mounting details of a General Motors Hydro-boost power brake booster (other power steering fluid-powered boosters similar)

1	Nuts	3	Booster unit	5	Gasket
2	Master cylinder	4	Nut (4)		

steering pump will cease.

Check the operation of the brakes. The brake pedal should not feel "hard" when depressed, and should feel smooth, not jerky.

Electro-mechanical type booster (General Motors Powermaster unit and similar designs)

Warning: *Failure to fully depressurize the hydraulic unit before starting this procedure could result in bodily injury and damage to the painted surfaces of the vehicle. Wear eye protection when performing this procedure.*

Make sure the ignition key is in the Off position, then disconnect the cable from the negative terminal of the battery. Depressurize the system by firmly pushing the brake pedal down a minimum of 10 times. Cover the fender and the area surrounding the brake master cylinder/power booster unit.

Disconnect the electrical connectors at the pressure switch and the hydraulic pump. Using a flare-nut wrench, disconnect the brake lines from the unit **(see illustration).**

Working under the dash, disconnect the pushrod from the brake pedal. Remove the mounting nuts that secure the unit to the firewall.

Carefully remove the booster /master cylinder unit from the engine compartment. **Note:** *The master cylinder portion of the unit should be bench-bled before installing the unit.*

Installation is the reverse of removal, but the system must be bled of air. Fill both sides of the reservoir to the Full marks with the recommended brake fluid. Turn the ignition switch to the On position. With the booster pump running, the brake fluid level in the booster side should decrease as the accumulator is pressurized.

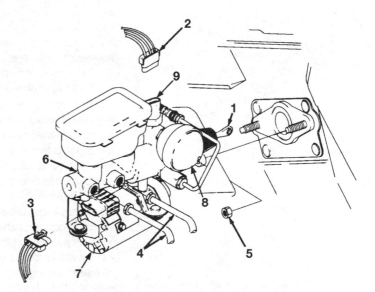

9.54 Installation details of the General Motors Powermaster power brake unit

1	Pushrod	6	Powermaster unit
2	Electrical connector	7	Electro-hydraulic pump
3	Electrical connector	8	Accumulator
4	Brake lines	9	Pressure switch
5	Nut		

Watch the level and don't let the reservoir run dry.

Note: *The pump must shut off after 20 seconds. If not, turn the ignition key off after 20 seconds. Perform the following steps if the fluid in the booster side of the reservoir does not drop:*

a) Loosen the booster line fitting at the casting boss, directly below the accumulator.

b) Wait for brake fluid to gravity bleed from the loosened fitting.

c) When fluid starts to flow, tighten the fitting. Check for leaks or flow back to the reservoir.

Install the reservoir cover. Make sure the ignition is Off, then depress the pedal ten times. Remove the reservoir cover and add fluid, if necessary. Repeat the bleeding procedure again and check the fluid level. Check the operation of the brakes before returning the vehicle to normal service.

Brake caliper

The following procedure depicts the removal, overhaul and installation procedures for the brake caliper. If an overhaul is indicated (usually because of fluid leaks, a stuck piston or broken bleeder screw) explore all options before beginning this procedure. New and factory rebuilt calipers are available on an exchange basis, which makes this job quite easy, and much less time consuming. Before beginning work, check with auto parts stores and dealer parts departments for parts availability and prices. In some cases, rebuilt calipers are actually cheaper than rebuild kits.

Removal

Warning: *Dust created by the brake system may contain asbestos, which is harmful to your health. Never blow it out with compressed air and don't inhale any of it. An approved filtering mask should be worn when working on the brakes. Do not, under any circumstances, use petroleum-based solvents to clean brake parts. Use brake system cleaner only.*

Note: *Always rebuild or replace the calipers in pairs - never rebuild just one of them. Loosen the front wheel lug nuts, raise the vehicle and support it securely on jackstands. Remove the front wheels.*

On models where the brake hose threads directly into the caliper, disconnect the brake line fitting from the brake hose (refer to *Brake hoses and lines* in this Chapter, if necessary), then unscrew the hose from the caliper. (If the caliper is only being removed for access to other components, don't disconnect the hose.) On brake hoses with banjo fittings, unscrew the fitting bolt and detach the hose. Discard the two copper sealing washers on each side of the fitting and use new ones during installation. Wrap a plastic bag around the end of the hose to prevent fluid loss and contamination.

To remove a fixed caliper, remove the bolts that attach the caliper assembly to the steering knuckle **(see illustration 5.101** in Chapter 5**)**. To remove a floating or sliding caliper, refer to the first few steps of the appropriate brake pad replacement sequence in Chapter 5 (caliper removal is the first part of the brake pad replacement procedure for those calipers).

Remove the brake pads from the caliper, also by referring to the appropriate sequence in Chapter 5.

Clean the caliper assembly with brake system cleaner. DO NOT use kerosene, gasoline or petroleum-based solvents. And while you've got the caliper off, be sure to check the pads as well and replace them if necessary.

Overhaul

Fixed caliper

On some fixed calipers the pistons are kind of a "loose" fit in their bores. Try to work the pistons out by hand. If you can't, you'll have to use air pressure to force the pistons out. On most calipers this can be accomplished by removing the bolts holding the caliper halves together, separate the halves, then setting the pistons face-down on the workbench and applying compressed air to the fluid inlet fitting. However, some manufacturers expressly state NOT to remove the bolts that hold the caliper halves together. These calipers are designed with short pistons which can be removed without separating the halves.

To find out if your caliper is the kind that should not be disassembled, place several rags in the middle of the caliper and apply air pressure to the fluid inlet port. **Warning:** *Wear eye protection, and never place your fingers in front of the pistons in an attempt to catch or protect them when applying compressed air, as serious injury could occur. Apply only enough air pressure to ease the pistons out.* If the pistons can be removed without separating the caliper halves, go ahead and remove them (also, ignore any references in this procedure related to disassembling the caliper halves). If they are too long and can't be removed, you have the kind of caliper that is OK to disassemble.

Remove the transfer tube, if equipped. Use a flare-nut wrench, if available **(see illustrations)**.

Mount the caliper assembly in a bench vise lined with wood or equipped with brass protector jaws. **Caution:** *Clamp on the caliper mounting lugs, NOT the caliper itself!* Remove the bridge bolts that hold the two halves of the caliper together. Separate the two caliper halves.

Peel the dust boot out and away from the caliper housing and out of the piston groove **(see illustration)**.

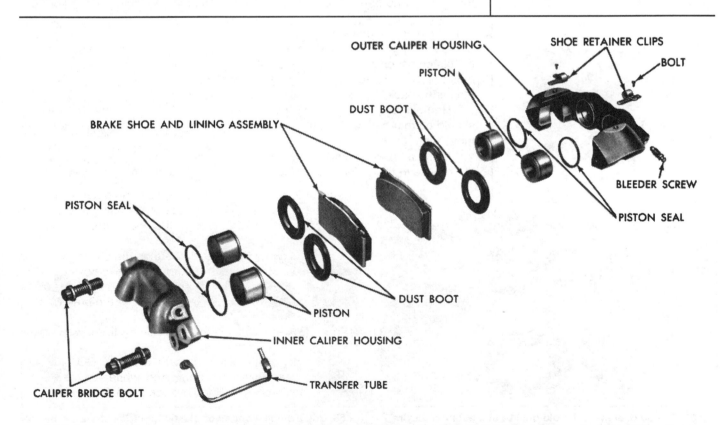

9.55 An exploded view of a fixed brake caliper - this type uses a transfer tube to deliver fluid from one half of the caliper to the other (Chrysler)

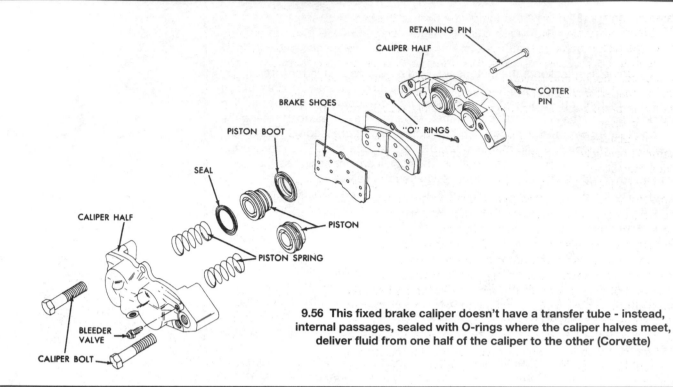

9.56 This fixed brake caliper doesn't have a transfer tube - instead, internal passages, sealed with O-rings where the caliper halves meet, deliver fluid from one half of the caliper to the other (Corvette)

Place each caliper-half face down (pistons facing down) on top of a block of wood and, while holding the caliper-half down against the wood with one hand, plug the hole for the brake hose, apply a brief burst of compressed air through the hole for the transfer tube and pop out the pistons. **Warning:** *Keep your fingers away from the pistons while doing this.* The outer caliper-half also has another hole, for the bleeder screw, which must be installed during this procedure, or the air will go in one hole and out the other, instead of pushing out the pistons. If a piston becomes cocked in its bore during removal, don't try to jerk it out with a pair of pliers, or you will damage the piston and the bore. Instead, carefully tap the piston back into its bore until it's square and try again.

Remove the pistons, noting which bore each of them belongs to - they should be returned to the same bores. If there are springs behind the pistons, remove them.

Using a small pointed wooden or plastic tool, remove the piston seals from the groove in the cylinder bore **(see illustration)**. Discard the old seals. And be

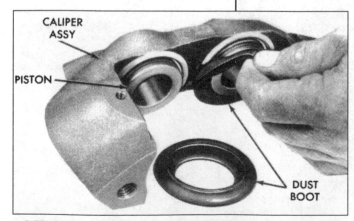

9.57 As you peel away the old dust boots, not how they're installed over both the caliper housing retainer and into the piston groove - that's exactly how the new boots must be installed during reassembly

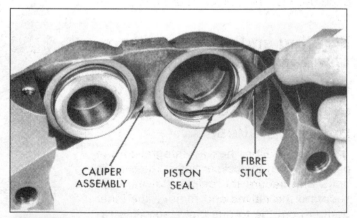

9.58 Use a small wooden or plastic tool to peel the old piston seals out of the piston bores - don't use a metal tool or you might damage the bore walls

careful! Don't scratch or gouge the piston bore or the seal groove.

Clean everything with brake system cleaner and allow the parts to dry. Inspect the piston bores in both housings for scratches, scoring and pitting. Black stains in the bore walls are caused by the piston seals and are harmless. Light scratches, scoring and pitting can be cleaned up with crocus cloth. If the damage is deeper, the caliper should be replaced (rebuilt units are available). Light honing is permissible on cast iron calipers, but NEVER attempt to hone an aluminum caliper.

The pistons should be similarly inspected and cleaned up as necessary. If a piston is severely damaged - pitted, scored or the chrome plating worn off - replace the caliper.

Clamp the mounting lugs of the inner half of the caliper in a bench vise with protector jaws. Dip the new piston seals in clean brake fluid and install them in the caliper grooves. Position each seal into its groove at one spot and gently work it around the piston bore with your finger until it's properly seated. **Warning:** *Do NOT use old seals!*

Coat the outside diameter of the pistons with clean brake fluid and install them in the cylinder bores with the open end of each piston facing away from the caliper **(see illustration)**. If the caliper is equipped with springs behind the pistons don't forget to install them first. Position the piston squarely in the bore and apply a slow, steady pressure until the piston is fully seated. If you encounter resistance, remove the piston and make sure the seal is properly installed.

Install the new dust boots into their grooves in the caliper and the pistons. Make sure the dust boots are properly seated.

Remove the inner caliper-half from the bench vise and install the outer caliper-half the same way. Install the seals, pistons and dust boots using the same method.

If your caliper does not have a transfer tube, install new O-ring seals in the counterbore(s) around the fluid passage(s) **(see illustration 9.56)**.

Install the inner caliper-half onto the outer caliper-half that's still clamped into the vise, install the bridge bolts and tighten them to the torque listed in the Specifications at the end of this manual. Install the transfer tube (if applicable) and tighten the fittings securely. Install the bleeder screw, if removed. Refer to page 9-40 for the caliper installation procedure.

Floating and sliding calipers (except opposed-piston calipers) - front (and rear on models with "drum-in-disc" parking brakes)

If your caliper has an integral torque plate on which the caliper floats, remove the bolts that secure the caliper halves, then separate the halves and remove the inner portion from the torque plate **(see illustrations)**. Some examples of vehicles that use this design are the Hyundai Excel, Dodge Colt and Mitsubishi Precis.

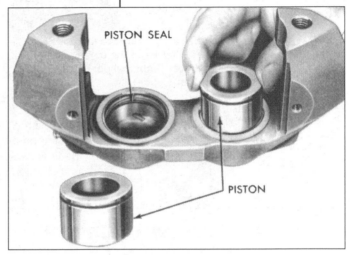

9.59 When you install the pistons into the caliper halves, coat them with fresh brake fluid, double-check the new seals to make sure they're properly seated into their grooves in the piston bore walls and make sure the pistons are square to the bores

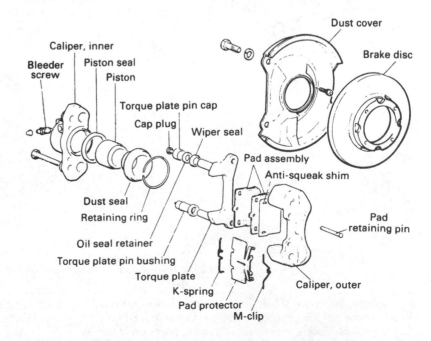

9.60 Exploded view of a floating caliper with an integral torque plate (mount)

9.61 With the caliper held in a vise, remove the bolts that secure the inner and outer halves of the caliper frame together, then separate them (caliper with integral torque plate) . . .

9.62 . . . then slide the inner half of the caliper off the torque plate

If the piston dust boot is held in place by a retaining ring, remove the ring and pull the dust boot out **(see illustrations).**

Place several shop towels or a block of wood in the center of the caliper to act as a cushion (unless you have the kind of caliper shown in illustration 9.60),

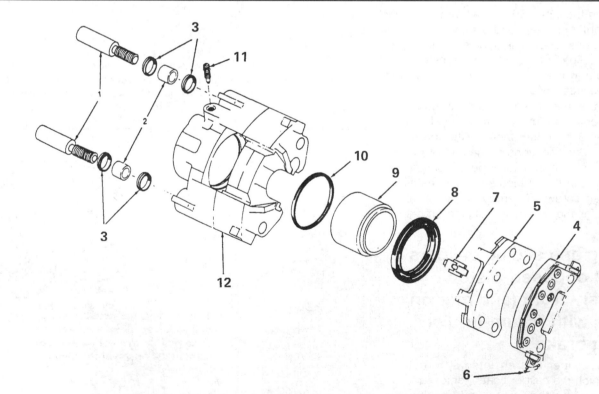

9.63 Exploded view of a floating brake caliper (General Motors)

1	Guide pin (mounting bolt)	5	Inner pad	9	Piston
2	Sleeve	6	Wear sensor	10	Piston seal
3	Bushing	7	Pad retainer	11	Bleeder screw
4	Outer pad	8	Dust boot	12	Caliper housing

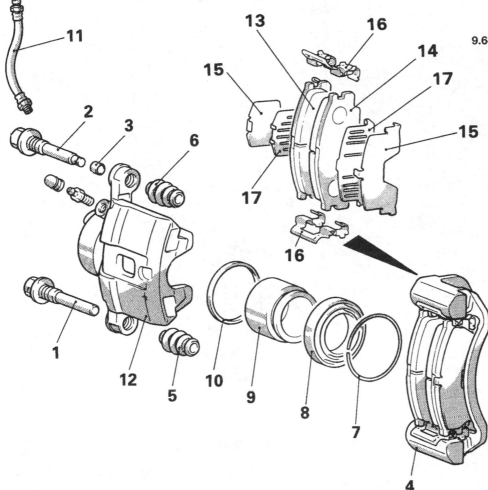

9.64 Exploded view of a single-piston floating caliper (Mitsubishi)

1   Guide pin (mounting bolt)
2   Lock pin (mounting bolt)
3   Bushing
4   Torque plate (caliper mounting bracket)
5   Guide pin boot
6   Lock pin boot
7   Retaining ring
8   Dust boot
9   Piston
10 Piston seal
11 Brake hose
12 Caliper body
13 Inner brake pad
14 Outer brake pad
15 Shim
16 Anti-rattle clip
17 Shim

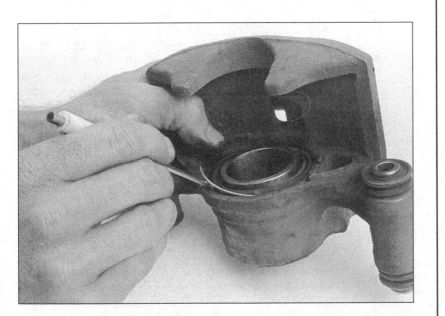

9.65 Remove the boot retaining ring with a small screwdriver - be extremely careful not to gouge or scratch the piston or the piston bore

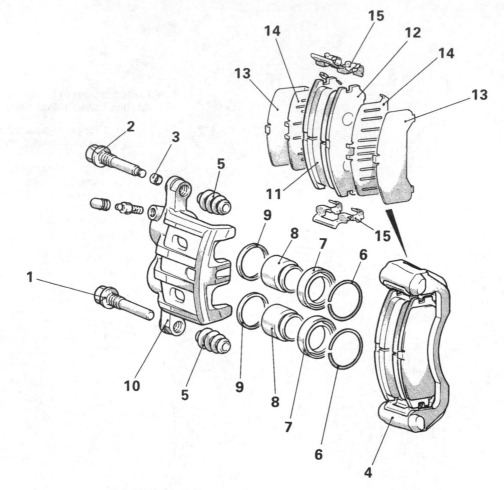

9.66 Exploded view of a dual-piston floating brake caliper (Mitsubishi)

1	Guide pin (mounting bolt)	6	Retaining ring	12	Outer brake pad
2	Lock pin (mounting bolt)	7	Piston dust boot	13	Shim
3	Bushing	8	Piston	14	Shim
4	Torque plate (caliper mounting bracket)	9	Piston seal	15	Anti-rattle clip
5	Dust boot	10	Caliper body		
		11	Inner brake pad		

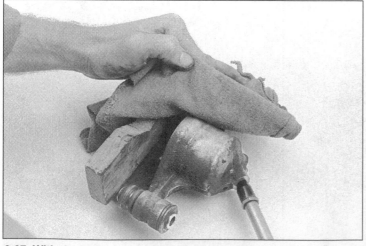

9.67 With the caliper padded to catch the piston, use compressed air to force the piston out of its bore - make sure your fingers aren't between the piston and the caliper

then use compressed air, directed into the fluid inlet, to remove the piston **(see illustration)**. If you separated the caliper halves to remove the torque plate (as shown in illustrations 9.61 and 9.62), place the caliper on the workbench, piston facing down, and direct the compressed air into the fluid inlet **(see illustration)**. Use only enough air pressure to ease the piston out of the bore. If the piston is blown out, even with the cushion in place, it may be damaged. **Warning:** *Never place your fingers in front of the piston in an attempt to catch or protect it when applying compressed air, as serious injury could occur.*

Remove the dust boot from the caliper bore. Some types of dust boots have to be pried out **(see illustration)**. Now, using a wood or plastic tool, remove the piston seal from the groove in the caliper bore **(see illustration)**. Metal tools may cause bore damage.

9.68 On calipers with integral torque plates, place the caliper piston-side down and direct compressed air into the brake hose inlet port to eject the piston

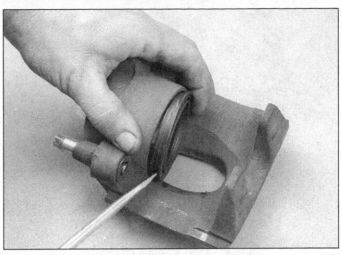

9.69 If the dust boot has a rigid casing, carefully pry it out of the caliper

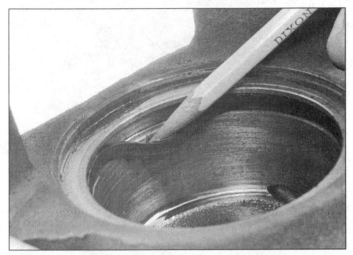

9.70 To remove the seal from the caliper bore, use a plastic or wooden tool - a pencil will do the job

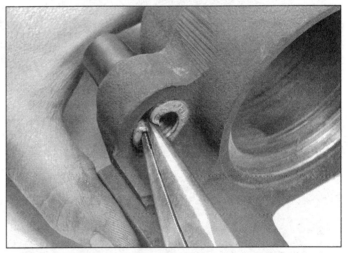

9.71 To remove a caliper pin boot, grab it with a pair of needle-nose pliers, twist it and push it through the caliper frame

Remove the bleeder screw, then remove and discard the caliper pin (mounting bolt) bushings, sleeves and boots, if equipped **(see illustration)**.

Clean the remaining parts with brake system cleaner. Allow the parts to air dry. Inspect the surfaces of the piston for nicks and burrs and loss of plating. **Note:** *Some pistons are made of a plastic-like material called phenolic. If your caliper contains this kind of piston, be sure to check it for cracks, chips and other surface irregularities* **(see illustration)**. If surface defects are present, the caliper must be replaced. Check the caliper bore in a similar way. Light polishing with crocus cloth is permissible to remove slight corrosion and stains. Light honing is also permissible on cast iron calipers, but NEVER attempt to hone an aluminum caliper. Discard the caliper pins if they're severely corroded or damaged.

Lubricate the new piston seal with clean brake fluid and

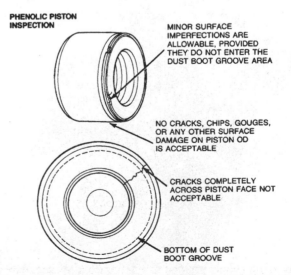

PHENOLIC PISTON INSPECTION

MINOR SURFACE IMPERFECTIONS ARE ALLOWABLE, PROVIDED THEY DO NOT ENTER THE DUST BOOT GROOVE AREA

NO CRACKS, CHIPS, GOUGES, OR ANY OTHER SURFACE DAMAGE ON PISTON OD IS ACCEPTABLE

CRACKS COMPLETELY ACROSS PISTON FACE NOT ACCEPTABLE

BOTTOM OF DUST BOOT GROOVE

9.72 Phenolic pistons must be carefully inspected for signs of damage, as shown

9.73 Position the new seal in the cylinder groove - make sure it isn't twisted

9.74 On calipers with dust boots that have rigid casings (or boots that use a retaining ring), slip the boot over the piston . . .

9.76 If you don't have a boot installation tool, gently seat the boot with a drift punch

position the seal in the cylinder groove using your fingers only **(see illustration)**.

If the dust boot on your caliper is held in place by a retaining ring, or if it has a hard casing around its outer diameter which seats in the caliper, install the new dust boot in the groove in the end of the piston **(see illustration)**. Dip the piston in clean brake fluid and insert it squarely into the cylinder. Depress the piston to the bottom of the cylinder bore **(see illustration)**. Seat the boot in the caliper counterbore using a boot installation tool **(see illustration 9.103)** or a blunt punch **(see illustration)**, or install the retaining ring **(see illustration),** depending on design.

If the dust boot on your caliper has a lip or flange at the bottom which seats in the upper groove in the caliper bore, install the boot on the bottom (closed end) of the piston **(see illustration)**. Lower the piston and boot assembly into the caliper bore and work the flanged portion (the lip) into its

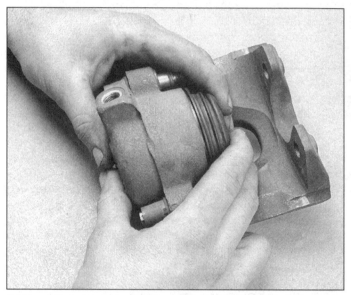

9.75 . . . then push the piston straight into the cylinder - make sure it doesn't become cocked in the bore

9.77 On calipers that use retaining rings, seat the boot in its groove, then install the retaining ring

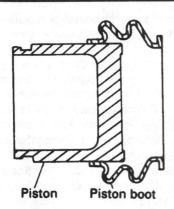

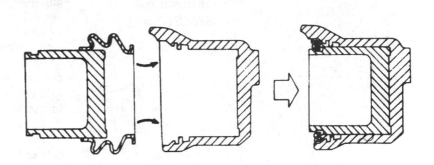

Piston   **Piston boot**

9.78 On calipers where the flange of the boot fits into a groove in the caliper, place the piston boot onto the bottom of the piston . . .

9.79 . . . then tuck the fluted portion of the boot into the upper groove of the caliper bore and bottom the piston in the bore

groove in the bore **(see illustration)**. If you have trouble seating the lip of the dust boot in its groove, try this alternative method:

a) Take the boot off the piston and install the flanged portion of the boot into its groove in the caliper bore **(see illustration)**. Coat the cylinder bore and the walls of the piston with clean brake fluid.

b) Place the caliper in a vise, with the jaws of the vise clamping onto one of the caliper mounting ears. Place the piston against the end of the new dust boot, apply a small burst of compressed air (make sure it's filtered and unlubricated) to the caliper bore to inflate the boot, then push the piston through the boot opening as the boot inflates **(see illustration)**. **Warning:** *Once the piston has been inserted into the boot, stop applying compressed air. Also, don't place your fingers between the piston and the caliper frame.* Although it's possible to push the piston through the boot without applying compressed air, it's extremely difficult and may damage the boot.

c) Continue to push the piston into the bore, then seat the dust boot into its groove near the outer end of the piston **(see illustration)**.

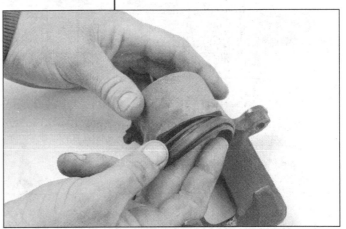

9.80 If you have trouble installing the piston and boot as described in illustrations 9.78 and 9.79, install the big end of the dust boot into the caliper, making sure the fluted portion is completely seated in its groove in the caliper bore . . .

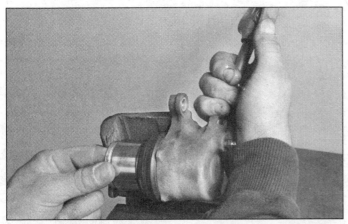

9.81 . . . then place the piston against the open end of the new dust boot, apply a small burst of filtered, unlubricated compressed air to the fluid inlet to inflate the boot. Push the piston through the small end of the boot as the boot inflates . . .

9.82 . . . as soon as the piston is in the boot, release the compressed air and push the piston far enough into the bore to seat the ridge on the boot into its groove in the piston. Push the piston the rest of the way into the bore.

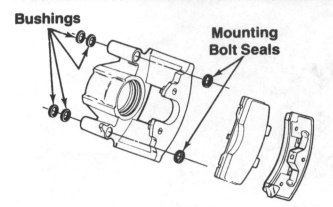

Bushings

Mounting Bolt Seals

9.83 Caliper bushing and seal details (General Motors)

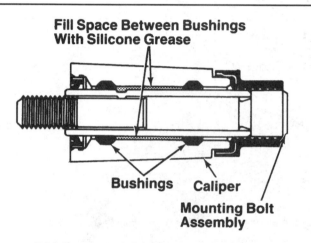

Fill Space Between Bushings With Silicone Grease

Bushings   **Caliper**

Mounting Bolt Assembly

9.84 Be sure to lubricate the caliper bushings

Install the new mounting pin bushings and sleeves **(see illustrations)**. Be sure to lubricate the sleeves with silicone grease.

If your caliper has an integral torque plate on which the caliper floats, lubricate the torque plate pins with silicone grease and install the torque plate to the inner half of the caliper **(see illustration)**. Attach the outer half of the caliper to the inner half, install the bolts and tighten them to the torque listed in the Specifications at the end of this manual.

Install the bleeder screw.

On sliding calipers, clean the sliding surfaces of the caliper and the caliper adapter. Refer to page 9-40 for the caliper installation procedure.

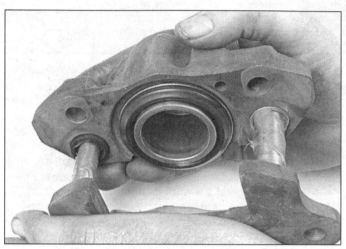

9.85 On calipers with integral torque plates, lubricate the torque plate pins with silicone-based grease, then assemble the plate and the inner portion of the caliper - the caliper halves can now be bolted together

Opposed-piston calipers - front

Remove the two bolts that attach the caliper to the yoke, then pry the yoke holder off the yoke **(see illustrations)**. This disengages the piston from the yoke.

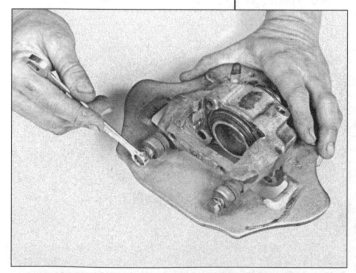

9.86 Remove the yoke-to-caliper bolts . . .

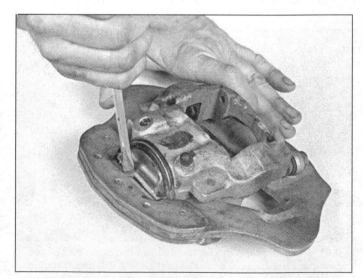

9.87 . . . then, using a screwdriver, pry the yoke holder off the yoke

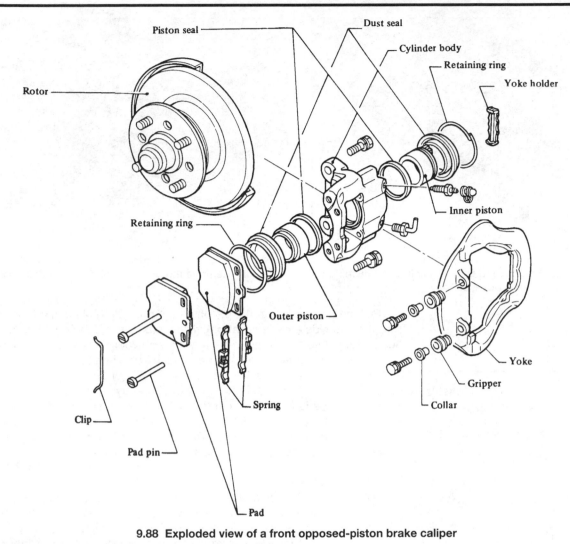

9.88 Exploded view of a front opposed-piston brake caliper

The caliper can now be detached from the yoke.

Remove the yoke holder from the inner piston, then remove the dust boot retaining rings from each piston **(see illustrations)**.

9.89 Remove the retaining ring from the inner piston dust boot . . .

9.90 . . . then remove the retaining ring from the outer piston dust boot

9.91 Both pistons are removed from the caliper by pushing on the outer piston

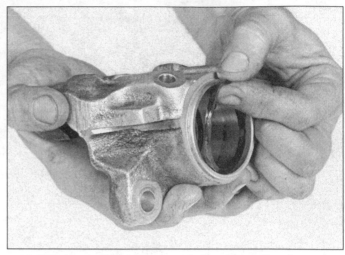

9.92 Remove the piston seals from the caliper bore

Pull the dust boots off the pistons, then push both pistons out of the caliper from the pad side **(see illustration)**. The piston seals can now be removed **(see illustration)**. If you can't remove the seals with your fingers, use a wood or plastic tool - metal tools can cause bore damage.

Remove the grippers and collars from the yoke.

Clean everything with brake system cleaner and allow the parts to dry. Inspect the piston bore for scratches, scoring and pitting. Black stains in the bore walls are caused by the piston seals and are harmless. Light scratches, scoring and pitting can be cleaned up with crocus cloth. If the damage is deeper, the caliper should be replaced (rebuilt units are available). NEVER attempt to hone an aluminum caliper.

The pistons should be similarly inspected and cleaned up as necessary. If a piston is severely damaged - pitted, scored or the chrome plating worn off - replace the caliper.

Check the yoke for cracks, excessive wear or other damage and replace it if necessary.

Lubricate the piston seals with clean brake fluid and install them into their grooves in the caliper bore, making sure they seat correctly and aren't twisted. Lubricate the walls of the pistons with clean brake fluid (or silicone grease, if it was supplied in the rebuild kit). Install the pistons *into their respective ends of the bore*. Don't install them like you removed them (both pistons through the same side of the caliper).

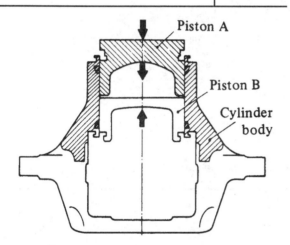

9.93 The pistons should be positioned like this when installing the dust boots

Piston A

Piston B

Cylinder body

Press the outer piston into the bore so that the inner edge of the piston seal groove is in line with the inner edge of the caliper seal grooves **(see illustration)**. Apply silicone grease to the inside of the dust boots, then install the dust boots and retaining rings **(see illustrations)**. Make sure the boots seat properly into the grooves in the pistons and the grooves in the caliper.

Press the yoke holder into its groove on the inner piston face, then position the caliper body in the yoke, making sure the yoke holder is aligned with the yoke **(see illustration)**. Place the entire assembly so the outer piston is facing up, then apply pressure to the outer piston to press the yoke holder onto the yoke **(see illustration)**.

Install the grippers and collars into their holes in the yoke **(see illustration)**. Be sure to lubricate the collars with silicone grease.

9.94 Apply silicone grease to the inside of the dust boots prior to installation

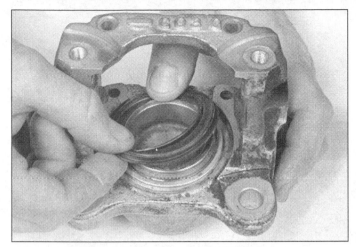

9.95 Install the outer piston dust boot . . .

9.96 . . . followed by the dust boot retaining ring

9.97 Install the yoke holder on the inner piston

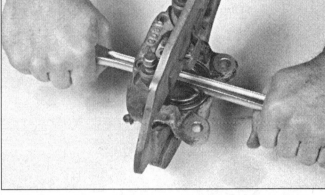

9.98 To seat the yoke holder on the yoke, place a prybar or similar tool across the outer piston, then push down

Once the yoke holder is pressed onto the yoke, install the two mounting bolts that retain the yoke to the caliper housing, tightening them securely.

Refer to page 9-40 for the caliper installation procedure.

Floating and sliding calipers (except opposed-piston calipers) - rear (calipers with integral parking brake mechanisms)

Note: *If you're not sure what type of rear caliper you have, refer to the component descriptions in Chapter 1. Also, the following procedures are typical for the type of caliper designs they represent. There may be some differences between the calipers shown and the ones on your vehicle. Whenever disassembling a rear brake caliper, make sketches of the relationship of the components and lay the parts out in order as they are removed. As stated in the beginning of this Section, it may actually be more cost effective to purchase factory-rebuilt calipers rather than buy a rebuild kit and overhaul them yourself.*

9.99 Install the bushings and boots (collars and grippers), lubricating them with silicone grease

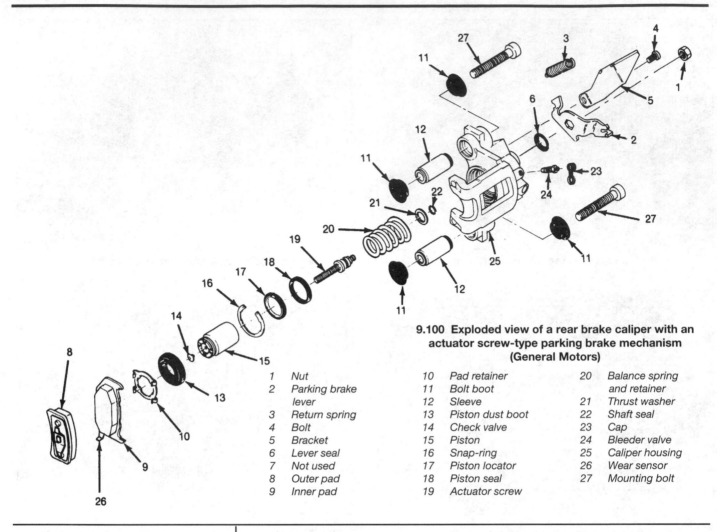

9.100 Exploded view of a rear brake caliper with an actuator screw-type parking brake mechanism (General Motors)

1	Nut	10	Pad retainer	20	Balance spring
2	Parking brake	11	Bolt boot		and retainer
	lever	12	Sleeve	21	Thrust washer
3	Return spring	13	Piston dust boot	22	Shaft seal
4	Bolt	14	Check valve	23	Cap
5	Bracket	15	Piston	24	Bleeder valve
6	Lever seal	16	Snap-ring	25	Caliper housing
7	Not used	17	Piston locator	26	Wear sensor
8	Outer pad	18	Piston seal	27	Mounting bolt
9	Inner pad	19	Actuator screw		

Calipers with an actuator *screw*-type parking brake mechanism

Remove the sleeves and bolt boots from the ends of the caliper, then remove the pad retainer (if equipped) from the end of the piston by rotating the retainer until the inside tabs line up with the notches in the piston **(see illustration)**.

Remove the parking brake lever nut, lever and seal. Place rags between the piston and the caliper frame, then turn the actuator screw with a wrench to work the piston out of the bore **(see illustration)**. Remove the balance spring then push on the parking brake lever end of the actuator screw to remove it. Remove the shaft seal and thrust washer from the actuator screw.

While being careful not to scratch the housing bore, remove the boot **(see illustration)**.

Using snap-ring pliers, remove the snap-ring, followed by the piston locator **(see illustration 9.100).** Using a plastic or wooden tool, remove the piston seal from the caliper bore **(see illustration 9.70).**

Remove the bleeder valve. Clean all parts not included in the repair kit with brake system cleaner. Allow the parts to dry. Inspect all parts for wear, damage and corrosion. **Note:** *It is okay to remove minor corrosion from the caliper bore using crocus cloth.*

Lubricate all of the internal parts with clean brake fluid. Install a new piston seal into its groove in the caliper bore **(see illustration 9.73).**

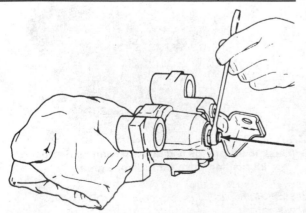

9.101 Rotate the actuator screw to remove the piston

Install a new piston locator on the piston. Install the thrust washer on the actuator screw with the grayish side toward the caliper housing and the copper side toward the piston assembly. Install the shaft seal on the actuator screw, then install the actuator screw onto the piston assembly.

Install the piston assembly with the actuator screw and balance spring into the lubricated bore of the caliper. Push the piston into the caliper bore so the locator is past the retainer groove in the caliper bore. **Note:** *A pair of adjustable pliers might be necessary to press the piston into the caliper.* Install the snap-ring using snap-ring pliers.

With the inside of the boot in the piston groove and boot fold toward the end of the piston that contacts the brake pad, install the caliper boot onto the piston. Use a seal driver to seat the caliper boot into the cylinder counterbore **(see illustration)**. If a seal driver isn't available, use a blunt punch and carefully tap around the outer circumference of the seal to seat it **(see illustration 9.76)**. Install the pad retainer on the piston.

Lubricate the outside diameter of the mounting bolt sleeves and the caliper sleeve cavities with silicone grease. Install one sleeve boot into a groove of the sleeve cavity **(see illustration)**. Install a sleeve through the opposite side of the sleeve cavity and continue pushing the sleeve until the boot lip seats in the sleeve groove.

Install the boot in the opposite side of the sleeve cavity groove and push the sleeve through the cavity far enough for that boot to slip into the remaining groove of the sleeve. Repeat the sleeve boot installation procedure to the remaining sleeve.

Refer to page 9-40 for the caliper installation procedure.

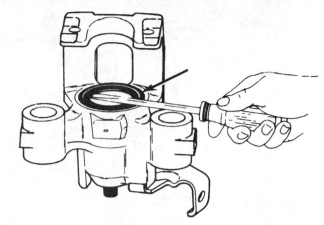

9.102 Carefully pry the dust boot out of the housing, taking care not to scratch the surface of the bore

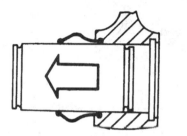

A

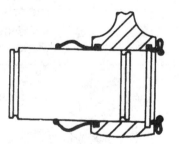

B

C

9.104 Sleeve boot installation procedure

A Install the sleeve boot into the groove in the sleeve bore, then push the sleeve into the boot
B Install the other boot into the groove in the other side of the bore
C Push the sleeve back and seat both boots into the grooves in the ends of the sleeve

9.103 Use a seal driver, if available, to seat the boot in the caliper housing

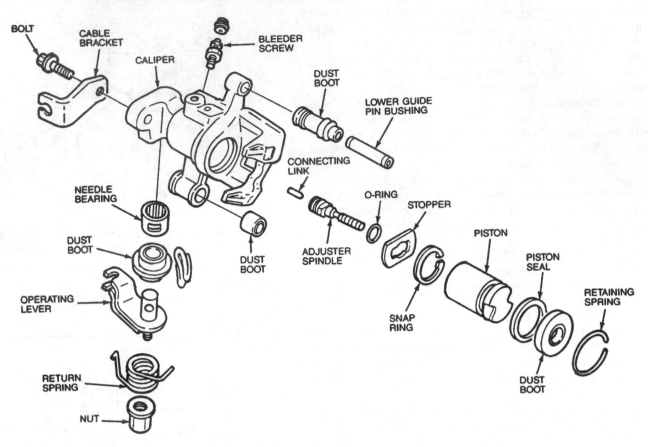

9.105 Exploded view of a rear caliper with an actuator-spindle parking brake mechanism

Calipers with an actuator *spindle*-type parking brake mechanism

Remove the caliper guide bushing and dust boots **(see illustration)**. Discard all rubber parts. Clean the exterior of the caliper with brake system cleaner.

Pry the retaining ring off the dust boot with a screwdriver. Discard the boot.

Rotate the piston counterclockwise using a brake piston turning tool (see Chapter 2) or a pair of needle-nose pliers, and remove the piston from the adjuster spindle.

Using a wood or plastic tool, remove the piston seal from the caliper bore groove **(see illustration 9.70)**. Metal tools may cause bore damage.

Remove the caliper bleeder screw.

Using a pair of snap-ring pliers, remove the snap-ring above the stopper **(see illustration 9.105)**. Remove the adjuster spindle, stopper and connecting link. Separate the adjuster spindle from the stopper. Remove the O-ring from the adjuster spindle. Discard the O-ring.

Remove the return spring from the operating lever of the parking brake mechanism **(see illustration)**. Remove the operating lever nut and lockwasher. Mark the relationship between the operating lever and the shaft. Remove the operating lever from the shaft.

Remove the seal from the caliper housing. Also remove the shaft and the needle bearing.

Carefully examine the piston for nicks and burrs and loss of plating. If surface defects are evident, replace the

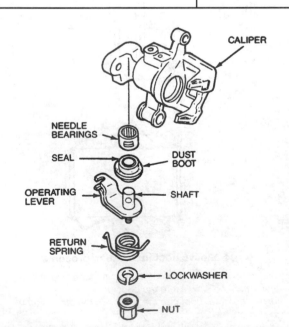

9.106 Exploded view of the parking brake lever assembly

caliper assembly. Inspect the caliper bore for similar damage and wear. You can lightly polish the bore with crocus cloth to remove light corrosion and stains. If that fails to clean up the damage, replace the caliper. Inspect the mounting bolt - if it's corroded or damaged, discard it. Also inspect the guide pin bushings for wear. If they're worn or corroded, replace them.

Lubricate the needle bearings with multi-purpose grease. Align the opening in the needle bearings with the bore in the caliper housing **(see illustration)**. Install the needle bearings.

Install the operating shaft into the caliper housing. Install the operating lever **(see illustration)**. Align the marks made during removal. Install the lockwasher nut.

Insert the connecting link into the operating lever shaft. Install the O-ring onto the adjuster spindle. Position the stopper onto the adjuster spindle so the pins will align with the caliper housing **(see illustration)**. Install the adjuster spindle and the stopper into the caliper.

Install the stopper retaining snap-ring. Be sure the operating lever and adjuster spindle move freely. Install the parking brake return spring.

Before reassembling the caliper, lubricate the piston bore and O-ring with clean brake fluid. Install the O-ring into its groove in the caliper bore **(see illustration 9.73)**. Make sure it isn't twisted.

Lubricate the piston with clean brake fluid, then insert it into the caliper bore. Rotate the piston clockwise using a brake piston turning tool or a pair of needle-nose pliers, to install it onto the adjusting spindle. Screw the piston in fully. Align the notches in the piston with the opening in the caliper.

Install a new dust boot in the piston groove with the fold toward the open end of the piston. Seat the boot in the caliper bore. Install the retaining ring.

Install the dust boots for the guide pin bushings **see illustration)**. Install the caliper guide pin bushing.

Refer to page 9-40 for the caliper installation procedure.

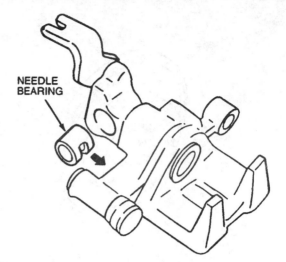

9.107 Align the opening in the needle bearings with the bore in the caliper housing

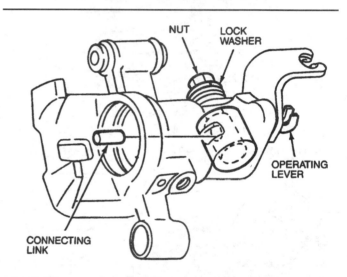

9.108 When you install the operating lever, be sure to align the marks you made before disassembly

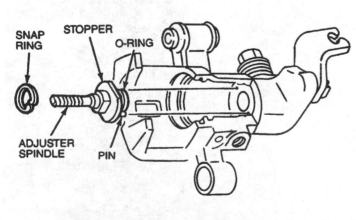

9.109 Exploded view of the adjuster spindle and stopper assembly

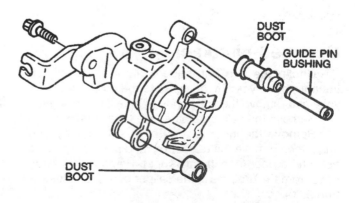

9.110 Exploded view of the guide pin (mounting bolt) bushings and dust boots

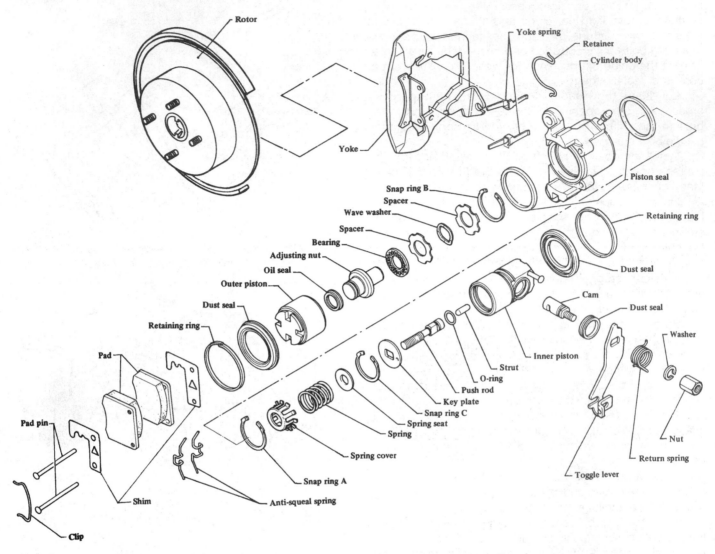

9.111 Exploded view of a rear opposed-piston brake caliper

Opposed-piston calipers - rear

Unclip the two anti-squeal clamps and detach them from the yoke, then separate the caliper from the yoke and remove the yoke springs **(see illustrations)**.

Remove the caliper retainer clip **(see illustration)**.

Remove the inner and outer dust boot retaining rings, then remove the dust boots and push the pistons from the caliper **(see illustrations)**. Remove the piston seals from the grooves in the caliper bore **(see illustration 9.70)**.

Remove the outer piston from the inner piston by

9.112 Remove the anti-squeal clamps from the yoke

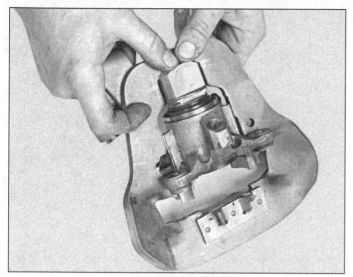

9.113 Separate the caliper from the yoke by pushing outward on the inner piston

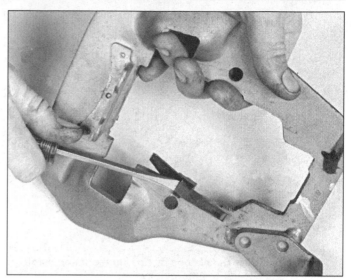

9.114 Pry the yoke springs from the yoke

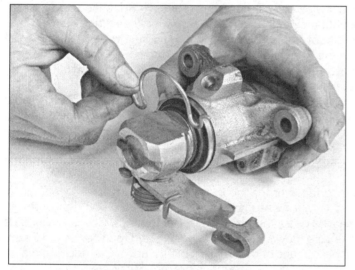

9.115 Remove the caliper retaining clip

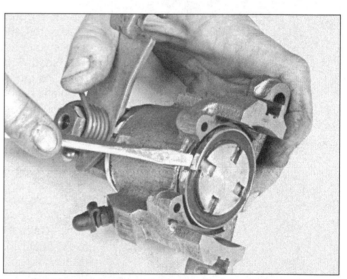

9.116 Remove the retaining rings from both piston dust boots . . .

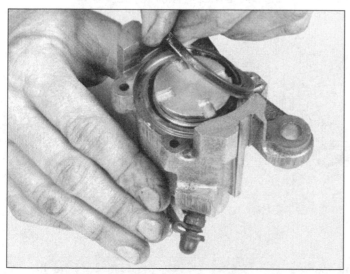

9.117 . . . peel the dust boots from the caliper

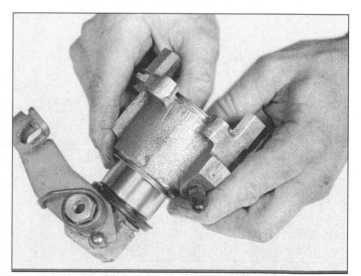

9.118 Push the pistons out of the caliper

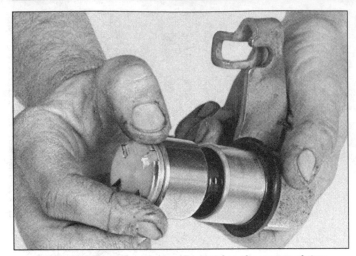

9.119 Separate the pistons by turning the outer piston counterclockwise

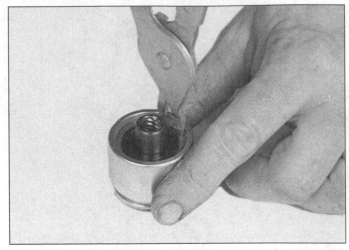

9.120 Using snap-ring pliers, remove the snap-ring from the outer piston . . .

rotating it counterclockwise **(see illustration)**. Remove the snap-ring from the outer piston, then lift out the components for inspection **(see illustrations)**.

Remove the dust boot from the inner piston **(see illustration)**. Remove the outer snap-ring from the inner piston. Lift off the spring cover and spring. Remove the inner snap-ring from the inner piston. Turn the inner piston upside-down and remove the spring seat and key plate.

Pull the pushrod out of the piston and remove the strut and O-ring. Disengage the return spring from the toggle lever **(see illustration)**. Remove the toggle lever and cam assembly by pulling it out of the inner piston. Remove the toggle lever lip seal from the inner piston.

Use a screwdriver to carefully pry out the seal and bearing retainer from the inner piston.

Check the outer surfaces of the inner and outer pistons for scoring, nicks, rust or other damage. If light rust or scoring is present, it can be removed with crocus cloth. If damage is deep, the piston must be replaced. Check the threads on the pushrod. Make sure they are in good condition and replace the rod if necessary. Inspect the inner surface of the caliper bore for any nicks, scoring, rust or other damage. Again, if light rust or scoring is present, it can be removed using crocus cloth, but the caliper will have to be replaced if the damage is deeper.

Inspect the needle bearing inside the outer piston for freedom of movement. If any of the rollers are binding, the bearing will have to be replaced by prying it out.

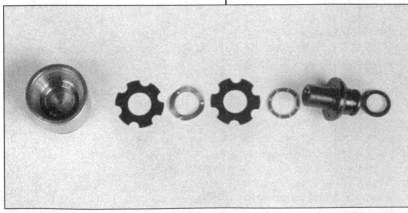

9.121 . . . then remove the contents of the piston and lay the components out in order

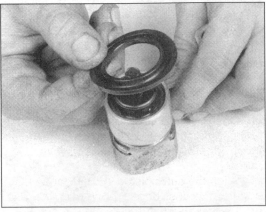

9.122 Remove the dust boot from the inner piston

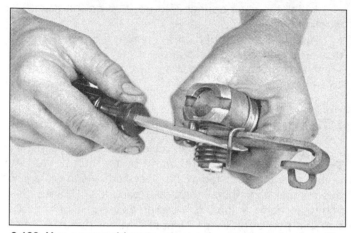

9.123 Use a screwdriver to remove the return spring from the toggle lever

9.124 The seal on the outer piston adjusting nut should be installed like this

Apply a light coat of silicone-based grease to the needle bearings.

Install the seal and bearing retainer by gently tapping it into place with a plastic-head hammer. Install a new lip seal onto the retainer.

Carefully insert the toggle lever and cam assembly into the outer piston.

Lubricate both ends of the strut with silicone-based grease and install it into the pushrod. Lightly lubricate the pushrod O-ring with silicone-based grease and install it into its groove on the pushrod. Insert the strut and pushrod into the piston.

Install the key plate over the pushrod. Be sure the pushrod is aligned so that the locating pin in the key plate is properly fitted into its hole in the piston. Install the inner snap-ring into its groove in the piston.

Install the spring seat, spring and spring cover then place an appropriately sized socket over the pushrod, so that it sits on the spring cover. Place the assembly in a vise and tighten the vise until the spring is compressed enough for the outer snap-ring to be installed in its groove. Be careful that the spring cover is properly centered in the bore so that the tangs don't get damaged.

Install a new lip seal onto the adjusting screw. Be sure the lip is facing in the proper direction **(see illustration)**. Lubricate the lip seal with silicone-based grease.

Place the adjusting screw into the bore in the outer piston. Place the bearing spacer, wave washer and spacer, in that order, over the adjusting nut. Install the snap-ring into the outer piston. Install the outer piston onto the pushrod.

Apply a light coat of silicone-based grease to the cylinder bore. Install the new piston seals into their grooves inside the bore **(see illustration 9.73)**. Lightly lubricate the seals with silicone-based grease.

Apply a light coat of silicone-based grease to the seal groove on the inner piston. Wipe off any residual grease that may have gotten onto the exterior surface of the piston. Install a new dust boot into position on the inner piston, then install the retaining ring.

Apply a light coat of silicone-based grease to the outer surfaces of both pistons. Apply a light coat of silicone-based grease into the two dust boot grooves on either side of the caliper housing. Carefully insert the pistons into the caliper bore. Align the inner piston so the inner edge of its boot groove is in line with the inner edge of the caliper boot groove, then pull the boot into position on the caliper.

Align the outer piston in a similar fashion to the inner piston so that the inner edge of the dust boot groove is in line with the inner edge of the caliper seal groove **(see illustration)**. Install the outer piston dust boot and retaining ring.

Place the caliper retainer into its groove.

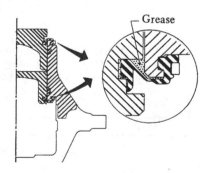

9.125 The pistons should be positioned like this when installing the dust boots. Apply silicone grease to the boots before installing them

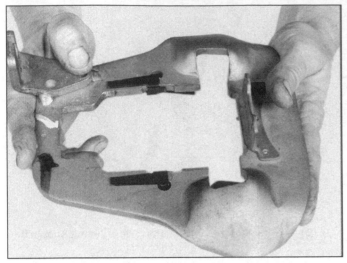

9.126 The yoke springs should be installed prior to attaching the caliper to the yoke

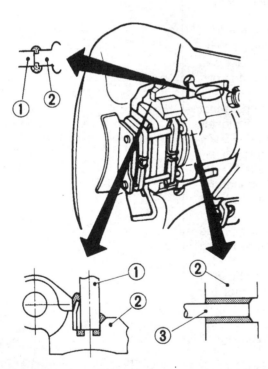

9.127 The frictional surfaces of the yoke and caliper and the pad pin holes should be lubricated with high-temperature grease prior to installation of the caliper to the yoke

1	Yoke	3	Pad pin
2	Caliper body		

Install the yoke springs into position on the yoke **(see illustration)**.

Apply a light film of high-temperature grease to the indicated surfaces **(see illustration)**.

Install the caliper into the yoke, then install the anti-squeak clamps onto the yoke.

Installation

If you're servicing a fixed caliper, install the caliper assembly and the caliper mounting bolts, then tighten the bolts to the torque listed in the Specifications at the end of this manual, then install the brake pads (see Chapter 5). If you're servicing a floating or sliding caliper, caliper installation is a part of brake pad installation in Chapter 5. Make sure you tighten the caliper pins/bolts to the torque listed in the Specifications at the end of this manual.

On models where the brake hose threads directly into the caliper, connect the brake hose to the caliper and tighten it securely. Position the other end of the hose in its bracket and connect the brake line fitting. Tighten the fitting securely. On models that use banjo fittings, connect the hose to the caliper using new sealing washers. Tighten the banjo bolt securely.

Bleed the brakes as outlined in later in this Chapter.

Install the wheels and lug nuts. Lower the vehicle and tighten the lug nuts securely.

After the job has been completed, firmly depress the brake pedal a few times to bring the pads into contact with the disc. Check the operation of the brakes before driving the vehicle in traffic.

Wheel cylinders

Note: *If an overhaul is indicated (usually because of fluid leakage or sticky operation) explore all options before beginning the job. New wheel cylinders are available, which makes this job quite easy. If you decide to rebuild the wheel cylinder, make sure a rebuild kit is available before proceeding. Never overhaul only one wheel cylinder. Always rebuild both of them at the same time.*

Removal

Remove the brake shoes (see Chapter 6). Unscrew the brake line fitting from the rear of the wheel cylinder **(see illustration)**. If available, use a flare-nut wrench to avoid rounding off the corners on the fitting. Don't pull the metal line out of the wheel cylinder - it could bend, making installation difficult.

Remove the bolt(s) or clip securing the wheel cylinder to the brake backing plate **(see illustration)**. Remove the wheel cylinder. Plug the end of the brake line to prevent the loss of brake fluid and the entry of dirt.

9.128 Unscrew the brake line fitting from the rear of the wheel cylinder with a flare-nut wrench, then remove the bolts (arrows) securing the wheel cylinder to the brake backing plate

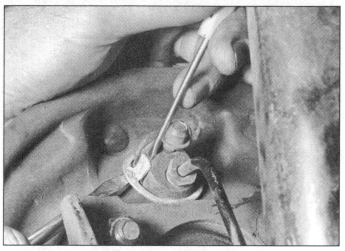

9.129 On wheel cylinders that aren't retained by bolts, a pair of screwdrivers can be used to remove the wheel cylinder retainer

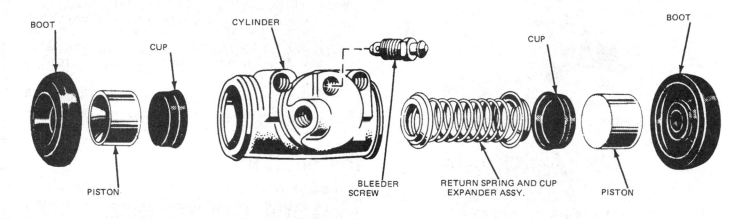

9.130 Exploded view of a typical wheel cylinder

BOOT — CUP — CYLINDER — BOOT — CUP — BLEEDER SCREW — RETURN SPRING AND CUP EXPANDER ASSY. — PISTON — PISTON

Overhaul

To disassemble the wheel cylinder, remove the rubber dust boot from each end of the cylinder, then push out the two pistons, the cups (seals) and the expander/spring assembly **(see illustrations)**. Discard the rubber parts and use new ones from the rebuild kit when reassembling the wheel cylinder.

Inspect the pistons for scoring and scuff marks. If present, the pistons should be replaced with new ones. Examine the inside of the cylinder bore for score marks and corrosion. If these conditions exist, the cylinder can be honed slightly to restore it, but replacement is recommended. If the cylinder is in good condition, clean it with brake system cleaner. **Warning:** *DO NOT, under any circumstances, use gasoline or petroleum-based solvents to clean brake parts!*

Remove the bleeder screw and make sure the hole is clean.

Lubricate the cylinder bore with clean brake fluid, then

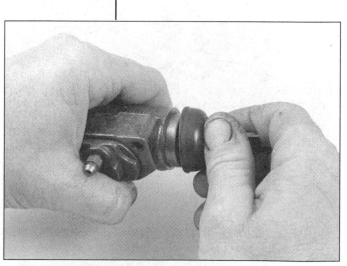

9.131 Remove the wheel cylinder dust boots

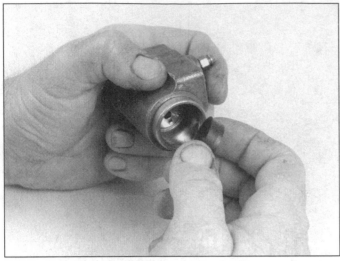

9.132 Install the piston cups with their open ends facing in, toward the cup expanders

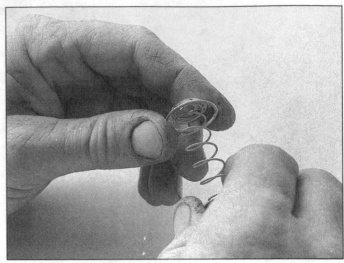

9.133 Attach the cup expanders to the spring . . .

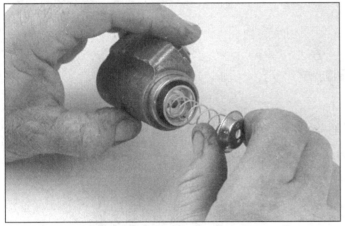

9.134 . . . then insert the expander/spring assembly into one end of the wheel cylinder housing (some wheel cylinders don't use expanders - they only have springs)

insert one of the new rubber cups into the bore **(see illustration)**. Make sure the lip on the rubber cup faces in.

Install the cup expanders, if equipped, onto the ends of the spring, insert the expander/spring assembly into the opposite end of the bore and push it in until it contacts the rear of the rubber cup **(see illustrations)**. Install the remaining cup in the cylinder bore.

Install the pistons **(see illustration)**. Insert the pushrods (if equipped) into the boots **(see illustration)**, then install the boots and pushrods.

Installation

Installation is the reverse of removal. Attach the brake line to the wheel cylinder before installing the mounting bolt(s) or clip and tighten the line fitting after the wheel

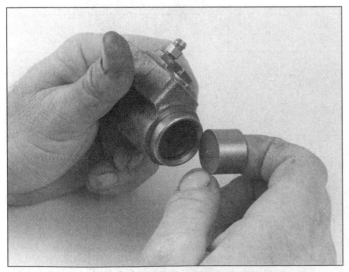

9.135 Install a piston into each end of the wheel cylinder housing

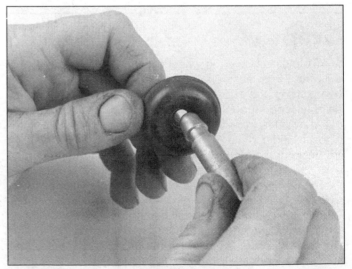

9.136 Insert the wheel cylinder pushrods into the new dust boots, then install the dust boots and pushrods onto both ends of the wheel cylinder (not all wheel cylinders have pushrods)

cylinder mounting bolt(s) have been tightened or the clip has been installed. If available, use a flare-nut wrench to tighten the line fitting.

Install the brake shoes and brake drum, (see Chapter 6).

Bleed the brakes following the procedure described later in this Chapter. Don't drive the vehicle in traffic until the operation of the brakes has been thoroughly tested.

Hydraulic control valves

All vehicle have either a pressure differential warning valve that incorporates a brake warning switch to warn if one half of the hydraulic system fails or a fluid level sensor in the master cylinder reservoir, and most vehicles have a proportioning valve which restricts - at a given ratio - hydraulic pressure to the rear wheels when system hydraulic pressure reaches a certain point. Models with disc brakes in the front and drum brakes in the rear also have a metering valve (or hold-off valve) which limits the front brake pressure until the rear brake shoes have overcome the return springs and contacted the drums. The proportioning valve on some vehicles (notably station wagons and pick-up trucks) is load-sensitive - it will allow more pressure to the rear brakes when there is a considerable load in the rear of the vehicle, but reduces the pressure to the rear brakes when the rear of the vehicle is light. The proportioning valves on some vehicles are screwed right into the outlet ports of the master cylinder.

Check

Pressure differential warning valve

The brake warning light normally comes on during the engine start sequence, then goes off again when the engine starts. It can be checked functionally by opening one bleeder screw slightly while an assistant presses down on the brake pedal. The light should come on with the ignition switch On.

Proportioning or metering valve

The testing procedures for these valves require special gauges for checking, as well as individual specifications for each different vehicle. If a problem is suspected with one of these valves, have it diagnosed by a dealer service department or other qualified repair shop, or simply replace the valve. Refer to Chapter 1 for descriptions of the proportioning and metering valves, and Chapter 3 for problems associated with these components.

Resetting

If the light for the brake warning system comes on, a leak or problem has occurred in the system and must be corrected. Once the leak or problem has been repaired, the pressure differential warning valve must be centered. Once the valve is centered, the brake light on the dash will go out.

To center the valve, first fill the master cylinder reservoir and make sure the hydraulic system has been bled. Turn the ignition switch to On or Accessory. Slowly press the brake pedal down and the piston will center itself, causing the light to go out. This procedure will work on some valves. If this does not cause the light to go out, have your assistant depress the brake pedal again and loosen a bleeder screw in the hydraulic circuit that did NOT have the problem. Close the bleeder screw as soon as the light goes out, or you'll have to reset the switch again, opening a bleeder screw in the other hydraulic circuit.

Check the brake pedal for firmness and proper operation.

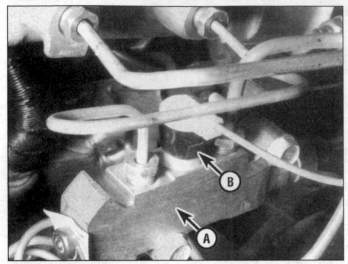

9.137 On most vehicles, the combination valve is located just under the master cylinder, and the pressure differential warning switch is screwed into the center of the valve

9.138 Typical load sensing proportioning valve mounting details

Replacement

Caution: *Brake fluid will damage paint. Cover all painted surfaces and avoid spilling fluid during this procedure.*

If a hydraulic system control valve is defective or if it is leaking, it must be replaced. Most are non-serviceable units anyway, so there is no point in disassembling one.

Remotely mounted valve (proportioning, metering or combination)

Disconnect the brake warning light connector from the warning light switch **(see illustration)**. Disconnect the brake line fittings from the valve assembly using a flare nut wrench to avoid rounding off the corners of the fittings. Plug the ends of the lines to prevent loss of brake fluid and the entry of dirt.

Remove the bolts and nuts securing the valve assembly to the chassis or bracket. If you're removing a load sensing proportioning valve **(see illustration)**, disconnect the linkage from the rear suspension. Remove the valve assembly. Installation is the reverse of removal. Bleed the system by following the procedure described later in this Chapter. If necessary, center the valve as described above (pressure differential valve only).

Master cylinder-mounted valve

If you're replacing a master cylinder-mounted proportioning valve, simply unscrew the line fitting from the valve (using a flare-nut wrench), then unscrew the valve from the master cylinder. Installation is the reverse of removal, but be sure to use a new sealing washer if one was present.

Hoses and lines

Note: *Refer to Chapter 4 for the brake hose and line inspection procedures.*

Flexible hose replacement

Clean all dirt away from the ends of the hose. Disconnect the metal brake line from the hose fitting **(see illustration)**. You'll need a back-up wrench on some fittings. Be careful not to bend the frame bracket or line. If necessary, soak the connections with penetrating oil. Remove the U-clip (lock) from the female fitting at the bracket and remove the hose from the bracket.

On models that use banjo fittings, unscrew the banjo bolt and disconnect the hose from the caliper, discarding the sealing washers on either side of the fitting. On models where the brake hose threads directly into the caliper, unscrew the hose from the caliper.

Using new sealing washers, attach the new brake hose to the caliper (models that use banjo fittings). Tighten the banjo bolt to the torque listed in the Specifications at the end of this manual. On models where the brake hose threads directly into the caliper, tighten the hose fitting securely.

Pass the female fitting through the frame or frame bracket. With the least amount of twist in the hose, install the fitting in this position. **Note:** *The weight of the vehicle must be on the suspension, so the vehicle should not be raised while positioning the hose. If this is not possible, raise the suspension with a floor jack to simulate normal ride height.*

Install the U-clip (lock) in the female fitting at the frame bracket. Attach the brake line to the hose fitting using a back-up wrench on the fitting. Tighten the fitting securely.

Carefully check to make sure the suspension or steering components don't make contact with the hose. Have an assistant push on the vehicle and also turn the steering wheel lock-to-lock during inspection.

Bleed the brake system as described later in this Chapter.

9.139 To disconnect a brake hose from the brake line fitting, place a backup wrench on the hose fitting (A) and loosen the tube nut (B) with a flare nut wrench. Remove the U-clip (C) to detach the hose from the bracket.

Metal brake lines

When replacing brake lines, be sure to use the correct parts. Don't use copper tubing for any brake system components. Purchase steel brake lines from a dealer parts department or auto parts store. Prefabricated brake line, with the tube ends already flared and fittings installed, is available at auto parts stores and dealer parts departments. These lines are also sometimes bent to the proper shapes.

When installing the new line make sure it's well supported in the brackets and has plenty of clearance between moving or hot components. Make sure you tighten the fittings securely. After installation, check the master cylinder fluid level and add fluid as necessary. Bleed the brake system as outlined in Section 11 and test the brakes carefully before placing the vehicle into normal operation.

Brake system bleeding

Warning: *Wear eye protection when bleeding the brake system. If the fluid comes in contact with your eyes, immediately rinse them with water and seek medical attention. Never use old brake fluid. It contains moisture and contaminants which will deteriorate the brake system components and could cause the fluid to boil when braking temperatures rise. This can lead to complete loss of pressure in the hydraulic system.*

Note: *Bleeding the brake system is necessary to remove any air that's trapped in the system when it's opened during removal and installation of any hydraulic component.*

Models without an Anti-lock Brake System (ABS)

Standard (two-person) bleeding procedure

It will probably be necessary to bleed the system at all four brakes if air has entered the system due to low fluid level, or if the brake lines have been disconnected at the master cylinder. If a brake line was disconnected only at a wheel, then only that caliper or wheel cylinder must be bled. If a brake line is disconnected at a fitting located between the master cylinder and any of the brakes, that part of the system served by the disconnected line must be bled. If you suspect air in the master cylinder, begin by bleeding the master cylinder on the vehicle **(see illustration 9.47)**.

Remove any residual vacuum from the brake power booster (if equipped) by applying the brake several times with the engine off. Remove the master cylinder reservoir cover and fill the reservoir with brake fluid. Reinstall the cover. **Note:** *Check the fluid level often during the bleeding operation and add fluid as necessary to prevent the fluid level from falling low enough to allow air bubbles into the master cylinder.*

Have an assistant on hand, as well as a supply of new brake fluid, an empty clear plastic container, a length of 3/16-inch plastic, rubber or vinyl tubing to fit over the bleeder valve and a wrench to open and close the bleeder valve.

The order of bleeding depends on the design of the brake system. Generally, the caliper or wheel cylinder farthest away from the master cylinder is bled first, then the other caliper or wheel cylinder, in the same half of the circuit just bled, would be bled of air.

If the hydraulic system in the vehicle being bled is split front-to-rear, as is the case with most rear-wheel drive vehicles, you would bleed the brakes in the following order:

> *Right rear*
> *Left rear*
> *Right front*
> *Left front*

If the hydraulic system is diagonally split, as is the case with most smaller front-wheel drive vehicles, you would bleed the brakes in this order:

> *Right rear*
> *Left front*
> *Left rear*
> *Right front*

Warning: *Regardless of the order used to bleed the brake hydraulic system, the vehicle should NOT be driven unless a satisfactory, firm brake pedal is obtained.*

Beginning at the first wheel cylinder or caliper to be bled, loosen the bleeder screw slightly, then tighten it to a point where it's snug but can still be loosened quickly and easily. Place one end of the tubing over the bleeder screw fitting and submerge the other end in brake fluid in the container **(see illustration)**.

Have the assistant depress the brake pedal *slowly* to get pressure in the system, then hold the pedal firmly depressed. While the pedal is held depressed, open the bleeder screw just enough to allow a flow of fluid to leave the valve. Watch for air bubbles to exit the submerged end of the tube. When the fluid flow slows after a couple of seconds, tighten the screw and have your assistant release the pedal.

Repeat this procedure until no more air is seen leaving the tube, then tighten the bleeder screw and proceed to the next wheel cylinder or caliper in the bleeding sequence, and perform the same procedure. Be sure to check the fluid in the master cylinder reservoir frequently.

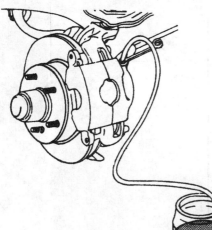

9.140 When bleeding the brakes, a hose is connected to the bleeder screw at the caliper or wheel cylinder and then submerged in brake fluid - air will be seen as bubbles in the tube and container (all air must be expelled before moving to the next wheel)

Note: *On the rear brakes of some drum brake systems, only the wheel cylinder on one side of the vehicle is equipped with a bleeder screw. Both wheel cylinders are bled through the cylinder that has the bleeder screw.*

Refill the master cylinder with fluid at the end of the operation. Check the operation of the brakes. The pedal should feel solid when depressed, with no sponginess. If necessary, repeat the entire process. Bleed the height (or load) sensing proportioning valve, if equipped. **Warning:** *Do not operate the vehicle if you are in doubt about the effectiveness of the brake system.*

Alternative methods

Vacuum bleed method

Several companies manufacture brake bleeder kits that use hand-held vacuum pumps to suck the air out of the hydraulic circuit. The pump is connected to a reservoir, which is connected to the bleeder screw. When the bleeder screw is opened and the vacuum pump is operated, fluid (and any air bubbles) will collect in the reservoir **(see illustration)**. The pump itself does not come in contact with the fluid. The normal bleeding sequence can be used.

This is a very convenient method which only requires one person, but it has a couple of potential drawbacks. Since the cups in the wheel cylinder depend on pressure to seal against the walls of the cylinder, negative pressure can cause air to be drawn around the lips of the seal and into the hydraulic circuit. This may not happen in every case, but is more likely if the cups in the wheel cylinder are old or the hydraulic system has not been maintained properly. Air can also be drawn past the bleeder screw threads when the screw is loosened.

One-person method

In addition to the vacuum bleed method, which is also a one-person method, another single-handed bleeding method exists. A number of one-person bleeder kits are available, and your local auto parts store will almost surely stock one of them.

The most common type of one-person bleeder kit is really nothing more than a hose with a check valve in it, and a container to collect the fluid **(see illustration)**. All you have to do is connect the bleeder hose to the bleeder screw, put the other end of the hose into the container (which must be partially filled with brake fluid) loosen the bleeder screw and slowly pump the brake pedal.

If you just can't seem to bleed all of the air out of a circuit, there are a couple of things you can try. Always keep in mind that air rises to the high points in an enclosed system.

a) Make sure the master cylinder is free of air. Re-bleed it if necessary.

b) Try raising the end of the vehicle being bled. This will help any air trapped in the system to find its way to the bleeder screw, which should be located at the highest point in its caliper or wheel cylinder.

c) Lightly tap the caliper or wheel cylinder with a hammer. This can dislodge air bubbles that cling to the walls of the hydraulic cylinder.

d) Follow the brake lines from the master cylinder, bleeding each line at each connection by loosening the fitting. Work from the closest fitting to the master cylinder to the farthest.

e) Try one of the alternative bleeding methods.

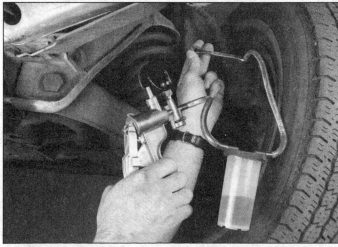

9.141 The vacuum-operated brake bleeding method is a convenient way to bleed the brakes, and you don't need a helper to do it! Just connect the hose from the container to the bleeder valve, connect the hand-held vacuum pump to the container, open the bleeder valve and apply vacuum. Fluid and air will be sucked out and will collect in the container.

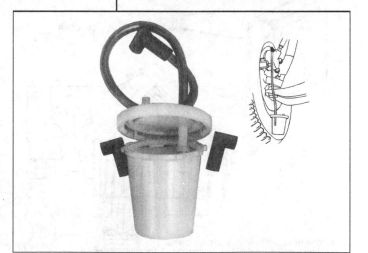

9.142 This type of brake bleeding setup also allows you to bleed the brakes alone. Just attach the hose to the bleeder screw and insert the other end into the container partially filled with fluid. Open the bleeder valve and slowly pump the brake pedal - fluid and air will be expelled from the caliper, but the check valve in the hose prevents anything from being sucked back in.

9.143 Another type of one-person brake bleeder is this bag-type. In operation it works in a similar manner to the setup shown in illustration 9.142, but it doesn't have a check valve. Air bubbles rise up into the bag, but they can't flow back into the hose. It is important to pump the pedal *slowly* in order for this method to be effective.

Fluid and air will be forced out of the circuit, past the check valve and into the container. The check valve prevents old fluid or air from being sucked back into the system.

Another type of one-man bleeding kit consists of a hose connected to a bag, which must be suspended above the level of the bleeder screw (see illustration). Connect the hose to the bleeder screw, open the bleeder screw and slowly pump the brake pedal. The old fluid and air will be forced from the caliper or wheel cylinder and into the bag. Since air rises, the bubbles that have been pushed into the bag can't be sucked back into the system.

Gravity bleeding

Gravity bleeding is exactly what it sounds like - the force of gravity pulls the brake fluid through the lines to the bleeder screw, and any air trapped in the lines flows out of the bleeder screw with it. This procedure can't be used in systems incorporating residual pressure check valves, because the check valves prevent the flow of fluid. When using this procedure you have to be sure to check the fluid level from time-to-time, since the procedure can take an hour or even more. This method might, however, be the answer to a hydraulic system that just can't seem to be purged of air.

To start the gravity bleeding procedure, raise both ends of the vehicle and support it securely on jackstands. Attach a length of clear plastic hose to the bleeder screw at each wheel and direct the end of each hose into a container. Open each bleeder screw one full turn, making sure fluid flows from each hose. **Note:** *If fluid does not flow, it may be necessary to start a siphon.* **Warning:** *Don't do this by mouth - use a hand-held vacuum pump to start the siphon!* Remember, don't let the fluid in the master cylinder reservoirs drop too low.

Allow the fluid to drain until the fluid in each hose is free of bubbles. Tighten the bleeder screws and check the operation of the brakes. Top up the fluid level in the master cylinder reservoir.

Pressure bleeding

This is the method most automotive repair shops bleed brake systems. It's quick, easy, and probably the most effective way to bleed a hydraulic system.

With this method, an adapter is attached to the top of the master cylinder reservoir. A hose is connected between the adapter and the pressure bleeder unit. Air pressure (from an air compressor) is connected to the bleeder unit also. This forces pressurized fluid through the system - all you have to do is open the bleeder valve at each wheel until the flow of fluid is free of air bubbles.

When using a pressure bleeder on some vehicles you'll have to use a metering valve override tool, which defeats the hold-off function of the metering valve so pressure can be directed to the front brakes (see illustration). On some metering valves a plunger must be pulled out and held in the extended position - on others a button on the valve must be depressed. Special spring tools are available for this purpose. **Caution:** *Never use a rigid clamp or wedge to depress or pull out the valve plunger, since this can damage the metering valve internally.* Either way, the tool must be removed after the bleeding process is complete. **Note:** *It isn't necessary to defeat the metering valve when bleeding brakes manually, since the pressure created in the system is*

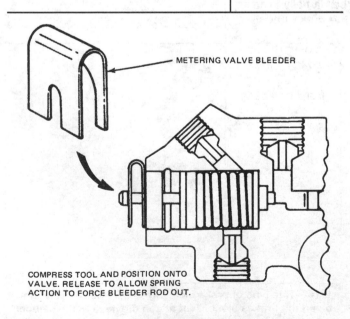

METERING VALVE BLEEDER

COMPRESS TOOL AND POSITION ONTO VALVE. RELEASE TO ALLOW SPRING ACTION TO FORCE BLEEDER ROD OUT.

9.144 When using a pressure bleeder, you'll have to install a metering valve bypass tool to defeat the metering valve. Otherwise, no fluid will flow to the front brakes.

greater than the hold-off point of the valve (approximately 150 to 165 psi). It also isn't necessary to defeat the valve when using vacuum or gravity bleeding methods, since there basically is no pressure generated in the system.

There are drawbacks, though. Pressure bleeding outfits are very expensive and require special adapters to fit different master cylinder reservoirs. They can usually be purchased only through specialized tool dealers. Also, extreme care must be taken to ensure that all pressure in the fluid reservoir is bled off before removing the adapter. If it isn't, brake fluid will "explode" everywhere, possibly causing damage to the vehicle's paint and injury to anyone nearby.

Models with an Anti-lock Brake System (ABS)

Warning: *Anti-lock Brake Systems can be bled **at the wheels** by using the following procedure. Some ABS systems, however, require special tools to cycle the modulator valves in the ABS hydraulic unit to enable the hydraulic unit to be bled. Therefore, any service work that requires opening the hydraulic system at the master cylinder/hydraulic unit should be performed by a dealer service department or other repair shop. If, after following this procedure you still can't obtain a satisfactory (firm) brake pedal, DO NOT drive the vehicle. Have it towed to a dealer service department or other repair shop.*

The front brakes of an ABS system can be bled in the normal manner. The rear brakes must be bled with a pressure bleeder or with the hydraulic accumulator fully charged. On most models this can be achieved by simply turning the ignition key to the On position. Some models require the engine to be running. Open the hood and have an assistant turn the ignition key to the On position while you listen to the ABS hydraulic unit. If the pump runs as soon as the key is turned On, it is unnecessary to start the engine.

To bleed the rear calipers or wheel cylinders, attach a length of clear plastic tubing to the bleeder screw at the right rear brake. Place the other end of the hose in a container partially filled with clean brake fluid.

Have an assistant depress the brake pedal and hold it in the applied position. SLOWLY loosen the bleeder valve and allow the fluid to flow for a few seconds, then close the valve. **Warning:** *The brake fluid is under extremely high pressure, and careless opening of the bleeder valves may cause the fluid to shoot out with great force.*

Have your assistant pump the brake pedal several times, then repeat the operation until the stream of fluid is free of air bubbles. Check the fluid level in the master cylinder fluid reservoir and add fluid, if necessary.

Repeat the procedure to the other rear wheel, then bleed the front brakes (if necessary) in the normal manner.

Notes

10 Anti-lock Braking Systems (ABS)

General information

Note: *The following system and component descriptions offer a general overview of the ABS system, how it works, and the components that comprise a typical ABS system. It is not possible to cover the specifics of all ABS systems nor include in-depth troubleshooting and repair procedures in a manual of this size.*

What is ABS?

When a vehicle is rolling down the road and the driver applies the brakes, the energy of the moving vehicle is converted to heat through the brakes, which slows down the vehicle. The coefficient of friction between the road and the tires increases too, as the tires "grab" at the road. If the coefficient of friction exceeds that which the tire can effectively deal with, the tire begins to slide along the road surface. This is an example of wheel slip.

Wheel slip can be defined as the difference between the speed of the vehicle and the speed of the wheel against the ground. When a wheel is locked up and sliding, wheel slip of 100-percent is present. If the wheel is rolling along with the vehicle and the brakes are not applied, zero wheel slip is taking place. Normal braking takes place somewhere between these extreme percentages.

Unfortunately, not all instances of slowing down a vehicle fall under the "normal" category. During panic stops or hard braking on slippery surfaces, it's easy to lock up the wheels. When the wheels are locked up, it is impossible to maintain control of the vehicle - the wheels must be rolling in order to have directional control. Besides that, engineers have discovered that a rolling wheel with a slip factor of 8 to 35-percent under braking generates a much greater coefficient of friction between the tire and the road than a sliding tire does. When the wheels stop rolling but the vehicle keeps moving, it will follow the path of least resistance. To the dismay of many a motorist, this path of least resistance has lead right into an object of *great* resistance - like a telephone pole or the proverbial brick wall!

Anti-lock Brake Systems work to avoid such situations by monitoring the rotational speed of the wheels. When the system senses a wheel turning too slowly in relation to the others, the hydraulic pressure to the brake at that wheel is reduced, preventing it from locking up.

All current anti-lock systems are failsafe. They all have some form of mechanical override - in the event of a malfunction such as failure of the Electronic Control Unit (ECU), the brakes will function just like a standard (non-ABS) system.

Advantages

Anti-lock Brake Systems have three distinct advantages over a non-ABS brake system:
a) Since brake systems with ABS don't allow the wheels to lock up, tire wear is improved. It is impossible for the tires to get flat-spotted with an ABS system.
b) The driver of a vehicle equipped with ABS has a much greater degree of control over the vehicle then the driver of a vehicle not equipped with ABS. Even under hard braking, a vehicle with ABS can be steered.
c) A vehicle with ABS has a reduced tendency to hydroplane when braking in wet road conditions.

Limitations

Although a vehicle equipped with ABS will stop in the shortest possible distance achievable under the design parameters of the system (how much wheel slip the system will allow), there is certainly no guarantee that all accidents will be avoided. Tires are still a limiting factor, and if the brakes are applied hard during cornering there is a good possibility that the tires will lose traction. Remember, the system monitors *rotational* speed of the wheels, not lateral motion against the road (although some systems incorporate a lateral acceleration sensor or switch which, when activated, tells the control unit to reduce braking pressure so the tires don't get overloaded and break traction).

If the driver applies the brakes too late, the ABS system will do its job just as it was designed to do, but the vehicle won't come to a complete stop before it hits the object that the driver is trying to avoid.

When braking on snowy or icy road surfaces, or in dirt or gravel, an Anti-lock Brake System will not bring the vehicle to a stop in the shortest possible distance. A locked or nearly locked wheel is preferred under these conditions, because a wedge of snow or dirt will build up under the front of the tires and help stop the vehicle. This can't happen with ABS, so many vehicles equipped with ABS have an override switch on the dash to defeat the ABS system during snowy conditions or when driving on unpaved roads.

ABS facts

Here are some important things to know regarding ABS brakes:
a) When the ABS is activated (when a wheel begins to lock up), it is normal for the driver to feel pedal pulsations. This is the result of the modulator valves cycling to reduce pressure in the brake lines. During normal braking no pulsations should be felt.
b) The amount of force on the brake pedal that is required to activate the ABS system depends on to road surface. If the surface of the road is very slippery it won't take much to slide the tires, in which case the ABS will kick in.
c) If the system experiences a malfunction, it will revert to a normal hydraulic brake system. You won't lose your brakes completely on account of a problem with the ABS portion of the brake system.
d) Some service operations to the ABS system should not be performed by the do-it-yourselfer. These include anything other than brake pad replacement or work to the hydraulic portion of the wheel brakes (calipers and wheel cylinders or brake hoses). This is because special tools are required to bleed the hydraulic modulator assembly of air.
e) If you replace a brake disc, brake drum, driveaxle, axleshaft or differential assembly (depending on system design) and the component being replaced is equipped with a toothed signal rotor, make sure the new part is equipped with a signal rotor also. In some cases the signal rotor is pressed onto the component, in which case it will have to be removed from the old part and pressed onto the new one.

f) ABS systems aren't just installed on vehicles with four-wheel disc brakes. Some vehicles with disc brakes in the front and drum brakes in the rear are equipped with ABS.

g) The presence of an ABS system does not affect normal brake service. Brake pads or shoes must still be checked at the recommended intervals. They don't last any longer, either. Calipers, wheel cylinders, hoses and parking brake mechanisms are also subject to the same amount of wear as in a non-ABS system.

h) There was a time when some domestic automobile manufacturers were hesitant to even install ABS systems on their vehicles, due to potential lawsuits if the system didn't operate properly and an ABS-equipped vehicle was involved in an accident. Even if the system did operate properly the manufacturers were afraid of being sued, because the ABS system was thought of by the general public as an accident-avoidance system, which it isn't.

i) Finally, a vehicle equipped with ABS can't out-brake an identical non-ABS vehicle driven by an expert driver on a dry, straight road. This is because a typical ABS system allows the amount of tire slip to drop as low as 5-percent, which is below the point at which maximum traction is achieved. A skilled driver can use his or her judgment to increase the amount of tire slip without locking the wheels. An Anti-lock Brake System has no "judgment" whatsoever.

Types of systems

There are two basic types of ABS systems - two-wheel anti-lock systems and four-wheel anti-lock systems. Within these two groupings are two sub-groupings - component systems and integral systems.

Two-wheel anti-lock systems

Two-wheel ABS systems are designed to improve vehicle stability and prevent the vehicle from spinning out during heavy braking. These systems operate on the rear wheels only and are popular on light trucks and vans, because the rear of these vehicles, when not carrying a cargo, are very light and tend to lock up the rear wheels much easier than on a passenger car **(see illustrations)**.

Two-wheel ABS systems have no effect on the front wheels at all, and can't prevent the loss of steering due to lock-up of the front wheels. A proportioning

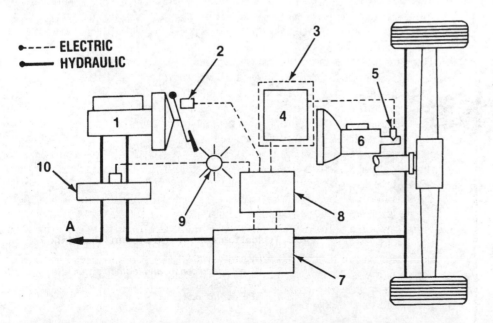

●----- ELECTRIC
●——— HYDRAULIC

10.1 Here's a schematic of a typical two-wheel anti-lock brake system as fitted to a late-model Chevy truck. On this vehicle it's called the Rear Wheel Anti-Lock (RWAL) system

A	To front brakes
1	Master cylinder
2	Brake light switch
3	Instrument cluster
4	Digital ratio adapter
5	Speed sensor
6	Transmission
7	Isolation/dump valve
8	RWAL control module
9	Brake warning lamp
10	Combination valve

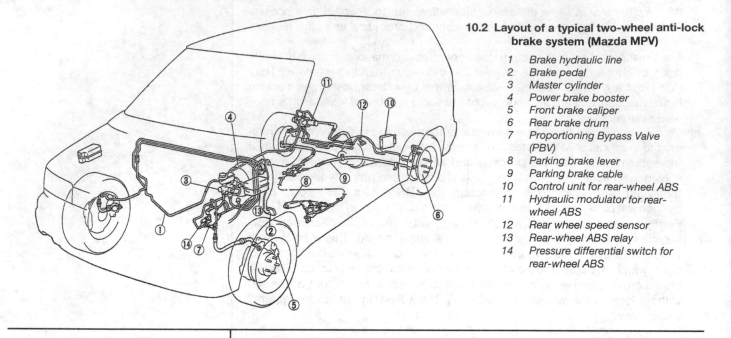

10.2 Layout of a typical two-wheel anti-lock brake system (Mazda MPV)

1   Brake hydraulic line
2   Brake pedal
3   Master cylinder
4   Power brake booster
5   Front brake caliper
6   Rear brake drum
7   Proportioning Bypass Valve (PBV)
8   Parking brake lever
9   Parking brake cable
10 Control unit for rear-wheel ABS
11 Hydraulic modulator for rear-wheel ABS
12 Rear wheel speed sensor
13 Rear-wheel ABS relay
14 Pressure differential switch for rear-wheel ABS

valve is unnecessary in a brake system with two-wheel ABS (although on some designs the proportioning valve is retained). Both rear wheels are controlled through the same hydraulic circuit. On two-wheel ABS systems with individual wheel speed sensors, hydraulic pressure regulation is based on the needs of the wheel with the least amount of traction (this is called the *select-low* principle).

Four-wheel anti-lock systems

Four wheel anti-lock systems monitor the rotational speeds of all four wheels **(see illustrations)**. This allows the driver to brake as hard as possible *and* steer the vehicle at the same time.

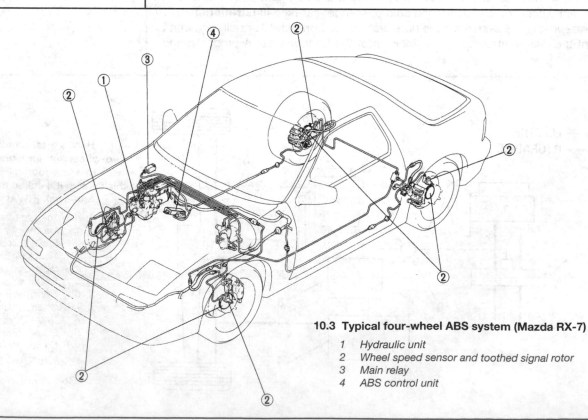

10.3 Typical four-wheel ABS system (Mazda RX-7)

1   Hydraulic unit
2   Wheel speed sensor and toothed signal rotor
3   Main relay
4   ABS control unit

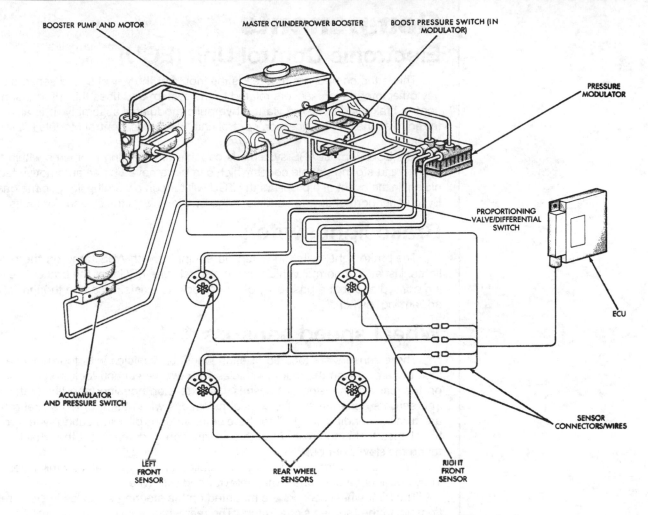

BOOSTER PUMP AND MOTOR

MASTER CYLINDER/POWER BOOSTER

BOOST PRESSURE SWITCH (IN MODULATOR)

PRESSURE MODULATOR

PROPORTIONING VALVE/DIFFERENTIAL SWITCH

ECU

ACCUMULATOR AND PRESSURE SWITCH

LEFT FRONT SENSOR

REAR WHEEL SENSORS

RIGHT FRONT SENSOR

SENSOR CONNECTORS/WIRES

10.4 Schematic of a typical four-wheel ABS system (Jeep Cherokee)

As in a two-wheel anti-lock system both rear wheels are controlled through a single hydraulic circuit. The front wheels are controlled through individual circuits.

Component systems

The component type system is sometimes referred to as an "add on" system. These types of systems are used on vehicles equipped with a standard master cylinder/power booster combination. The hydraulic control unit is installed downstream of the master cylinder (**see illustration 10.2**).

Although referred to as an add on system, retrofit kits are not available to convert a non-ABS brake system into an ABS system (at least not at the time of writing). Much more work would be involved than simply installing a hydraulic modulator assembly in the brake lines between the master cylinder and the wheels.

Integral systems

Integral anti-lock systems are self-contained units which combine the master cylinder, power booster and hydraulic modulator into a single assembly (**see illustration**).

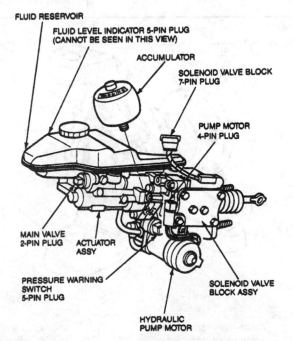

FLUID RESERVOIR

FLUID LEVEL INDICATOR 5-PIN PLUG (CANNOT BE SEEN IN THIS VIEW)

ACCUMULATOR

SOLENOID VALVE BLOCK 7-PIN PLUG

PUMP MOTOR 4-PIN PLUG

MAIN VALVE 2-PIN PLUG

ACTUATOR ASSY

PRESSURE WARNING SWITCH 5-PIN PLUG

HYDRAULIC PUMP MOTOR

SOLENOID VALVE BLOCK ASSY

10.5 Details of an integral ABS actuation assembly

Components

Electronic Control Unit (ECU)

This is the component that reads the inputs of the wheel speed sensors and any other information sensors related to the system and uses this information to control brake line pressure via the hydraulic modulator assembly. It is also referred to as the microprocessor, control unit, Electronic Control Module (ECM) or "brain."

The ECU in most ABS systems is capable of detecting problems within the system and storing trouble codes which can be retrieved by an automotive technician. In the event of a problem, the ECU will turn on a warning light on the dash. Even under normal driving conditions the ECU monitors the system for faults.

Brake light switch

The brake light switch in a vehicle equipped with ABS turns on the brake lights, just as in a normal vehicle, but it also tells the ECU that the brakes are being applied and that a possibility of wheel lock up exists. It can be thought of as an "arming" switch.

Wheel speed sensors

These sensors are located at each wheel on vehicles with four-wheel anti-lock systems, or on the rear of vehicles with two-wheel anti-lock systems, in or on the rear axle housing or transmission extension housing (see illustrations). The sensors generate small electrical pulsations when the toothed signal rotors are turning, sending a signal to the electronic controller indicating wheel rotational speed. The controller uses this information to determine if the wheel is rotating too slowly or not.

Regardless of the vehicle they're installed on, they are all of similar design. They consist of a permanent magnet wrapped with wire.

The front wheel sensors are mounted on the steering knuckles in close relationship to the toothed signal rotors. The rear wheel sensors are mounted to the rear hub carriers, the rear brake backing plate or to the axle housing or transmission output shaft.

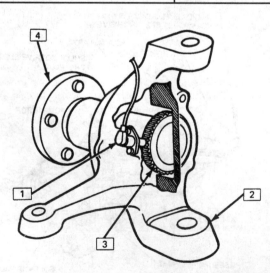
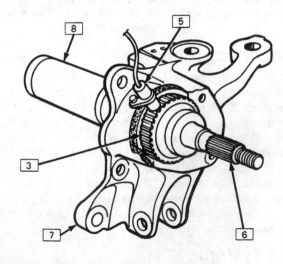

10.6 Typical wheel speed sensors

1	Front-wheel speed sensor	4	Front hub and bearing assembly
2	Steering knuckle	5	Rear-wheel speed sensor
3	Toothed signal rotor (part of 4, 6)	6	Rear driveaxle spindle

7	Rear knuckle
8	Rear driveaxle

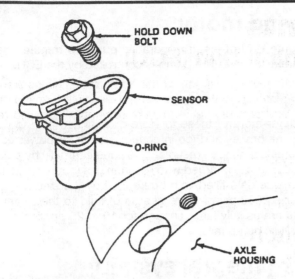

10.7 On some vehicles the rear-wheel speed sensor mounts on the rear axle housing - the toothed signal rotor is integral with the differential ring gear or carrier

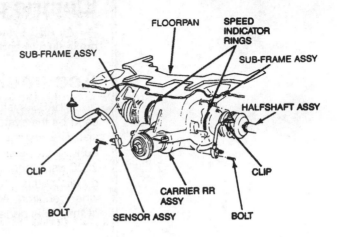

10.8 On some vehicles the rear-wheel speed sensors are attached to the rear axle housing, but the toothed signal rotors are integral with the outer races of the inner CV joints (Ford Thunderbird)

Toothed signal rotors

Also called sensor rings, these devices turn with the wheel hubs, driveaxles, rear axle or transmission output shaft **(see illustration)**. As the teeth of the ring pass the sensor, an alternating current voltage is produced in the sensor, which is read by the control unit. The control unit is able to count the frequency of this voltage signal and equates this to vehicle speed. The control unit then decides if it should or shouldn't enter ABS mode based on these calculations.

Lateral acceleration switch

This component is not present on all ABS systems. As stated previously, the lateral acceleration switch activates when a certain lateral G-force is attained. The switch sends the control unit this signal and the control unit commands the hydraulic modulator assembly to reduce braking pressure. It does this because braking increases the coefficient of friction between the wheel and the road, but so does cornering. When heavy braking and heavy cornering are combined, it's easy to exceed the coefficient of friction that the tires can handle, which will result in a slide.

10.9 On this Saturn the toothed signal rotors are pressed onto the ends of the outer CV joints (lower arrow) - the wheel speed sensor is mounted on the steering knuckle (upper arrow)

Modulator assembly

The hydraulic modulator assembly regulates line pressure to the brakes based on information it receives from the ECU. The modulator can maintain braking pressure or, if necessary, reduce it to prevent the wheels from locking up. It can't apply the brakes all by itself and it can't increase brake pressure past the point that was originally applied before being reduced to prevent wheel lock up.

The modulator houses a set of solenoid valves which, depending on the system, are capable of cycling at a rate of ten times per second. It is the cycling of these valves which reduces braking pressure.

In an integral ABS system, the modulator portion of the unit is called the *valve block*.

Electric pump and motor

All ABS systems use some kind of electro-hydraulic pump to transfer the brake fluid to and from the brake circuits. The pump is controlled by the ECU.

Accumulator

All ABS systems use one or more of these to store pressurized fluid to be used by the system to apply the brakes and (on integral systems) for power assist. The accumulators are spherical-shaped reservoirs, divided internally by a diaphragm. One side of the diaphragm is charged with nitrogen gas (approximately 2000 psi or more). The other side gets filled with brake fluid by the electro-hydraulic pump. This allows for a 2000 psi or more charge of fluid to the system when required. Accumulators are usually large enough for 10 to 20 applications of the brake pedal if the hydraulic pump fails.

Master cylinder (integral system)

The master cylinder in an integral ABS system works like a regular non-ABS master cylinder, but it doesn't send pressure to the wheel brakes like a non-ABS master cylinder does. Each hydraulic circuit from the integral ABS master cylinder pressurizes the front brakes only (each circuit serves one wheel). The rear brakes are controlled by pressure from the hydraulic power booster.

Hydraulic power booster (integral system)

The hydraulic power booster in an integral ABS system is kind of like a General Motors Powermaster brake booster (see Chapter 1). It uses hydraulic pressure to assist in applying the brake master cylinder. It also supplies and regulates all of the pressure for the rear brakes.

Operation

Although operation of the ABS system has been described in terms of how the individual components work, lets see how they all work together.

When the driver slams on the brakes in a panic stop situation, the system must decide to do one of the three following things:

a) Maintain the same pressure that the driver has applied
b) Reduce pressure to one or more of the hydraulic circuits
c) Increase pressure back to that which the driver has applied after it has been reduced to avoid wheel lock up.

The electronic control unit constantly monitors the inputs from the wheel speed sensors. If one (or more) of the sensors relays a higher rate of wheel deceleration than the other sensors, the control unit knows that a wheel lock-up is likely to take place. Initially, nothing happens - the pressure to the wheel in question is maintained. This is called the *pressure holding* phase. The hydraulic modulator is told by the control unit to partially close the solenoid valve, preventing any increase in brake pressure to that wheel. Fluid return is also blocked off by the solenoid, so the pressure stays the same. The fluid return delivery pump begins to run, but its outlet check valve stays shut unless the pump pressure exceeds the amount of pressure being created by the master cylinder.

If the wheel continues to decelerate at an abnormal rate, the control unit signals the hydraulic modulator to reduce line pressure to the brake at the wheel. The modulator closes the solenoid valve for the wheel, which prohibits any more pressure from being applied. In addition, the valve is positioned in such a way as to open a passage back to the fluid reservoir. This bleeds off line pressure and allows the wheel to turn. **Note:** *Each front wheel has its own solenoid valve. The rear wheels share a single solenoid valve which is controlled by the ECU using the select-low principle.*

At this point, the wheel accelerates again because it isn't being braked as hard as it was. The control unit will only let it gain a certain amount of speed, however. Once it has been determined that the wheel is no longer in danger of locking up, the solenoid valve stops cycling and normal braking resumes. The modulator allows pressure to build up to the point it was at before the pressure reduction took place.

This entire chain of events takes place in less than one second. In fact, this sequence can occur from *four to ten times per second,* depending on how slippery the surface of the road is.

Precautions

There are several precautions that must be taken when working on or around the components of an anti-lock brake system. If the following points of advice are not taken, you run the risk of incurring serious injury, either through careless servicing techniques or disabling of the anti-lock system due to improper servicing techniques.

a) Portions of the ABS system operate at pressures over 2,600 psi. Before working on any part of the hydraulic system, always depressurize the system by pumping the brake pedal a number of times (some system require as many as 20 pumps) with the ignition in the Off position. When bleeding the rear brakes on an ABS system, open the bleeder screw *slowly* (see Chapter 9 for more information on brake system bleeding).

b) Do not unplug or connect electrical connectors related to the ABS system while the ignition is in the On position. This could cause a voltage spike severe enough to damage the microprocessor in the Electronic Control Unit.

c) If it becomes necessary to replace a brake hose or line, purchase only genuine replacement parts or parts of original equipment quality and design. DO NOT fabricate your own lines.

d) Never strike a speed sensor or a sensor ring with a hammer (don't drop them either). This could cause the component to become polarized or demagnetized, which would lead to inaccurate signals being sent to the ECU.

e) If the vehicle is equipped with four-wheel ABS, never install larger diameter tires on one end of the vehicle. It's okay to install wider tires, as long as they are the same diameter as the original ones. All four tires must be of the same diameter, or the ECU will become confused.

f) Don't get grease on the wheel speed sensors or the toothed signal rotors.

g) Don't overtighten the wheel lug nuts, as this could warp the brake disc or drum. This can cause inaccurate speed signals.

h) Whenever a speed sensor is disturbed or replaced, the gap between the sensor and the toothed signal rotor must be checked and, if necessary, adjusted (not all sensors are adjustable).

i) Never use silicone-based brake fluid in an ABS hydraulic system. Silicone fluid has a tendency to trap tiny air bubbles in the fluid, which are difficult to bleed out. Normal ABS operation can agitate silicone fluid, which may promote the formation of air bubbles in the system. Only use the recommended fluid in the ABS system (consult your owners manual).

j) If you install a cellular telephone or citizen's band radio (or any other transmitting device), don't route the wires or install the antenna near any of the ABS system components or wiring harnesses. Certain frequencies can interfere with system operation.

k) Before performing any welding on the vehicle, disconnect the cables from the battery (negative first, positive last) and unplug the electrical connectors from the ECU.

l) Most manufacturers recommend against repairing ABS system wiring harnesses. If one of the wires in an ABS wiring circuit becomes damaged, replace that portion of the harness. Even slight changes in resistance can lead to inaccurate speed readings.

Troubleshooting

All vehicles equipped with ABS have an amber "ANTI-LOCK" warning light on the dash. This light alerts the driver when the ECU has detected a malfunction in the system. The light will also come on whenever the ignition key is turned to the On position, as a bulb check feature. If the light doesn't come on, you should check for a burned-out bulb as soon as possible. Otherwise, you won't know if a problem arises in the ABS system until you need the system to work correctly and find out it doesn't.

In some systems the red BRAKE light also will come on in the event of a problem. In some cases the light(s) will flash on and off. The flashing of these lights is part of a much more involved diagnostic procedure, and interpreting the flashing sequence must be left to a technician trained in ABS system diagnosis.

If a problem with the ABS system arises, there are a few general checks that can be made. If these checks don't reveal the source of the problem, take the vehicle to a dealer service department or other repair shop qualified to service ABS systems. Some vehicles will flash trouble codes through the warning light on the instrument panel when a diagnostic mode is entered, much like a CHECK ENGINE light for an engine management system. Unfortunately, most ABS systems require expensive electronic test equipment, pressure gauges and adapter fittings to successfully perform diagnostic procedures. Some of this equipment is proprietary to the manufacturers and is not available through auto parts stores or tool dealers. Furthermore, the ABS diagnosis chapters in a manufacturer's factory service manual frequently exceed one hundred pages or so, making it impossible to provide a comprehensive ABS troubleshooting guide within the covers of this manual. Remember, vehicles equipped with anti-lock braking systems are also susceptible to the same problems as a vehicle that isn't equipped with ABS.

a) Always begin diagnosis to any electrical system with a check of the fuses. If you find a blown fuse, try to figure out why the fuse blew. Never replace a fuse with one of a higher amperage rating or with a piece of foil.

b) Check the battery and make sure it has a full charge. ABS systems depend on an accurate supply of voltage. Batteries should put out 12 or 13 volts without the engine running, and approximately 14.2 volts with the engine running.

c) Make sure all electrical connectors related to the ABS system are securely connected. Remember, however, don't disconnect or connect any of the connectors with the ignition in the On position.

d) Check the gap between the wheel speed sensors and the toothed signal rings. While specifications vary, they should be somewhere in the neighborhood of 0.015 to 0.040-inch. Also make sure the sensors or rings are not damaged.

Haynas Automotive Brake Manual

11 Modifications

Introduction

While it is safe to say that all passenger vehicles on the road come equipped from the factory with adequate brakes that do their job well, there is always the driver who enjoys attacking remote canyon roads, putting himself and his vehicle to the ultimate test. Speed is craved by a certain few, and there will always be those who just can't leave their cars alone. They simply must extract every ounce of performance from it.

But it doesn't do any good to be able to go fast if you can't slow down. As a matter of fact, you *must* be able to slow down rapidly if you want to go fast. This can be accomplished through *a)* increasing the amount of friction that the brakes are able to develop, or *b)* getting rid of the heat created by the brakes more rapidly, or *c)* both of the above.

Some of the ways to achieve these goals are relatively simple, inexpensive replacement of parts. Other methods are very costly, require much fabrication, and some shouldn't even be performed on a car intended for street use.

11.1 The addition of high-performance brake pads is a very effective and relatively inexpensive way to increase braking performance. These Carbon Metallic® pads by Performance Friction Corp. are said to last longer than most brake pads, are highly resistant to brake fade, and don't wear out brake discs like some semi-metallic pads do.

Lining material

When brakes are subjected to extreme conditions such as heavy, repeated braking, the most noticeable result is a reduction in braking efficiency. This is commonly referred to as brake fade. One way to reduce this problem is to experiment with different brake lining materials.

For disc brakes you might want to try replacing your standard organic pads with semi-metallic linings. These pads can tolerate more heat before brake fade is noticed, which means repeated hard deceleration becomes safer. Drawbacks to semi-metallic brake pads include higher cost and noisy operation. Some semi-metallic pads also have a tendency to wear out brake discs much more rapidly than stock brake pads.

An exception to this is the Carbon Metallic® pad made by Performance Friction Corporation (Clover, South Carolina) **(see illustration)**. This company claims their pads aren't noisy and do not accelerate brake disc wear like some semi-metallic pads. These pads last longer than standard pads and don't contain asbestos either, which is a plus.

There are many other manufacturers who offer high-performance brake pads - check with your local speed shop or auto parts store. Alternative drum brake linings are also available. Different compositions and brake shoes with more lining material can be experimented with.

Wheels

If you were a brake assembly tucked away behind a stock wheel and hubcap, you'd probably feel quite claustrophobic. Not only that, you would get very warm and probably wouldn't feel like working very hard after a while. Aftermarket wheels could solve this dilemma, improving your view and reducing your temperature considerably.

Modern high-performance wheels not only look good, they open up the area around the wheel hub for improved airflow to the brakes. Some wheels even feature "blade" construction. When they rotate, they operate like the blades of a fan, sucking air around the brake which cools the braking surface and keeps the fluid temperatures down.

If you care about the appearance of your vehicle, you'll probably spend a little more time when washing it after installing aftermarket wheels, due to the fact that you'll most likely notice more ugly, black brake dust covering the wheels. This is something you'll have to learn to live with. Several companies manufacture covers that fit behind the wheels and eliminate or greatly reduce this residue. The addition of these covers, however, defeats the improved ventilation benefits of the wheels, and really aren't recommended if you're interested in high performance.

Disc brake modifications

Remove the dust shield

This is a modification that should not be performed if you intend to operate your vehicle on the street (or in the dirt, for that matter). These shields have been installed for a reason - to protect the brakes from dirt, gravel and other road debris.

If you plan to run your vehicle only on controlled surfaces, than this modification might be considered, as it will allow more cooling air to reach the brake assembly.

Install vented brake discs

The addition of vented brake discs enhance braking in three ways. They are usually heavier, which means they can soak up more heat, there are more surfaces exposed to the air from which the heat can dissipate, and the cooling vanes direct air over these surfaces which helps to carry the heat away.

11.2 Vented brake discs drastically reduce braking temperatures and increase brake performance. As you can see, these discs are also drilled to further improve cooling. These discs are fabricated by Neuspeed, a manufacturer of high-performance aftermarket Volkswagen parts, and are a bolt-on replacement for various water-cooled VW models.

Depending on the vehicle, the addition of vented brake discs may be an easy bolt-on conversion. Many vehicles, when originally introduced, were equipped with solid discs. After being in production a couple of years, the manufacturer may have equipped the same vehicle with vented discs. This information can be gleaned from factory service manuals or your local salvage yard. You'll probably obtain all of the necessary parts from the salvage yard anyway, so it would be wise to consult with the people who operate the yard for advice on this conversion.

For some vehicles that are considered enthusiast cars or "cult" cars, the conversion procedure is made easy. There are many aftermarket companies that manufacture high-performance brake discs and all the necessary equipment to install them **(see illustration)**.

Two items concerning ventilated discs should be brought up. One, some ventilated discs are directional - the way the vanes are positioned or curved act to greatly increase the airflow between the plates of the disc. If the disc is installed on the wrong side of the vehicle it won't work as it should. Two, the addition of a ventilated disc may also require caliper replacement, since ventilated discs are thicker.

The first step to take when considering a modification such as this one is to look through as many enthusiast magazines you can get your hands on. Often they will feature this very conversion in one of their tech articles. You can consult the publishers of these magazines to obtain back-issues, if necessary. Also, look in the classified section at the rear of the magazine. There might be a company advertising exactly what you are looking for. Or, you could write to a car club or organization that is devoted to the kind of car you are customizing. They too are usually listed in these magazines and would be more than happy to offer assistance.

Install bigger brakes

Along the same lines as installing ventilated discs is the addition of *bigger* brakes. A larger diameter brake is a more powerful brake. Think of it as a lever - a longer lever will enable you to move a heavier object, right? Well, a bigger brake will soak up more torque and slow the vehicle down quicker. It will also operate at lower temperatures than a standard diameter brake.

Again, follow the same path as described in the previous section. Enthusiast magazines are a valuable source of information. Many of the same companies that manufacture vented discs also offer big brake kits **(see illustration)**.

You can also consult your local wrecking yard for information regarding factory brake replacements. Sometimes a manufacturer will equip certain versions of the same carline with heavy-duty brakes. Parts interchange manuals or even factory service manuals can provide the necessary information. Look for terms like Heavy Duty, Police, Taxi or even Station Wagon.

11.3 Bigger brakes equal more powerful brakes. In some cases this job will involve a little more work than others, but a kit like this makes it easy. Also marketed by Neuspeed, this Big Brake kit is designed for Golfs and Jettas and features larger diameter, thicker discs. It contains everything necessary to complete the conversion except for 15-inch wheels with the proper offset.

Drum brake modifications
Install finned brake drums

Certain vehicles of years ago were equipped with finned brake drums. These cooling fins increase the surface area of the outside of the drum, enabling the drum to cast off heat better. In many instances these drums can be installed on a similar model vehicle with little or no modifications required. However, when performing such a modification it may also be necessary to replace the brake backing plates, wheel cylinders and brake shoes.

The addition of finned *aluminum* brake drums is even better, as the aluminum can get rid of heat faster than cast iron can. Aluminum drums have a cast iron friction surface inside.

Consult with experts at your local salvage yard, refer to parts interchange manuals (which the people at the wrecking yard will undoubtedly have) or refer to factory parts manuals to find out if this modification is an option for you.

Ventilate the backing plates

Although brakes need heat to work at their greatest potential (they work best within a certain heat range), excessive heat is their worst enemy. Brake backing plates do a good job of sealing in much of this heat. Some of this heat radiates from the backing plate. The rest of it is absorbed by the drum and the other components.

One way to reduce the temperatures inside a drum brake is to make holes in the brake backing plate. This can be done with a large drill bit or a hole saw. If you choose to undertake this modification, do not remove any material near the anchor pin or anchor plate, wheel cylinder, shoe hold-down pins or any area where contours or bends have been formed in the metal, as these points are important to the structural integrity of the backing plate.

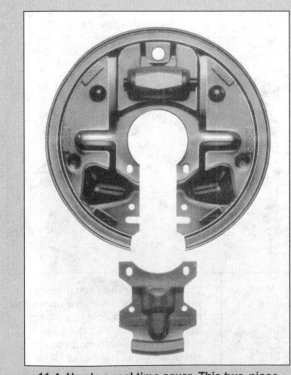

11.4 Here's a real time saver. This two-piece backing plate manufactured by Jones Brake Technologies, Inc. eliminates the need for removing the axle to replace a damaged brake backing plate.

Two-piece backing plate

Not all modifications are intended to increase performance. Some are designed to enhance serviceability or reduce the amount of work necessary to perform a certain job.

Brake backing plates live in a harsh environment of heat, cold, torque, vibration, moisture mud, snow rocks and road salt. If a human were to be subjected to such a sentence it would be considered cruel and unusual punishment. Backing plates take the punishment, but rarely complain.

After years of this abuse, however, rust holes or other damage can occur, requiring replacement of the backing plate. This job entails more work than you may think. Besides removing the brake components, the axle has to be removed. On some vehicles this means the differential cover must also be removed and the pinion shaft and axle retainer clip taken out, which is probably a lot more work than you bargained for.

You don't necessarily have to go through all of this trouble if you own a 1978 or later General Motors vehicle with 9-1/2 inch drum brakes. A company by the name of Jones Brake Technologies, Inc. (St. Thomas, Ontario, Canada) manufactures a two-piece backing plate which greatly simplifies this job **(see illustration)**. All you have to do is remove the brake drum, brake shoes and wheel cylinder, cut the old backing plate off and slip the new, two-piece unit into place. All of the necessary hardware is supplied. These are available singularly or in pairs and will save you much time and unnecessary expense if the need to replace a backing plate arises.

Convert from drum brakes to disc brakes

Depending on what kind of vehicle you have, this can be a relatively easy job to an ordeal that you probably shouldn't attempt. The key is in the research and, like the other modifications mentioned, should be gone about in the same manner. Check with enthusiast magazines and junkyards first. As with the modifications listed under *Disc brake modifications*, many companies offer complete conversion packages, which makes the job quite easy.

Some older cars, when introduced, were only available with drum brakes. Many of these same models continued through to years when disc brakes became common. In cases like this, disc brake conversion may be as simple as a direct swap, and you'll be able to obtain all the parts you need from a like vehicle at a junkyard. Since disc brakes require a considerable amount of effort to apply, the addition of a power booster, if one isn't already installed, is advisable. Also check with factory service manuals for the diameter of the pistons in the master cylinder that is on your car, and compare this with the diameter of the master cylinder pistons in the version with disc brakes. It may be necessary to install a different master cylinder, too. You'll also need a metering valve, and perhaps a different proportioning valve, which can be taken from the same vehicle you're getting the other parts from.

Other modifications

Other changes to the brake system to increase pressure at the brakes (which will increase friction) are:

a) Install larger wheel cylinders or calipers with larger pistons. Larger piston surface area (on the receiving end) equates to more pressure applied to the shoes or pads. Check with factory parts manuals and/or parts interchange manuals to see if this is a viable option. If you do this, be reminded that brake pedal travel will also increase. Make sure there is adequate reserve distance between the pedal and the floor when the brakes are fully applied.

b) Add a power brake booster, if the vehicle is not equipped with one.

c) Install braided steel, Teflon coated brake lines. This will give you a reduced amount of deflection (by eliminating "ballooning" of the hoses) in the brake system, and a firmer pedal feel. Special adapter fittings will probably be required. Consult with the experts at your local speed shop - they'll most likely be able to set you up with everything you need for this conversion.

High-performance/racing components

If you're building a street rod or a race car and have a healthy bank account, your options are almost boundless. Shown here is just a fraction of the neat componentry that is available when the sky is the limit.

Calipers

From lightweight high-performance units that will bolt right in place of your existing calipers to all-out, composite four-piston fixed calipers for race use only, improvements made in this area are about the most effective way to generate higher braking forces **(see illustrations)**. Depending on the calipers being installed, it may also be necessary to upgrade the brake discs, as well.

11.5 Wilwood Engineering (Camarillo, CA) makes disc brake components for just about every requirement, from high-quality single-piston calipers like this . . .

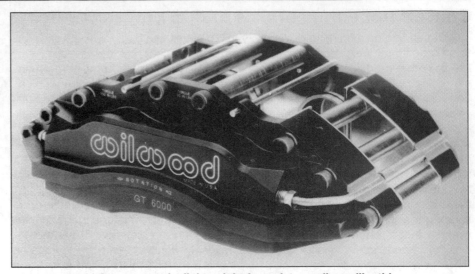

11.6 . . . to exotic, lightweight four-piston calipers like this

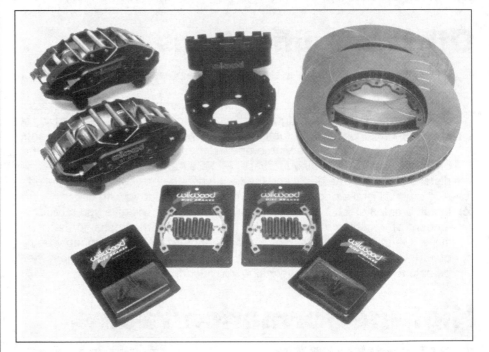

11.7 Wilwood sells kits which contain all of the necessary hardware for installation

Brake pedal/master cylinder arrangements

Thanks to the many different pedal assemblies and master cylinder arrangements available, what was once a difficult problem for car constructors is now much easier. There are pedal/master cylinder setups readily available for virtually every need. Whether you're building a hot rod, sandrail or a race car, there is surely a pedal assembly which will fit the bill **(see illustrations)**.

Many of these pedal setups have two single-piston master cylinders connected by a balance bar. This allows the driver to fine-tune front-to-rear brake balance as track conditions and/or car setup require. With the installation of a remote balance bar adjuster, these adjustments can be made by a driver during race conditions **(see illustration 11.11)**.

Another way to change front-to-rear brake balance is with the addition of an adjustable proportioning valve **(see illustration)**. This device allows a driver to increase or decrease pressure to the rear brakes, in increments.

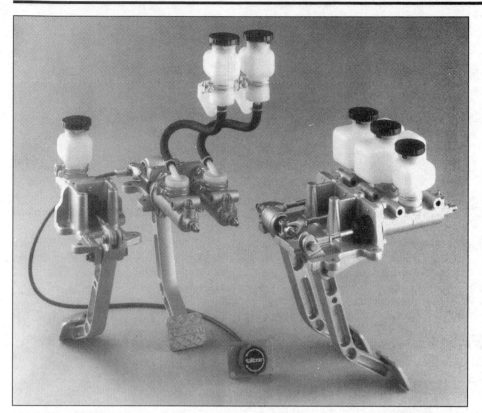

11.8 Tilton Engineering, Inc. is the manufacturer of these brake pedal/clutch pedal and master cylinder assemblies. They make pedal assemblies to satisfy every fabricators' need. Notice the brake balance bars on the two units on the right.

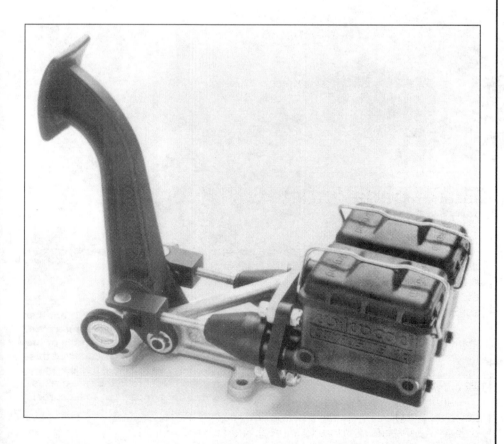

11.9 This Wilwood floor-mounted brake pedal/master cylinder assembly also is equipped with a balance bar

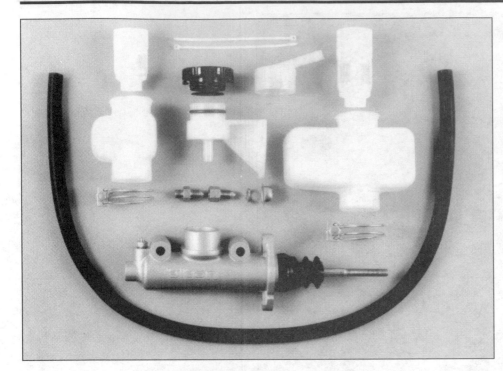

11.10 Aftermarket master cylinders are available for just about any arrangement needed. This Tilton universal master cylinder can be fitted with different shaped reservoirs or even a remote reservoir, as installation requirements dictate.

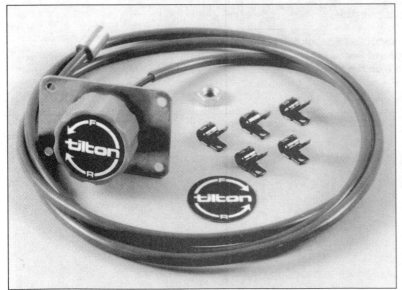

11.11 Here's a remote balance bar adjuster, used for adjusting front-to-rear brake bias from the driver's seat

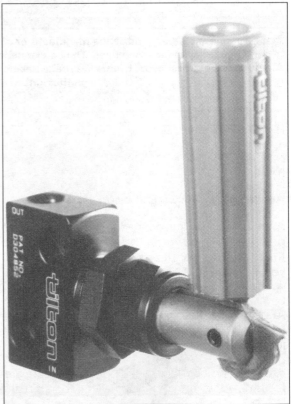

11.12 An adjustable proportioning valve is another way to alter front-to-rear brake bias during a race. As the tires go away, car setup deteriorates or the track conditions worsen, the driver can click this valve up a notch or two (this one has seven positions) to reduce pressure to the rear brakes which can "tighten up" a loose car, to a degree.

Glossary

A

ABS: Anti-lock Brake System.

Accumulator: A vessel that stores hydraulic fluid under pressure.

Adapter: See *Caliper mounting bracket*.

Air-suspended power booster: A type of power booster that contains atmospheric pressure in both chambers of the booster when the brake pedal is at rest. When the pedal is applied, the front chamber is opened to manifold vacuum, causing the diaphragm of the booster to move toward the master cylinder which assists the driver in the application of the brakes.

Anchor: The stationary portion of a leading/trailing drum brake on which the heels of the brake shoes ride.

Anchor plate: See *Caliper mounting bracket*.

Anchor pin: The stationary portion of a duo-servo drum brake on which the tops of the brake shoes rest. The secondary shoe bears against the anchor pin when the brakes are applied and the vehicle is moving forward. Conversely, when the vehicle is backing up and the brakes are applied, the primary shoe bears against it.

Anti-lock brake system: A brake control system that monitors the rotational speeds of the wheels and reduces hydraulic pressure to any wheel it senses locking up.

Arcing: A process where the brake shoes are ground to the proper curvature for the drums they are to be used with. Modern brake shoes are pre-arced.

Asbestos: A fibrous mineral used in the composition of brake friction materials. Asbestos is a health hazard and the dust created by brake systems should never be inhaled or ingested.

Asbestosis: An incurable lung disease caused by the inhalation of asbestos fibers.

Automatic adjusters: Brake adjusters that are actuated by the application of the parking brake or by normal brake operation, to compensate for lining wear.

Haynes Automotive Brake Manual

B

Backing plate: The part of a drum brake to which the wheel cylinder(s) and the brake shoes are attached.

Banjo fitting: A type of hydraulic fitting, shaped like a banjo, through which a hollow bolt passes, allowing fluid transfer from a hydraulic line to a hydraulic component.

Bleeder screw: The hollow screw that is loosened to open a bleeder valve, allowing fluid and air bubbles to pass through during a bleeding procedure.

Bleeder valve: A valve on a wheel cylinder, caliper or other hydraulic component that is opened to purge the hydraulic system of air.

Bonded linings: Brake linings that are affixed to the shoe or backing plate with high-temperature adhesive and cured under pressure and heat.

Brake adjuster: A mechanism used to adjust the clearance between the brake linings and the brake drum.

Brake balance: The ratio of front-to-rear braking force.

Brake caliper: The component of a disc brake that converts hydraulic pressure into mechanical energy.

Brake disc: The component of a disc brake that rotates with the wheels and is squeezed by the brake caliper and pads, which creates friction and converts the energy of the moving vehicle into heat.

Brake drum: The component of a drum brake that rotates with the wheel and is acted upon by the expanding brake shoes, which creates friction and converts the energy of the moving vehicle into heat.

Brake dust: The dust created as the brake linings wear down in normal use. Brake dust usually contains dangerous amounts of asbestos.

Brake fade: The partial or complete loss of braking power which results when the brakes are overheated and can no longer generate friction.

Brake lathe: The machine used to resurface the friction surfaces of brake discs or drums.

Brake lines: The rigid steel and flexible rubber hoses composing the portion of the hydraulic system that transfers brake fluid from the master cylinder to the calipers and/or wheel cylinders.

Brake lining: The friction material that is either riveted or bonded to the brake backing plates or brake pads.

Brake pads: The components of a disc brake assembly that are surfaced with brake lining and clamped against the brake disc to generate friction.

Brake shoes: The components of a drum brake assembly that are surfaced with brake lining and forced against the brake drum to generate friction.

Brake system cleaner: A type of solvent designed exclusively for cleaning brake system components. It will not destroy plastic, rubber or synthetic rubber components and it dries quickly, without leaving a residue.

Breather port: The small passage between the master cylinder fluid reservoir and the area behind the primary cups of the pistons. This port allows fluid from the reservoir to fill the area behind the primary cups of the pistons, preventing a vacuum from being formed behind the cups when the brakes are applied, which prevents air bubbles from traveling around the lips of the primary cups as the brakes are released.

Bridge bolts: High-strength bolts used to fasten together the halves of a split brake caliper.

Burnish: The process of "breaking-in" new brake pads or shoes so the linings conform to the disc or drum friction surfaces.

C

Caliper: See *Brake caliper*.

Caliper mounting bracket: The component that connects a brake caliper to the steering knuckle, hub carrier or rear axle.

Cam: An eccentric shaped device mounted on a shaft that raises and lowers the component in contact with it. Some brake adjuster designs use a cam (or cams) to set the clearance between the brake shoes and the brake drum.

Clevis: A U-shaped device with a pin or bolt passing through it used for attaching the master cylinder or power booster pushrod to the brake pedal. Clevises are sometimes used in other parts of the brake system, like attaching the parking brake cable to the parking brake lever at the rear brakes.

Coefficient of friction: A numerical value indicating the amount of work required to slide one surface against another. The coefficient of friction equals the side force acting on the object divided by the weight of the object.

Combination valve: A hydraulic valve usually incorporating a pressure differential warning switch, a metering valve and a proportioning valve. Not all combination valves contain all of these control valves.

Compensating port: The small passage between the master cylinder fluid reservoir and the pressure chamber (the area in front of the primary seals of both pistons) that allows fluid to flow in or out as necessary, depending on requirements.

Component Anti-lock Brake System: A type of Anti-lock Brake System in which the hydraulic control unit is not part of the master cylinder/power booster assembly.

Cup: A type of lip seal used on hydraulic pistons.

D

Diaphragm: A flexible partition used to separate two chambers or elements.

Discard diameter: The diameter at which a worn brake drum should be replaced.

Discard thickness: The thickness at which a brake disc should be replaced.

Disc: See *Brake disc*.

Disc brake: A brake design incorporating a flat, disc-like rotor onto which brake pads containing lining material are squeezed, generating friction and converting the energy of a moving vehicle into heat.

Dished brake disc: A disc that has worn thinner at the inner part of its friction surface. This is an abnormal form of wear.

Double anchor drum brake: See *Leading/trailing* drum brake.

Drum brake: A brake design incorporating a drum with brake shoes inside that expand to contact the drum, creating friction and converting the energy of the moving vehicle into heat.

Dual circuit brake system: A brake hydraulic system composed of two separate hydraulic circuits.

Duo-servo drum brake: A type of self-energizing drum brake that has servo action in both forward and reverse.

Dust boot: A rubber diaphragm-like seal that fits over the end of a hydraulic component and around a pushrod or end of a piston, not used for sealing fluid in but keeping dust out.

E

Electro-hydraulic pump: An electrically powered hydraulic pump used to create pressure in certain portions of the brake system. Typically found in General Motors Powermaster brake boosters and in ABS hydraulic control units.

Electronic Control Unit (ECU): The "brain" of an ABS system. The ECU reads impulses from the wheel speed sensors to determine if anti-lock braking needs to take place. If so, the ECU controls the cycling of the solenoid valves in the hydraulic control unit.

Emergency brake: Another term for parking brake.

Equalizer: A bracket or cable connector which balances tension equally on the cables to the parking brakes.

F

Filler port: See *breather port*.

Firewall: The partition between the passenger compartment and the engine compartment. Sometimes referred to as a bulkhead.

Fixed caliper: A caliper containing one or more pistons on each side of the brake disc that is bolted to the steering knuckle, hub carrier or rear axle and is incapable of movement.

Flare-nut wrench: A wrench designed for loosening hydraulic fitting tube nuts (flare-nuts) without damaging them. Flare-nut wrenches are kind of like a six-point box-end wrench with one of the flats missing, which allows the wrench to pass over the tubing but still maintain a maximum amount of contact with the nut.

Floating caliper: A caliper that rides on bushings, has one or more pistons only on one side of the caliper and moves laterally as the piston pushes on the inner brake pad, which pulls the outer pad against the brake disc.

Four-wheel ABS: An anti-lock brake system that operates on all four wheels.

Freeplay: The amount of travel before any action takes place. In a brake pedal it is the distance the pedal moves before the pistons in the master cylinder are actuated.

Friction: Surface resistance to relative motion.

G

Glazed lining: A brake lining that has been overheated and become smooth and glossy.

Grinding: The process of resurfacing a brake disc or drum on a brake lathe using a power-driven abrasive stone.

Grommet: A round rubber seal which fits into a hole or recess, intended to seal or insulate the component passing through it.

Guide pin: A caliper mounting bolt used for fastening a floating caliper to its mounting plate.

H

Hard spots: Shiny bluish/brown glazed areas on a brake drum or disc friction surface, caused by extreme heat. Hard spots can usually be removed by resurfacing.

Hat: The portion of a detachable brake disc that comes in contact with the wheel hub.

Heat checking: Small cracks on a brake disc or drum friction surface caused by heat. Heat checks can usually be removed by resurfacing.

Hold-down pin, spring and retainer: The most common method of retaining a brake shoe to the backing plate. The pin passes through the backing plate and brake shoe. The spring and retainer are fastened to the pin, which holds the shoe against the backing plate.

Hold-off valve: See *Metering valve*.

Hydraulically operated power booster: A power booster that uses hydraulic pressure to assist the driver in the application of the brakes. This hydraulic pressure usually comes from the power steering pump or an electro-hydraulic pump.

Hydraulic control unit: The portion of an anti-lock brake system that houses the solenoid valves and electro-hydraulic pump.

Hydraulic modulator: See *Hydraulic control unit*.

Hydroplane: The action that takes place when an accumulation of water builds up in front of a tire, causing the tire to ride on a layer of water instead of on the pavement.

Hygroscopic: The tendency to attract or absorb moisture from the air.

I

Inboard disc brakes: Disc brakes not mounted out at the wheels, but near the differential.

Inertia: The tendency of a body at rest to remain at rest, and a body in motion to remain in motion.

Integral Anti-lock Brake System: An anti-lock system that incorporates the master cylinder and power booster into the hydraulic control unit.

K

Kinetic energy: The energy of a body in motion.

Knockback: The action of a brake disc with excessive runout pushing back the brake pads when the brakes are not applied.

L

Lateral runout: Side-to-side warpage of the brake disc friction surfaces.

Leading shoe: A shoe in a non-servo action drum brake assembly that self-energized by the rotation of the brake drum.

Leading/trailing drum brake: A drum brake design in which both brake shoes are attached to an anchor plate, and only one of the shoes is self-energized.

Load Sensing Proportioning Valve (LSPV): A hydraulic system control valve that works like a proportioning valve, but also takes into consideration the amount of weight carried by the rear axle.

M

Manual adjuster: A type of brake adjuster that must be adjusted from time-to-time, with the use of a hand tool.

Master cylinder: The component in the hydraulic system which generates pressure for the calipers and/or wheel cylinders.

Metering valve: A hydraulic control valve placed in the circuit to the front brakes, designed to restrict pressure to the front brake calipers until the rear brake shoes overcome the tension of the retracting springs.

N

Non-servo drum brake: A drum brake design in which the application of one shoe has no effect on the other.

O

Organic linings: Brake lining material using asbestos as its main ingredient.

Out-of-round: The condition of a brake drum when it has become distorted and is no longer perfectly round. In many cases an out-of-round brake drum can be salvaged by resurfacing on a brake lathe.

P

Pad wear indicators: Mechanical or electrical devices which warn the driver when the lining material on the brake pads has worn to the point that they should be replaced.

Parallelism: The relationship between one friction surface of a brake disc and the other.

Parking brake: The mechanically actuated portion of a drum brake or disc brake caliper, used to prevent the vehicle from rolling when it is parked, applied by a lever, pedal or rod.

Pascal's Law: The law of physics stating that "pressure, when applied to a confined liquid, is transmitted undiminished." Discovered by Blaise Pascal (1623 - 1662).

Power booster: A device using vacuum or hydraulic power to assist the driver in the application of the brakes.

Pressure bleeder: A device that forces brake fluid under pressure into the master cylinder, so that by opening the bleeder screws all air will be purged from the hydraulic system.

Pressure differential warning switch: A component of the brake hydraulic system that warns the driver of a failure in one of the circuits.

Primary shoe: The shoe in a duo-servo drum brake that transfers part of its braking force to the secondary shoe.

Proportioning valve: A hydraulic control valve located in the circuit to the rear wheels which limits the amount of pressure to the rear brakes to prevent wheel lock-up during panic stops.

R

RABS: Rear-wheel Anti-lock Brake System (Ford).

Replenishing port: See *Breather port*.

Reservoir: A container attached to the master cylinder, either directly or by hoses, that stores extra brake fluid for the hydraulic system.

Residual pressure: Pressure remaining in a hydraulic circuit after the brakes have been released.

Residual pressure check valve: A small valve, usually located in the outlet port(s) of the master cylinder, which maintains a certain amount of pressure in the hydraulic circuit(s) when the brakes are released. Used only in drum brake hydraulic circuits to keep the lips of the wheel cylinder cups sealed against the walls of the cylinder.

Resurfacing: The process of machining a brake drum or disc on a brake lathe to remove surface imperfections from the friction surface.

Riveted linings: Brake linings that are riveted to the pad backing plate or brake shoe.

Rotor: See *Brake disc*.

Runout: Side-to-side warpage of the brake disc friction surfaces.

RWAL: Rear Wheel Anti Lock (General Motors).

S

Scoring: Grooves or deep scratches on a friction surface caused by metal-to-metal contact (worn-out brake pads or shoes) or debris caught between the friction material and the friction surface.

Secondary shoe: The shoe in a duo-servo drum brake assembly that is acted upon by the primary shoe. It provides more stopping power than the primary shoe (about 70-percent).

Select-low principle: The method by which the rear brake application of an ABS brake system is monitored and controlled, based on the rear wheel with the least amount of traction.

Self-energizing action: The action of a rotating brake drum that increases the application pressure of the brake shoe(s).

Semi-metallic lining: Brake lining incorporating a high-percentage of metal in its composition.

Servo-action drum brake: See *Duo-servo drum brake*.

Sliding caliper: Similar to a *floating caliper*, but instead of riding on guide pins and bushings, the caliper slides on machined "ways" and is retained by keys or spring plates.

Solid brake disc: A brake disc that is solid metal between its friction surfaces.

Star wheel: The portion of a brake adjuster that turns the adjuster screw.

T

Tire slip: The difference between the speed of the vehicle and the speed between the tire and the ground, expressed in a percentage.

Toothed signal rotor: The component of an ABS system that rotates with the hub, driveaxle, axle or ring gear, used along with the wheel speed sensors for generating impulses to be rear by the ABS Electronic Control Unit (ECU). The ECU counts these impulses and determines if a wheel is decelerating too rapidly or not.

Torque: A turning or twisting force imposed on a rotating part.

Torque plate: See *Caliper mounting bracket*.

Traction: The amount of adhesion between the tire and ground.

Trailing shoe: A shoe in a drum brake assembly that is not self-energized.

Two-wheel ABS: An anti-lock brake system that only operates on the rear wheels.

U

Uni-servo drum brake: A servo-action drum brake that only has servo action when the vehicle is braked in a forward direction.

V

Vacuum: As an automotive term, vacuum is any pressure less than atmospheric pressure.

Vacuum-operated power booster: A power booster that uses engine manifold vacuum to assist the driver in the application of the brakes.

Vacuum-suspended power booster: A type of power booster that contains vacuum in both chambers of the booster when the brake pedal is at rest. When the pedal is applied, the rear chamber is vented to the atmosphere, causing the diaphragm of the booster to move toward the master cylinder which assists the driver in the application of the brakes.

Vapor lock: The abnormal condition that occurs when brake fluid contains too much moisture and is overheated, causing the moisture in the fluid to boil. Gas bubbles are formed in the fluid, which causes a spongy brake pedal or a complete loss of hydraulic pressure.

Vented brake disc: A brake disc that has cooling passages cast or drilled between its friction surfaces.

Viscosity: The property of a fluid that resists the force tending to cause the fluid to flow.

W

Ways: Machined abutments on which a sliding brake caliper rides.

Wheel cylinder: The component in a hydraulic system that converts hydraulic pressure into mechanical force to apply the brake shoe(s).

Wheel speed sensor: The component of an anti-lock brake system that picks up the impulses of the toothed signal rotor, sending these impulses to the ABS ECU.

Notes

Torque Specifications

Note: *The following brake torque specifications have been compiled from all of the Haynes Automotive Repair Manuals available at the time of this manual's publication. They are torque specifications considered to be the most critical for proper assembly of the brake system components. If you can't find your vehicle in the following list, consult a dealer service department or factory service manual for the specifications required.*

Haynes Automotive Brake Manual

Torque specifications	Ft-lbs
	(unless otherwise indicated)

Acura

Legend (1986 - 1990) and Integra (1986 - 1989)

Brake hose banjo fitting bolt	25
Front disc brake	
Caliper guide pin bolts	
Integra	33
Legend	24
Caliper mounting bracket bolts	
Integra	53
Legend	56
Rear disc brake	
Caliper flange (guide pin) bolts	
Integra	17
Legend	20
Caliper mounting bracket bolts	28

AMC/Renault

Alliance and Encore (1983 through 1987)

Caliper mounting bolts	26
Caliper carrier bolts	74

Mid-Size Models (1970 through 1983)

Caliper guide pin (1970 through 1974)	35
Caliper support key screw (1975 through 1981)	15
Caliper mounting pin (1982)	26
Caliper anchor plate bolts (1982 and earlier)	80

Audi

4000 (1980 through 1987)

Caliper-to-suspension strut	
Hex bolts	37
Serrated bolts	52
Guide pin bolts (Girling)	26

5000 (1977 through 1983)

Caliper-to-suspension strut bolts	83
Rear caliper carrier bolt (1980 on)	47
Rear caliper mounting bolt (1980 on)	25

5000 (1984 through 1988)

Caliper guide pin bolts (Girling)	26
Caliper guide pins (Teves)	18
Front caliper carrier bracket-to-strut bolts	52
Rear caliper carrier bracket-to-rear axle bolts	48

Austin Healy

See entry under MG/Austin Healy

Austin Marina

Austin Marina (1971 through 1975)

Caliper mounting bolts	50

Torque specifications	Ft-lbs
	(unless otherwise indicated)

BMW

1602 and 2002 (1959 through 1977)
Includes 1500, 1502, 1600 and 2000 Touring

Caliper mounting bolts	70
Disc-to-hub bolts	48

2500, 2800, 3.0 and Bavaria (1969 through 1976)

Caliper mounting bolts	
Front	58 to 69
Rear	44 to 48
Caliper bridge bolts	
Front	33 to 40
Rear	16

320i (1975 through 1983)

Caliper mounting bolts	58 to 69
Disc-to-hub bolts	43 to 48

528i and 530i (1975 through 1980)

Caliper mounting bolts	
Front	58 to 68
Rear	43 to 48
Disc-to-hub bolts	
M6 bolt	36 to 42 in-lbs
M8 bolt	22 to 24
Disc-to-hub nut	43 to 48

3 and 5 Series (1982 to 1992)

Front caliper	
Guide (mounting) bolts	22 to 26
Carrier-to-strut housing bolts	80 to 80
Rear caliper	
Guide (mounting) bolts	22 to 25
Carrier-to-trailing arm bolts	44 to 49
Brake hose banjo fitting bolt	120 to 144 in-lbs

Buick

Mid-Size Rear-Wheel Drive (1974 through 1987)
Regal
Century
Wagons

Caliper mounting bolt	35
Brake hose banjo fitting bolt	
1974 through 1976	22
1977 and 1978	32
1979 and later	18

Skylark X-Cars (1980 through 1985)

Caliper mounting bolt	28
Brake hose banjo fitting bolt	22

Buick/Olds/Pontiac Full-Size Front-Wheel Drive models (1985 through 1993)

Caliper mounting bolts	38
Brake hose banjo fitting bolt	33

Buick/Oldsmobile/Pontiac Full-Size Rear Wheel Drive models (1970 through 1990)

Caliper mounting bolts	37
Brake hose banjo fitting bolt	32

Haynes Automotive Brake Manual

Torque specifications	Ft-lbs
	(unless otherwise indicated)

Cadillac

Rear-wheel Drive Models (1970 through 1992)
Caliper mounting bolts
    1970 to 1982 30
    1983 on .... 35 to 38
Brake hose banjo fitting bolt 30

Capri

Capri 2000 (1971 through 1975)
Caliper mounting bolts.. 50
Disc-to-hub bolts 34

Capri II 2800 (1975 through 1978)
Caliper mounting bolts.. 50
Brake disc-to-hub .... 30 to 34

Capri V6 2600, 2800 (1971 through 1975)
Caliper mounting bolts.. 45 to 50
Disc-to-hub bolts 30 to34

Chevrolet/GMC

Chevrolet and GMC Pick-Ups (1967 through 1987)
  Includes 1967 through 1991 Suburban, Blazer and Jimmy
Caliper mounting bolts.. 35
Brake hose banjo fitting bolt 22

Chevrolet and GMC Pick-ups (1988 through 1994)
Caliper mounting bolts.. 28
Brake hose banjo fitting bolt 32

Chevrolet and GMC S-10/S-15 Pick-Ups (1982 through 1993)
  Includes S-10 Blazer, S-115 Jimmy and Olds Bravada
Caliper mounting bolts.. 37
Brake hose banjo fitting bolt 13

Chevrolet Astro/GMC Safari Mini-Vans (1985 through 1993)
Caliper mounting bolts.. 37
Brake hose banjo fitting bolt 32

Camaro (1970 through 1981)
Caliper mounting bolts.. 35
Brake hose banjo fitting bolt 22

Camaro (1982 through 1992)
Caliper mounting bolts (single-piston front caliper)....................... 35
Caliper bracket bolts (twin-piston caliper) 137
Rear caliper guide pin bolt (1989 on)
    Allen head.. 16
    Hex head 26
Rear caliper bracket bolts 70
Brake hose banjo fitting bolt 32

Chevelle/El Camino (1969 through 1987)
Caliper mounting bolts.. 35
Brake hose banjo fitting bolt 22

Torque specifications	Ft-lbs (unless otherwise indicated)

Chevette/Pontiac T1000 (1976 through 1987)

Caliper bracket-to-steering knuckle bolts (1976 through 1982)	95
Caliper-to-bracket bolts (1983 on)	29 to 34
Brake hose banjo fitting bolt...	44

Citation X-Cars (1980 through 1985)

Caliper mounting bolt	28
Brake hose banjo fitting bolt...	22

Corsica and Beretta (1987 through 1992)

Caliper mounting bolts	38
Brake hose banjo fitting bolt...	33

Corvette (1968 through 1982)

Caliper mounting bolts	70
Caliper bridge bolts	
1974 and earlier..	130
1975 on	
Front caliper	130
Rear caliper	60

Corvette (1984 through 1991)

Caliper self-locking bolts	22 to 25
Caliper mounting bracket bolts (1984 w/o washer)	
Front	70
Rear	44
Caliper mounting bracket bolts (1985 and later w/washers)	
Front	133
Rear	70

Chevrolet Full Size Models (1969 through 1990)

Caliper mounting bolts	35
Brake hose banjo fitting bolt...	22

Luv Pick-Ups (1972 through 1982)

Caliper mounting bracket	64
Disc-to-hub bolts..	36
Brake hose banjo fitting bolt...	29

Monte Carlo (1970 through 1988)

Caliper mounting bolt	35
Brake hose banjo fitting bolt...	22

Nova (1969 through 1979)

Caliper mounting bolt	35
Caliper mounting bracket-to-steering knuckle bolts	
To steering knuckle (upper bolt)	140
Steering arm-to-knuckle bolt............................	70
Brake hose banjo fitting bolt...	22

Nova/Geo Prizm (1985 through 1992)

Caliper mounting bolts	
Front	18
Rear	14
Front caliper torque plate-to-steering knuckle bolts...	65
Rear caliper torque plate-to-axle carrier bolts	34
Brake hose banjo fitting bolt	
Nova	17
Prizm	22

Haynes Automotive Brake Manual

Torque specifications	Ft-lbs
	(unless otherwise indicated)

Sprint/Geo Metro (1985 though 1991)
Caliper mounting bolts	20
Brake hose banjo fitting bolt	15 to 18
Disc-to-hub bolts	37

Chevrolet/GMC Vans (1968 through 1992)
Caliper mounting bolt	35
Caliper banjo fitting bolt	22

Chrysler

Mid-S;ize Front-Wheel Drive Models (1982 through 1993)
Caliper guide pin(s)	
ATE	18 to 26
All others	
1984 through 1991	25 to 35
1992-on	30
Caliper mounting bracket-to-steering knuckle bolts	
1982 and 1983	70 to 100
1984 through 1991	130 to 190
1992 on	160
Brake hose-to-caliper inlet fitting bolt	
1984 through 1991	19 to 29
1992 on	35

Full-Size Front-Wheel Drive Models (1988 through 1993)
Caliper guide pins (front)	
1990 and earlier	25 to 35
1991 on	
Double-pin, attached to adapter	25 to 35
Double pin, attached to knuckle	18 to 25
Caliper attaching bolts (rear)	
1991 and earlier	20
1991 on	16
Caliper mounting bracket-to-steering knuckle bolts	160
Brake hose banjo fitting bolt	
Front	
1991 and earlier	19
1992 on	24

Datsun

1200 (1970 through 1973)
Caliper mounting botls	33 to 44
Disc-to-hub bolts	20 to 27

200SX (1977 through 1979)
Caliper mounting bolts	65
Disc-to-hub bolts	35
Brake hose banjo fitting bolt	144 in-lbs

200SX (1980 through 1983)
Caliper mounting bolts	
Front	62
Rear	33
Disc-to-hub bolts	33
Brake hose banjo fitting bolt	13

B-210 (1973 through 1978)
Caliper mounting bolt	53 to 72
Disc-to-hub bolts	40

Torque specifications	Ft-lbs
	(unless otherwise indicated)

F10 (1976 through 1979)
Caliper mounting bolts	40 to 47
Disc-to-hub bolts	18 to 25
Brake hose banjo fitting bolt	12 to14

210 (1979 through 1982)
Caliper mounting bolts	53 to 72
Disc-to-hub bolts	28 to 38
Brake hose banjo fitting bolt	12 to 14

240Z and 260Z (1970 through 1978)
Caliper mounting bolts	70
Disc-to-hub bolts	35

280ZX (1979 through 1983)
Caliper mount-to-steering knuckle bolts	
Front	62
Rear	33
Caliper pin bolts	20
Disc-to-hub bolts	47
Brake hose banjo fitting bolt	13

310 (1978 through 1982)
Caliper mounting bolts	40 to 47

510 (1968 through 1973)
Includes PL521 Pick-Up
Caliper mounting bolts	52 to 65
Disc-to-hub bolts	28 to 38

510 (1978 through 1981)
Caliper mounting bolts	70
Disc-to-hub bolts	28 to 38
Brake hose banjo fitting bolt	12 to 14

610 (1972 through 1976)
Caliper mounting bolts	53 to 71
Disc-to-hub bolts	28 to 38
Brake hose banjo fitting bolt	12 to14

810/Maxima (1977 through 1984)
Front caliper mounting bolts	
1979 and earlier	53 to 72
1980 on	
Torque plate-to-steering knuckle	53 to 72
Caliper-to-torque plate	12 to 15
Disc-to-hub bolt	
1979 and earlier	28 to 38
1980 on	36 to 51
Rear caliper pin bolt	16 to 23
Rear caliper mounting bolt	28 to 38

Sentra
See entry under Nissan

Dodge/Plymouth/Chrysler

Aries/Reliant (1981 through 1989)
Caliper guide pin	
ATE	18 to 24
Kelsey-Hayes	14 to 22
Caliper adapter-to-steering knuckle	70 to 100
Brake hose banjo fitting bolt	168 in-lbs

Haynes Automotive Brake Manual

Torque specifications	**Ft-lbs** (unless otherwise indicated)

Caravan/Voyager (1984 through 1993)

Caliper guide pin(s)	
ATE	18 to 26
Kelsey-Hayes	25 to 35
Caliper mounting bracket-to-steering	
knuckle bolts	130 to 190
Brake hose banjo fitting bolt	19 to 29

Challenger/Sapporo (1978 through 1983)

Caliper bracket bolts	51 to 65
Disc-to-hub bolts	30 to 35

Colt/Champ (1978 through 1987)

Caliper bridge bolts	60
Caliper mounting bolts	
1983 and earlier models	55
1984 and later models	16 to 23
Brake hose banjo fitting bolt	12

Dakota Pick-ups (1987 through 1993)

Caliper guide pins (mounting bolts)	18 to 26
Caliper adapter-to-steering knuckle bolts	95 to 125
Brake hose banjo fitting bolt	30 to 40

Dart/Valiant (1967 through 1976)

Caliper mounting bolts (fixed)	75
Caliper bridge bolts (fixed)	70 to 80
Caliper retaining plate bolts (sliding)	35
Caliper adapter mounting botls (sliding)	85

Daytona/Laser (1984 through 1989)

Caliper adapter bolts	130 to 190
Caliper guide pins	18 to 22

Omni/Horizon (1978 through 1990)

Caliper guide pins	
ATE caliper	
1978 through 1980	25 to 40
1981 on	18 to 26
Kelsey-Hayes	25 to 35
Caliper adapter-to-steering knuckle bolts	
1978 through 1982	70 to 100
1983 on	130 to 190
Brake hose banjo fitting bolt	19 to 29

Pick-Ups (1974 through 1991)

Caliper adapter-to-steering knuckle bolts	
1-ton, D-200 and 250 HD	160
All others	110
Caliper retaining screw key lock	15

Ram 50/D-50 Pick-Ups and Raider (1979 through 1993)

Caliper mounting bolts	
Sliding caliper	51 to 65
Floating caliper	
Upper	29 to 36
Lower	23 to 30
Caliper mounting bracket bolts	
2WD models	51 to 65
4WD models	58 to 72
Disc-to-hub bolts	
2WD	36
4WD	36 to 44
Brake hose banjo fitting bolt	120 to 144 in-lbs

Torque specifications	Ft-lbs
	(unless otherwise indicated)

Rear-Wheel Drive Models (1971 through 1989)

Brake hose banjo bolt-to-caliper	
1971 through 1987	25
1988 and 1989	30 to 40
Fixed caliper	
Caliper bridge bolts..	70 to 80
Caliper mounting bolts......................................	50 to 80
Floating caliper	
Caliper guide pins (mounting bolts)	30 to 35
Caliper adapter-to-steering knuckle bolts	95 to 125
Sliding caliper	
Caliper retaining clip bolts...............................	220 in-lbs
Caliper adapter-to steering knuckle bolts	
1973 through 1975..	180
1976 on	95 to 125

Shadow/Duster and Sundance (1987 through 1993)

Caliper guide pins	
1987 through 1991	25 to 35
1992 on	30
Caliper mounting bracket-to-steering knuckle bolts	
1987 through 1991	130 to 190
1992 on	160
Brake hose banjo fitting bolt	
1987 through 1991	19 to 29
1992 on	35

Spirit/Acclaim

Caliper guide pin(s)	
1989 through 1991	25 to 35
1992	30
Caliper mounting bracket-to-steering knuckle bolts	
1989 through 1991	130 to 190
1992	160
Brake hose banjo fitting bolt	
1989 through 1991	19 to 29
1992	35

Vans (1971 through 1991)

Caliper adapter-to-steering knuckle bolts	
1/2-inch	95 to 125
5/8-inch	140 to 180
Caliper retaining plate bolts.............................	14 to 22

Eagle

Talon

See entry under Mitsubishi

Fiat

124 Sport (1968 through 1978)

Brake caliper block retaining bolts	35
Caliper mounting plate bolts	43
Disc-to-hub bolts...	15

Strada (1979 through 1982)

Caliper mounting bolts	
1982 and earlier (single cam)	35
1982 twin cam..	65
Brake hose banjo fitting bolt...........................	20

Torque specifications	Ft-lbs
	(unless otherwise indicated)

Ford/Mercury

Crown Victoria/Grand Marquis (1988 through 1994)
Caliper mounting (front) bolts
    Front
        1988 through 1991 .... 40 to 60
        1992 on .... 45 to 65
    Rear (1992 on) .... 19 to 26
Brake hose banjo fitting bolts
    Front
        1988 through 1991 .... 20 to 30
        1992 on .... 30 to 40
    Rear (1992 on) .... 30 to 40

Escort/Lynx (1981 through 1990)
Caliper mounting pins .... 18 to 25
Brake hose banjo fitting bolt
    1981 .... 20 to 30
    1982 on .... 30 to 40

Escort/Tracer (1991 through 1993)
Caliper mounting bolts
    Front .... 29 to 36
    Rear .... 33 to 43
Brake hose banjo bolt .... 16 to 20

Fairmont/Zephyr (1978 through 1983)
Caliper bolt/locating pins .... 30 to 40

Fiesta (1977 through 1980)
Caliper bracket bolts 40
Caliper mounting pins (XR2 models only) 17

Full-Size Models (1975 through 1987)
Front caliper anchor plate-to-steering knuckle
    Upper .... 90 to 120
    Lower .... 55 to 75
Rear caliper anchor plate-to-axle .... 90 to 120
Rear caliper end retainer .... 75 to 95
Caliper key retaining screws .... 12 to 16
Brake hose-to-caliper .... 10 to 15

Granada/Monarch (1975 through 1980)
Front caliper anchor plate-to-steering knuckle
    Upper .... 90 to 120
    Lower .... 55 to 75
Rear caliper anchor plate-to-axle .... 90 to 120
Rear caliper end retainer .... 75 to 95
Caliper key retaining screws .... 12 to 16
Brake hose-to-caliper .... 10 to 15

Mid-Size Models (1975 through 1986)
Front caliper anchor plate-to-spindle bolts
    Upper .... 90 to 120
    Lower .... 55 to 75
Rear caliper anchor plate-to-axle bolt .... 90 to 120
Rear caliper end retainer .... 75 to 95
Brake hose-to-caliper .... 10 to 15

Mustang (1964-1/2 through 1973)
Front caliper anchor plate-to-steering knuckle
    Upper .... 90 to 120
    Lower .... 55 to 75

Torque specifications	Ft-lbs
	(unless otherwise indicated)

Mustang II (1974 through 1978)
Caliper anchor plate lower bolt	75
Caliper anchor plate upper bolt	100
Caliper banjo fitting bolt	25

Mustang (1979 through 1992)/Capri (1979 through 1986)
Front caliper locating pins	
All except SVO and 1987 and later 5.0L V8	30 to 40
Mustang SVO and 1987 and later 5.0L V8	40 to 60
Rear brake caliper end retainer	75 to 96
Rear brake caliper end retainer screw	16 to 22
Brake hose banjo fitting bolt	
Front	17 to 25
Rear	20 to 30

Pick-Ups and Bronco (1973 through 1979)
Caliper key retaining screw	12 to 20
Caliper bridge bolts (heavy duty)	155 to 185
Anchor plate-to-steering knuckle	
1973 through 1975	55 to 75
1976 through 1979	74 to 102
Brake hose-to-caliper	17 to 25

Pick-Ups and Bronco (1980 through 1994)
Caliper key retaining screw	
Light duty	12 to 20
Heavy duty	14 to 22
Anchor plate-to-spindle	74 to 102
Brake hose banjo fitting bolt	17 to 25

Pinto/Bobcat (1970 through 1980)
Caliper anchor plate lower bolt	75
Caliper anchor plate upper bolt	100
Caliper banjo fitting bolt	25

Probe (1989 through 1992)
Front caliper anchor-to-steering	
knuckle bolts	58 to 72
Rear caliper anchor-to-spindle bolts	33 to 49
Disc brake caliper mounting bolts	
Front	23 to 30
Rear	12 to 17
Brake hose banjo fitting bolt	16 to 22

Taurus/Sable (1986 through 1992)
Front brake caliper mounting pins	18 to 25
Rear brake caliper pinch bolts	23 to 26
Brake hose banjo fitting bolt	
Front	30 to 40
Rear	96 to 132 in-lbs

Tempo/Topaz (1984 through 1993)
Caliper mounting pins	18 to 25
Brake hose banjo fitting bolt	30 to 40

Thunderbird/Cougar (1983 through 1988)
Front caliper locating pins	
Non-turbo	30 to 40
Turbo	40 to 60
Rear caliper slider pin pinch bolt	18 to 24

Thunderbird/Cougar (1989 through 1993)
Front caliper locating pins	19 to 25
Rear caliper slider pin pinch bolt	23 to 26
Anchor plate bolts	45 to 65
Brake hose banjo fitting bolt	30 to 40

Haynes Automotive Brake Manual

General Motors

Buick Regal
Chevy Lumina
Pontiac Grand Prix
Olds Cutlass Supreme
(1988 through 1990)

Caliper mounting bolts	
Front	79
Rear	92
Front caliper mounting bracket-to-steering knuckle bolts	148
Brake hose banjo fitting bolt	32

Buick Skylark (1986 through 1991)
Buick Somerset (1985 through 1987)
Oldsmobile Calais (1985 through 1991)
Pontiac Grand Am (1985 through 1991)

Caliper mounting bolts	38
Brake hose banjo fitting bolts	33

Chevy Lumina APV
Oldsmobile Silhouette
Pontiac Trans Sport
(1990 through 1992)

Caliper mounting bolts	38
Brake hose banjo fitting bolt	33

A-Cars (1982 through 1993)

Buick Century
Chevy Celebrity
Olds Ciera, Cutlass Cruiser
Pontiac 6000

Front caliper mounting bolts	
1990 and earlier	25 to 28
1991 on	38
Rear caliper mounting bolts	
1988 and earlier	30 to 45
1989 and later	74
Rear caliper bridge bolts (1989 and later)	74
Brake hose banjo fitting bolt	
1990 and earlier	18 to 30
1991 on	33

J-Cars (1982 through 1994)

Buick Skyhawk
Cadillac Cimarron
Chevrolet Cavalier
Olds Firenza
Pontiac J2000 and Sunbird

Caliper mounting bolts	
1984 and earlier	21 to 25
1985 on	35 to 38
Brake hose-to-caliper banjo bolt	
1984 and earlier	18 to 30
1985 on	33

Geo

Metro

See entry under Chevrolet/GMC

Prizm

See entry under Chevrolet/GMC

Torque specifications	Ft-lbs (unless otherwise indicated)	

Storm

Caliper mounting bolts	36	
Caliper bracket-to-steering knuckle bolts	76	
Brake hose banjo fitting bolt	22	

Tracker

See entry under Suzuki

See entries under Chevrolet/GMC

GMC

Honda

Accord (1976 through 1983)

Caliper anchor plate bolts	58 to 66
Disc-to-hub bolts	36 to 43
Brake hose banjo fitting bolt	84 to 120 in-lbs

Accord (1984 through 1989)

Front brake caliper mounting bolts	
1984 and 1985	20
1986 on (except LX-i and SE-i models)	33
1988 on LX-i and SE-i models	24
Front brake caliper mounting bracket bolts	
1984 and 1985	56
1986 on (except LX-i and SE-i models)	53
1988 on LX-i and SE-i models	56
Brake hose-to-caliper banjo bolt (front or rear)	25
Rear brake caliper guide bolts	17
Rear brake caliper mounting bracket bolts	28

Accord (1990 through 1993)

Brake hose banjo bolt	25
Caliper mounting bolts	
Front	
Models with short bolts	36
Models with long bolts	54
Rear	17
Caliper mounting bracket bolts	
Front	80
Rear	28

Civic F1200 (1973 through 1979)

Caliper mounting bolts	36 to 43

Civic 1500 (1975 through 1979)

Caliper mounting bolts	43
Disc-to-hub bolts	40

Civic (1980 through 1983)

Caliper mounting bolts	56
Caliper guide pin bolts	
Hatchback	20
Sedan	13
Brake hose banjo fitting bolt	25

Civic (1984 through 1991)

Brake hose-to-caliper banjo bolt (front or rear)	25
Front disc brake	
Caliper guide pin (mounting bolts)	
1984 through 1987 (all models)	
Upper bolt	14
Lower bolt	13

Torque specifications	Ft-lbs
	(unless otherwise indicated)

1988	
Civic, CRX Std and Si, 2WD Wagon	
Upper bolt	40
Lower bolt	33
CRX HF	24
4WD Wagon	36
1989 on	
Civic, CRX DX and 2WD Wagon	
Upper bolt	25
Lower bolt	20
Civic EX	24
CRX HF	17
CRX Si and 4WD Wagon	36
Caliper mounting bracket bolts	
1984 through 1987	56
1988 on	53
Rear disc brake	
Caliper mounting bolts	16
Caliper mounting bracket bolts	28

Prelude (1979 through 1989)

Front caliper mounting bolts	
1979 through 1985	13
1986 and 1987	
Carbureted models	13
Fuel-injected models	33
1988 on	
S models	36
Si models	24
Caliper mount-to-steering knuckle bolts	55
Rear caliper guide bolts	
1984 and 1985	22
1986 and 1987	
Carbureted models	22
Fuel-injected models	17
1988 on	17
Rear caliper mount-to-hub carrier bolts	28
Brake hose banjo fitting bolts	25

Hyundai

Excel (1986 through 1993)

Caliper mounting bolts	
Caliper with integral torque plate (Sumitomo)	43 to 58
Tokico style caliper	16 to 23
Caliper bridge bolts (Sumitomo caliper with integral torque plate)	58 to 69
Torque plate-to-steering knuckle bolts (Tokico)	47 to 54
Brake hose banjo fitting bolt	18 to 22

Isuzu

Pick-ups (1981 - 1993) and Trooper (1984 through 1991)

Caliper mounting bolts	22 to 25
Caliper bracket bolts	
Front	
1981 through 1987	62 to 66
1988 on	103 to 126
Rear	69 to 84
Brake hose banjo fitting bolt	24 to 27
Disc-to-hub bolts	36

Torque specifications	Ft-lbs
	(unless otherwise indicated)

Jaguar

XJ6 (1968 through 1986)

Caliper mounting bolts	
Front	55
Rear	50
Disc-to-hub bolts	35

XJ12 and XJS (1972 through 1985)

Caliper mounting bolts	55
Disc-to-hub bolts	35
Master cylinder banjo bolts	23

Jeep

Cherokee/Comanche (1984 through 1993)

Caliper mounting pins	30
Caliper anchor plate bolts	77
Brake hose banjo fitting bolt	23

CJ (1949 through 1986)

Caliper support key retaining bolt	15 to 18
Caliper anchor bracket-to-steering knuckle	100
Caliper mounting pins (1982 on)	30
Caliper banjo fitting bolt	20 to 30

Wrangler

Caliper mounting pins	
Pre-1990 models	30
1990 models	35
Caliper anchor plate bolts	
Pre-1990 models only	77
Brake hose banjo fitting bolt	23

Mazda

626 (1979 through 1982)

Caliper slide pin	
1981 and earlier	37
1982	
Lower	36
Upper	62

626/MX-6 (1983 through 1991)

Disc brake caliper mounting bolts	
Upper	12 to 18
Lower	14 to 21
Caliper mounting bracket-to-steering	
knuckle bolts	36 to 55
Brake hose banjo fitting bolts	
Front	16 to 19
Rear	17 to 25

GLC (1981 through 1985)

Caliper torque plate-to-steering knuckle bolts	41 to 48
Caliper slide pin bolt	33 to 40
Brake hose banjo fitting bolt	16 to 22

MPV

Brake hose banjo fitting bolt	16 to 22
Brake caliper lock bolts	61 to 69
Caliper mount-to-steering knuckle bolts	65 to 80

Haynes Automotive Brake Manual

Torque specifications	Ft-lbs (unless otherwise indicated)

Pick-Ups (1972 through 1993)

Caliper mounting bolts	23 to 30
Caliper bracket-to-steering knuckle bolts	36 to 55
Brake hose banjo fitting bolt	16 to 19
Disc-to-hub bolts	33 to 42

RX7 (1979 through 1985)

Rear caliper mounting bolt	22 to 30
Brake hose banjo fitting bolt	16

RX-7 (1986 through 1991)

Front caliper lock pin (single-piston caliper)	22 to 30
Front caliper mounting bracket bolts	58 to 72
Front caliper mounting bolts (four-piston caliper)	58 to 72
Rear caliper lock pin	
1986 through 1988	22 to 30
1989 on	12 to 17
Rear caliper mounting bracket bolt	33 to 40
Brake hose banjo fitting bolt	16 to 22

Mercedes-Benz

190 (1984 through 1988)

Front caliper guide pins (bolts)	26
Front caliper carrier (mounting bracket)-to-steering knuckle bolts	85
Rear caliper-to-hub carrier bolts	37

230, 250 and 280 (1968 through 1972)

Caliper mounting bolts	82
Brake hose banjo bolts	144 in-lbs
Disc-to-hub bolts	82

280 (1977 through 1981)

Caliper mounting bolts	
Front	84
Rear	66
Disc-to-hub bolts	84

350, 450 (1971 through 1980)

Caliper mounting bolts	
Front	85
Rear	66
Disc-to-hub bolts	96 in-lbs

Diesel 123 Series (1976 through 1985)

Caliper mounting bolts	
Front	84
Rear	66
Disc-to-hub bolts	84

MG/Austin-Healey

Midget/Sprite (1958 through 1980)

Caliper mounting bolts	48
Disc-to-hub bolts	43

MGB (1962 through 1980)

Caliper mounting bolts	40 to 45

Torque specifications	Ft-lbs
	(unless otherwise indicated)

Mitsubishi

Cordia/Tredia/Precis/Mirage/Galant (1983 through 1993)

Caliper mounting bolts or pins	
8 mm diameter pin or bolt..	23
10 mm diameter pin	35
14 mm diameter pin	62
Caliper bridge bolts	58 to 69
Caliper torque plate bolts	
Front	
Caliper with integral torque	
plate (Sumitomo caliper)...	43 to 58
Tokico caliper	47 to 54
All others...	58 to 72
Rear	36 to 43
Brake hose banjo fitting bolt	
1990 and later mirage	132 to 156 in-lbs
All others	18 to 22

Eclipse/ Eagle Talon/Plymouth Laser (1990 through 1994)

Front brake caliper	
Caliper mounting bolts..	58 to 72
Guide pin pin bolt	
Single-piston caliper	
Up to April 1989	16 to 23
May 1989 on	46 to 62
Double-piston caliper	54
Rear brake caliper	
Caliper mounting bolts..	36 to 43
Guide pin and lock pin bolts	16 to 23
Brake hose-to-rear caliper banjo fitting............................	18 to 25

Pick-ups and Montero (1983 through 1993)

Caliper mounting bolts	
Sliding caliper...	51 to 65
Floating caliper	
Upper bolt..	29 to 36
Lower bolt..	23 to 30
Caliper mounting bracket bolts	
2WD	51 to 65
4WD	58 to 72
Brake hose banjo bolt..	120 to 144 in-lbs
Disc-to-hub bolts	
2WD	51 to 65
4WD	36 to 44

Nissan

300ZX (1984 thru 1989)

Brake hose connector-to-caliper.......................................	12 to 14
Caliper	
Torque member fixing bolt	
1984 thru 1986..	53 to 72
1987 on	28 to 38
Pin bolt	
1984 thru 1986..	16 to 23
1987 on	23 to 30

Maxima (1985 through 1991)

Caliper mounting bolts	16 to 23
Torque plate mounting bolts	
Front	53 to 72
Rear	28 to 38
Caliper banjo fitting bolt	12 to 14

Haynes Automotive Brake Manual

Torque specifications	Ft-lbs
	(unless otherwise indicated)

Pulsar (1983 through 1986)
Caliper mounting bracket-to-steering
    knuckle bolts .... 40 to 47
Guide pin bolts .... 16 to 23

Nissan/Datsun Pick-Ups and Pathfinder (1980 through 1993)
Caliper mounting bolts (1980 through 1983) .... 53 to 72
Yoke-to-cylinder body bolts (1980 through 1983) .... 12 to 15
Caliper slide pin bolts (1984 on) .... 16 to 23
Torque plate-to-steering knuckle bolts (1984 on) .... 53 to 72
Rear brake caliper guide pins (1988 on) .... 16 to 23
Torque plate-to-rear axle case (1988 on) .... 40 to 47
Disc-to-hub bolts
    1980 through 1983 33
    1984 on .... 36 to 51
Brake hose banjo fitting bolt .... 144 to 168 in-lbs

Sentra (1982 through 1990)
Caliper mounting bracket-to-steering
    knuckle 64
Caliper guide pin and lock pin 31
Disc-to-hub bolts 34
Brake hose-to-caliper 20

Stanza (1982 through 1990)
Caliper mounting bracket-to-steering
    knuckle bolts 90
Caliper guide and lock pins 35
Disc-to-hub bolts 50

Oldsmobile

Bravada
See entry under Chevrolet/GMC

Cutlass Rear-Wheel Drive (1974 through 1988)
Caliper mounting bolts 37
Brake hose banjo fitting bolt 32

Full-Size Front-Wheel Drive (1985 through 1993)
See entry under Buick

Full-Size Rear-Wheel Drive (1970 through 1990)
See entry under Buick

Omega
See entry under Pontiac

Peugeot

504 Diesel (1974 through 1983)
Caliper mounting bolts
    Front 51
    Rear 31
Disc-to-hub bolts 34

Torque specifications	Ft-lbs (unless otherwise indicated)	

Plymouth

Laser
See entry under Mitsubishi

Other models
See entry under Dodge/Plymouth/Chrysler

Pontiac

Fiero (1984 through 1988)
Caliper mounting bolts	40
Caliper parking brake lever nut	35
Caliper mounting knuckle assembly bolt	35
Front brake hose-to-caliper	
1984 and 1985	39
1986 and 1987	30

Firebird (1970 through 1981)
Caliper mounting bolts	35
Caliper banjo fitting bolt	22

Firebird (1982 through 1992)
Caliper mounting bolts (single-piston front caliper)	35
Caliper bracket bolts (twin-piston caliper)	137
Rear caliper guide pin bolt (1989 on)	
Allen head	18
Hex head	26
Rear caliper bracket bolts	70
Brake hose banjo fitting bolt	32

Full-Size Front-Wheel Drive Models (1985 through 1993)
See entry under Buick

Full-Size Rear-Wheel Drive Models (1970 through 1990)
See entry under Buick

Phoenix/Omega X-Cars (1980 through 1984)
Caliper mounting bolt	28
Brake hose banjo fitting bolt	22

Porsche

911
Caliper mounting bracket-to-strut	34
Caliper mounting bolts	
Front	51
Rear	43
Disc-to-hub bolts	17
Brake hose banjo fitting bolt	14

914 (four-cylinder)
Caliper mounting bolts	
Front	61
Rear	50
Caliper bridge bolt	16
Brake hose banjo fitting bolt	14

Torque specifications	Ft-lbs (unless otherwise indicated)
924 (1976 through 1982)	
Caliper mounting bolts	60
944 (1983 through 1989)	
Caliper-to-steering knuckle bolts	63
Disc-to-hub nut/bolt	17

Renault

5 LeCar (1976 through 1983)	
Caliper bracket bolts	67
Caliper guide bolts	35
Disc-to-hub bolts	26
8 and 10 (1962 through 1972)	
Caliper pin nut	120 in-lbs
Caliper mounting nuts	15
12 (1979 to 1980)	
Caliper mounting bolts	50
Disc-to-hub bolts	15
16 (1965 through 1978)	
Caliper mounting bolts	25
15 and 17 (1973 through 1979)	
Caliper bracket bolts	49
Disc-to-hub bolts	19
Fuego (1982 through 1984	
Caliper mounting bolt	42
Caliper bracket bolts	74
Caliper guide bolts (Girling)	26

Saab

99 (1969 through 1980)	
Caliper mounting bolts	60
900 (1979 through 1988)	
1983 and earlier models	
Caliper yoke bolts	
Front	81 to 96
Rear	52 to 66
Disc-to-hub bolts	22 to 36
1984 on	
Caliper guide pins	
Front	22 to 26
Rear	18 to 22
Caliper yoke bolts	
Front	52 to 82
Rear	30 to 49

Saturn

Saturn (1991 through 1994)	
Caliper lock/guide pins (front and rear calipers)	27
Caliper torque plate bolts	
Front	81
Rear	63
Brake hose banjo fitting bolt	36

Torque specifications	Ft-lbs	
	(unless otherwise indicated)	

Subaru

1600 and 1800 (1980 through 1989)

Caliper mounting bolt	36 to 51
Torque plate to steeering knuckle bolts	36 to 51
Brake hose banjo fitting bolt...	15

Suzuki

Samurai and Sidekick/Geo Tracker

Caliper mounting bolts	
Samurai	18 to 20
Sidekick and Tracker..	36 to 57
Caliper bracket bolts	29 to 43
Brake hose banjo fitting bolt...	14 to 18

Toyota

Camry (1988 through 1991)

Front caliper mounting bolts	
1986 and earlier..	18
1987 and later..	29
Front caliper torque plate-to-steering knuckle	
1986 and earlier..	65
1987 and later..	79
Rear caliper mounting bolts	14
Rear caliper torque plate-to-axle carrier	34
Brake hose banjo fitting bolt...	22

Carina (1971 through 1974)

Caliper bracket bolts	50 to 65

Celica Front-Wheel Drive (1986 through 1992)

Caliper mounting bolts	
Front	
1986 through 1988...	18
1989..	27
1990 and 1991	
13-inch wheels	18
14 and 15-inch wheels	29
1992 on..	29
Rear	14
Caliper torque plate bolts	
Front	
1986 through 1988...	69
1989..	73
1990 on..	79
Rear	34
Brake hose banjo fitting bolt...	22

Celica Rear-Wheel Drive (1971 through 1985)

Caliper-to-torque plate bolts	12 to 17
Torque plate-to-steering knuckle bolts	
1971 through 1979	40
1980 and 1981	40 to 54
1982 through 1985	58 to 75
Disc-to-hub bolts	
1971 through 1981	40
1982 through 1985	40 to 54
Rear torque plate mounting bolts..	34

Haynes Automotive Brake Manual

Torque specifications	Ft-lbs
	(unless otherwise indicated)

Celica Supra (1979 through 1992)

Front caliper mounting bolt(s)
1982 through 1986	14
1987 and later	27

Front caliper torque plate-to-steering knuckle bolts
1979 through 1986	34
1987 and later	77
Rear caliper mounting bolt(s)	14

Rear caliper torque plate-to-axle assembly bolts
1979 through 1981	48
1982 and later	34
Brake hose banjo fitting bolt	22

Corolla Rear-Wheel Drive (1975 through 1979)

Split-type floating caliper with integral torque plate
Caliper mounting bolts	38
Caliper bridge bolts	60
Sliding caliper mounting bolts	50
Disc-to-hub bolts	65

Corolla Rear-Wheel Drive (1980 through 1987)

Front brake caliper-to-torque plate
1980 through 1983	40 to 54
1984 on	14

Torque plate-to-steering knuckle bolts
1980 through 1983	29 to 39
1984 on	47
Caliper bridge bolts	58 to 68
Rear brake caliper-to-torque plate	14
Rear torque plate bolts	34

Corolla Front-Wheel Drive (1984 through 1992)

Caliper-to-torque plate bolts
Front	18
Rear	14

Front caliper torque plate-to-steering
knuckle bolts	65
Rear caliper torque plate-to-axle carrier bolts	34

Brake hose banjo fitting bolt
All except 1985 models	22
1985 models only	17

Corolla Tercel (1980 through 1982)

Caliper-to-torque plate bolt	11 to 15
Torque plate-to-steering knuckle	33 to 39
Brake hose banjo fitting bolt	120 to 156 in-lbs

Corona (1974 through 1982)

Caliper-to steering knuckle bolts
Fixed caliper	77
Sliding caliper	57

Cressida (1978 through 1982)

Caliper slide pins	62 to 68
Caliper mount-to-steering knuckle	68 to 86
Disc-to-hub bolts	29 to 39
Brake hose banjo fitting bolt	120 to 156 in-lbs

Hi-Lux Pick-Ups (1969 to 1978)

Caliper mounting bolts	80
Disc-to-hub bolts	35
Brake hose banjo fitting bolt	18

Land Cruiser (1968 through 1982)

Caliper mounting bolts	65
Disc-to-hub bolts	35

Torque specifications	Ft-lbs (unless otherwise indicated)

Mark II 6-cylinder (1972 through 1976)
Caliper mounting bolts	80
Disc-to-hub bolts	38

MR2 (1985 thru 1987)
Caliper mounting bolt	
Front	18
Rear	14
Brake hose banjo fitting bolt	11

Pick-ups and 4Runner (1979 through 1992)
Caliper-to-steering knuckle bolts	
1983 and earlier 2WD, 1984 and earlier 4WD	68 to 86
1984 and 1985 4WD models	55 to 75
1986 and later 4WD models	90
Caliper-to-torque plate bolts	
1984 and later 2WD (FS17 type)	62 to 68
1985 and later 2WD (PD60 and PD66 types through 1988)	29
1989 and later 2WD	47
1989 and later 4WD	90
Torque plate-to-steering knuckle bolts	73 to 86
Disc-to-hub bolts	
2WD	40 to 54
4WD	29 to 39

Previa (1991 through 1993)
Caliper mounting bolts	
Front	27
Rear	18
Caliper torque plate bolts	
Front	
1991	77
1992 on	
Single-piston type	65
Double-piston type	61
Rear	65
Brake hose banjo fitting bolt	22

Triumph

Spitfire (1962 through 1981)
Caliper mounting bolts	65
Disc-to-hub bolts	34

TR7 (1975 through 1981)
Caliper mounting bolts	74
Disc-to-hub bolts	32

Volvo

120 and 130 Series (1961 through 1973)
Includes 1800 Sports
Caliper mounting bolts	65 to 70

140 Series (1966 through 1974)
Caliper mounting bolts	
Front	65 to 70
Rear	45 to 50

Torque specifications	Ft-lbs
	(unless otherwise indicated)

240 Series (1974 through 1990)
Caliper mounting bolts
    Front .... 65 to 72
    Rear .... 38 to 46

260 Series (1975 through 1982)
Front brake caliper bolts .... 72 to 74
Rear brake caliper bolts .... 42 to 43
Caliper mounting bracket bolts .... 36 to 37

740 and 760 Series (1982 through 1988)
Front caliper bracket bolts 74
Rear caliper bracket bolts (use new bolts) 43
Upper guide pin-to-caliper bracket 18
Caliper buide pin bolts 25

Volkswagen

Beetle/Karmann Ghia (1954 through 1979)
Caliper mounting bolts 30
Caliper bridge bolts .... 14 to 18

Dasher (1974 through 19810
Caliper-to-steering knuckle bolts 45
Caliper guide bolts (vented discs) 29
Caliper guide self-locking bolts (solid discs) 26

Rabbit, Golf, Jetta Scirocco, Pick-Up (1975 through 1992)
Front caliper mounting bolts
    1975 through 1984
        Standard 43
        Self-locking 52
    1985 on 18
U-shaped pad retainer bolt (Girling) 15
Caliper guide pins (Kelsy-Hayes) 30
Rear caliper mounting bolts 26
Rear disc brake pad
    carrier-to-axle bolt 44

Rabbit/Jetta/Pick-Up Diesel (1977 through 1984)
Caliper-to-wheel bearing housing 43

Transporter 1600 (1968 through 1979)
Caliper mounting bolts 72

Transporter 1700 (1972 thorugh 1979)
Caliper-to-steering knuckle
    1972 72
    1973 on 116
Caliper bridge bolts 25
Disc-to-hub bolts 18
Brake hose banjo fitting bolt 14

Type 3 (1963 through 1973)
Caliper mounting bolts
    1971 and earlier 29
    1972 on .... 58 to 65

Vanagon (1980 through 1983)
Caliper-to-steering knuckle bolts 118

Index

Haynes Automotive Brake Manual

HAYNES AUTOMOTIVE MANUALS

NOTE: New manuals are added to this list on a periodic basis. If you do not see a listing for your vehicle, consult your local Haynes dealer for the latest product information.

ACURA
*1776 **Integra & Legend** '86 thru '90

AMC
Jeep CJ - *see JEEP (412)*
694 **Mid-size models,** Concord, Hornet, Gremlin & Spirit '70 thru '83
934 **(Renault) Alliance & Encore** '83 thru '87

AUDI
615 **4000** all models '80 thru '87
428 **5000** all models '77 thru '83
1117 **5000** all models '84 thru '88

AUSTIN
Healey Sprite - *see MG Midget (265)*

BMW
*2020 **3/5 Series** not including diesel or all-wheel drive models '82 thru '92
276 **320i** all 4 cyl models '75 thru '83
632 **528i** & 530i all models '75 thru '80
240 **1500 thru 2002** except Turbo '59 thru '77
348 **2500, 2800, 3.0 & Bavaria** '69 thru '76

BUICK
Century (front wheel drive) - *see GENERAL MOTORS (829)*
*1627 **Buick, Oldsmobile & Pontiac Full-size (Front wheel drive)** '85 thru '93
Buick Electra, LeSabre and Park Avenue; **Oldsmobile** Delta 88 Royale, Ninety Eight and Regency; **Pontiac** Bonneville
1551 **Buick Oldsmobile & Pontiac Full-size (Rear wheel drive)**
Buick Estate '70 thru '90, Electra'70 thru '84, LeSabre '70 thru '85, Limited '74 thru '79
Oldsmobile Custom Cruiser '70 thru '90, Delta 88 '70 thru '85,Ninety-eight '70 thru '84
Pontiac Bonneville '70 thru '81, Catalina '70 thru '81, Grandville '70 thru '75, Parisienne '83 thru '86
627 **Mid-size Regal & Century** all rear-drive models with V6, V8 and Turbo '74 thru '87
Regal - *see GENERAL MOTORS (1671)*
Skyhawk - *see GENERAL MOTORS (766)*
552 **Skylark** all X-car models '80 thru '85
Skylark '86 on - *see GENERAL MOTORS (1420)*
Somerset - *see GENERAL MOTORS (1420)*

CADILLAC
*751 **Cadillac Rear Wheel Drive** all gasoline models '70 thru '92
Cimarron - *see GENERAL MOTORS (766)*

CAPRI
296 **2000 MK I Coupe** all models '71 thru '75
Mercury Capri - *see FORD Mustang (654)*

CHEVROLET
*1477 **Astro & GMC Safari Mini-vans** '85 thru '93
554 **Camaro V8** all models '70 thru '81
866 **Camaro** all models '82 thru '92
Cavalier - *see GENERAL MOTORS (766)*
Celebrity - *see GENERAL MOTORS (829)*
625 **Chevelle, Malibu & El Camino** '69 thru '87
449 **Chevette & Pontiac T1000** '76 thru '87
550 **Citation** all models '80 thru '85
*1628 **Corsica/Beretta** all models '87 thru '92
274 **Corvette** all V8 models '68 thru '82
*1336 **Corvette** all models '84 thru '91
1762 **Chevrolet Engine Overhaul Manual**
704 **Full-size Sedans** Caprice, Impala, Biscayne, Bel Air & Wagons '69 thru '90
Lumina - *see GENERAL MOTORS (1671)*
Lumina APV - *see GENERAL MOTORS (2035)*
319 **Luv Pick-up** all 2WD & 4WD '72 thru '82
626 **Monte Carlo** all models '70 thru '88
241 **Nova** all V8 models '69 thru '79
*1642 **Nova/Geo Prizm** front wheel drive '85 thru '92
420 **Pick-ups '67 thru '87** - Chevrolet & GMC, all V8 & in-line 6 cyl, 2WD & 4WD '67 thru '87; Suburbans, Blazers & Jimmys '67 thru '91
*1664 **Pick-ups '88 thru '93** - Chevrolet & GMC, all full-size (C and K) models, '88 thru '93
*831 **S-10 & GMC S-15 Pick-ups** '82 thru '92
*1727 **Sprint & Geo Metro** '85 thru '91
*345 **Vans - Chevrolet & GMC,** V8 & in-line 6 cylinder models '68 thru '92

CHRYSLER
*2058 **Full-size Front-Wheel Drive** '88 thru '93
K-Cars - *see DODGE Aries (723)*
Laser - *see DODGE Daytona (1140)*
*1337 **Chrysler & Plymouth Mid-size** front wheel drive '82 thru '93

DATSUN
402 **200SX** all models '77 thru '79
647 **200SX** all models '80 thru '83
228 **B - 210** all models '73 thru '78
525 **210** all models '78 thru '82
206 **240Z, 260Z & 280Z** Coupe '70 thru '78
563 **280ZX** Coupe & 2+2 '79 thru '83
300ZX - *see NISSAN (1137)*
679 **310** all models '78 thru '82
123 **510 & PL521 Pick-up** '68 thru '73
430 **510** all models '78 thru '81
372 **610** all models '72 thru '76
277 **620 Series Pick-up** all models '73 thru '79
720 Series Pick-up - *see NISSAN (771)*
376 **810/Maxima** all gas models, '77 thru '84
368 **F10** all models '76 thru '79
Pulsar - *see NISSAN (876)*
Sentra - *see NISSAN (982)*
Stanza - *see NISSAN (981)*

DODGE
400 & 600 - *see CHRYSLER Mid-size (1337)*
*723 **Aries & Plymouth Reliant** '81 thru '89
*1231 **Caravan & Plymouth Voyager Mini-Vans** all models '84 thru '93
699 **Challenger/Plymouth Saporro** '78 thru '83
Challenger '67-'76 - *see DODGE Dart (234)*
236 **Colt** all models '71 thru '77
610 **Colt/Plymouth Champ (front wheel drive)** all models '78 thru '87
*1668 **Dakota Pick-ups** all models '87 thru '93
234 **Dart, Challenger/Plymouth Barracuda & Valiant** 6 cyl models '67 thru '76
*1140 **Daytona & Chrysler Laser** '84 thru '89
*545 **Omni & Plymouth Horizon** '78 thru '90
*912 **Pick-ups** all full-size models '74 thru '91
*556 **Ram 50/D50 Pick-ups & Raider and Plymouth Arrow Pick-ups** '79 thru '93
*1726 **Shadow/Plymouth Sundance** '87 thru '93
*1779 **Spirit & Plymouth Acclaim** '89 thru '92
*349 **Vans - Dodge & Plymouth** '71 thru '91

EAGLE
Talon - *see Mitsubishi Eclipse (2097)*

FIAT
094 **124 Sport Coupe & Spider** '68 thru '78
273 **X1/9** all models '74 thru '80

FORD
*1476 **Aerostar Mini-vans** '86 thru '92
788 **Bronco and Pick-ups** '73 thru '79
*880 **Bronco and Pick-ups** '80 thru '91
268 **Courier Pick-up** all models '72 thru '82
1763 **Ford Engine Overhaul Manual**
789 **Escort/Mercury Lynx** '81 thru '90
*2046 **Escort/Mercury Tracer** '91 thru '93
*2021 **Explorer & Mazda Navajo** '91 thru '92
560 **Fairmont & Mercury Zephyr** '78 thru '83
334 **Fiesta** all models '77 thru '80
754 **Ford & Mercury Full-size,** Ford LTD & Mercury Marquis ('75 thru '82); Ford Custom 500,Country Squire, Crown Victoria & Mercury Colony Park ('75 thru '87); Ford LTD Crown Victoria & Mercury Gran Marquis ('83 thru '87)
359 **Granada & Mercury Monarch** '75 thru '80
773 **Ford & Mercury Mid-size,** Ford Thunderbird & Mercury Cougar ('75 thru '82); Ford LTD & Mercury Marquis ('83 thru '86); Ford Torino,Gran Torino, Elite, Ranchero pick-up, LTD II, Mercury Montego, Comet, XR-7 & Lincoln Versailles ('75 thru '86)
*654 **Mustang & Mercury Capri** incl. Turbo Mustang, '79 thru '92; Capri, '79 thru '86
357 **Mustang V8** all models '64-1/2 thru '73
231 **Mustang II** 4 cyl, V6 & V8 '74 thru '78
649 **Pinto & Mercury Bobcat** '75 thru '80
1670 **Probe** all models '89 thru '92
*1026 **Ranger/Bronco II** gas models '83 thru '93
*1421 **Taurus & Mercury Sable** '86 thru '92
*1418 **Tempo & Mercury Topaz** '84 thru '93
1338 **Thunderbird/Mercury Cougar** '83 thru '88
*1725 **Thunderbird/Mercury Cougar** '89 and '90
*344 **Vans** all V8 Econoline models '69 thru '91

GENERAL MOTORS
*829 **Buick Century, Chevrolet Celebrity, Olds Cutlass Ciera & Pontiac 6000** all models '82 thru '93
*766 **Buick Skyhawk, Cadillac Cimarron, Chevrolet Cavalier, Oldsmobile Firenza Pontiac J-2000 & Sunbird** '82 thru '92
1420 **Buick Skylark & Somerset, Olds Calais & Pontiac Grand Am** '85 thru '91
*1671 **Buick Regal, Chevrolet Lumina, Oldsmobile Cutlass Supreme & Pontiac Grand Prix** front wheel drive '88 thru '90
*2035 **Chevrolet Lumina APV, Oldsmobile Silhouette & Pontiac Trans Sport** '90 thru '92

GEO
Metro - *see CHEVROLET Sprint (1727)*
Prizm - *see CHEVROLET Nova (1642)*
*2039 **Storm** all models '90 thru '93
Tracker - *see SUZUKI Samurai (1626)*

GMC
Safari - *see CHEVROLET ASTRO (1477)*
Vans & Pick-ups - *see CHEVROLET (420, 831, 345, 1664)*

HONDA
351 **Accord CVCC** all models '76 thru '83
1221 **Accord** all models '84 thru '89
2067 **Accord** all models '90 thru '93
160 **Civic 1200** all models '73 thru '79
633 **Civic 1300 & 1500 CVCC** '80 thru '83
297 **Civic 1500 CVCC** all models '75 thru '79
1227 **Civic** all models '84 thru '91
*601 **Prelude CVCC** all models '79 thru '89

HYUNDAI
*1552 Excel all models '86 thru '93

ISUZU
*1641 **Trooper & Pick-up,** all gasoline models Pick-up, '81 thru '93; Trooper, '84 thru '91

JAGUAR
*242 **XJ6** all 6 cyl models '68 thru '86
*478 **XJ12 & XJS** all 12 cyl models '72 thru '85

JEEP
*1553 **Cherokee, Comanche & Wagoneer Limited** all models '84 thru '93
412 **CJ** all models '49 thru '86
*1777 **Wrangler** all models '87 thru '92

LADA
*413 **1200, 1300. 1500 & 1600** '74 thru '91

MAZDA
648 **626** Sedan & Coupe (rear wheel drive) all models '79 thru '82
*1082 **626 & MX-6 (front wheel drive)** '83 thru '91
267 **B Series Pick-ups** '72 thru '93
370 **GLC Hatchback (rear wheel drive)** all models '77 thru '83
757 **GLC (front wheel drive)** '81 thru '85
*2047 **MPV** all models '89 thru '93
460 **RX-7** all models '79 thru '85
*1419 **RX-7** all models '86 thru '91

MERCEDES-BENZ
*1643 **190 Series** 4-cyl gas models, '84 thru '88
346 **230, 250 & 280** 6 cyl sohc '68 thru '72
983 **280 123 Series** gas models '77 thru '81
698 **350 & 450** all models '71 thru '80
697 **Diesel 123 Series** '76 thru '85

MERCURY
See FORD Listing

MG
111 **MGB** Roadster & GT Coupe '62 thru '80
265 **MG Midget & Austin Healey Sprite Roadster** '58 thru '80

MITSUBISHI
*1669 **Cordia, Tredia, Galant, Precis & Mirage** '83 thru '93
*2022 **Pick-up & Montero** '83 thru '93
*2097 **Eclipse, Eagle Talon & Plymouth Laser** '90 thru '94

MORRIS
074 **(Austin) Marina 1.8** all models '71 thru '78
024 **Minor 1000** sedan & wagon '56 thru '71

NISSAN
1137 **300ZX** all models incl. Turbo '84 thru '89
*1341 **Maxima** all models '85 thru '91
*771 **Pick-ups/Pathfinder** gas models '80 thru '93
876 **Pulsar** all models '83 thru '86
*982 **Sentra** all models '82 thru '90
*981 **Stanza** all models '82 thru '90

OLDSMOBILE
Bravada - *see CHEVROLET S-10 (831)*
Calais - *see GENERAL MOTORS (1420)*
Custom Cruiser - *see BUICK Full-size RWD (1551)*
*658 **Cutlass** '74 thru '88
Cutlass Ciera - *see GENERAL MOTORS (829)*
Cutlass Supreme - *see GM (1671)*
Delta 88 - *see BUICK Full-size RWD (1551)*
Delta 88 Brougham - *see BUICK Full-size FWD (1551), RWD (1627)*
Delta 88 Royale - *see BUICK Full-size RWD (1551)*
Firenza - *see GENERAL MOTORS (766)*
Ninety-eight Regency - *see BUICK Full-size RWD (1551), FWD (1627)*
Ninety-eight Regency Brougham - *see BUICK Full-size RWD (1551)*
Omega - *see PONTIAC Phoenix (551)*
Silhouette - *see GENERAL MOTORS (2035)*

PEUGEOT
663 **504** all diesel models '74 thru '83

PLYMOUTH
Laser - *see MITSUBISHI Eclipse (2097)*
Other PLYMOUTH titles, see DODGE

PONTIAC
T1000 - *see CHEVROLET Chevette (449)*
J-2000 - *see GENERAL MOTORS (766)*
6000 - *see GENERAL MOTORS (829)*
Bonneville - *see Buick Full-size FWD (1627), RWD (1551)*
Bonneville Brougham - *see Buick Full-size (1551)*
Catalina - *see Buick Full-size (1551)*
1232 **Fiero** all models '84 thru '88
555 **Firebird** V8 models except Turbo '70 thru '81
867 **Firebird** all models '82 thru '92
Full-size Rear Wheel Drive - *see BUICK Oldsmobile, Pontiac Full-size RWD (1551)*
Full-size Front Wheel Drive - *see BUICK Oldsmobile, Pontiac Full-size FWD (1627)*
Grand Am - *see GENERAL MOTORS (1420)*
Grand Prix - *see GENERAL MOTORS (1671)*
Grandville - *see BUICK Full-size (1551)*
Parisienne - *see BUICK Full-size (1551)*
551 **Phoenix/Oldsmobile Omega** '80 thru '84
Sunbird - *see GENERAL MOTORS (766)*
Trans Sport - *see GENERAL MOTORS (2035)*

PORSCHE
*264 **911** all Coupe & Targa models except Turbo & Carrera 4 '65 thru '89
239 **914** all 4 cyl models '69 thru '76
397 **924** all models incl. Turbo '76 thru '82
*1027 **944** all models incl. Turbo '83 thru '89

RENAULT
141 **5 Le Car** all models '76 thru '83
079 **8 & 10** 58.4 cu in engines '62 thru '72
097 **12** Saloon & Estate '70 thru '80
768 **15 & 17** all models '73 thru '79
081 **16** 89.7 cu in & 95.5 cu in engines '65 thru '72
Alliance & Encore - *see AMC (934)*

SAAB
247 **99** all models including Turbo '69 thru '80
*980 **900** including Turbo '79 thru '88

SUBARU
237 **1100, 1300, 1400 & 1600** '71 thru '79
*681 **1600 & 1800** 2WD & 4WD '80 thru '89

SUZUKI
*1626 **Samurai/Sidekick/Geo Tracker** '86 thru '93

TOYOTA
1023 **Camry** all models '83 thru '91
150 **Carina** Sedan all models '71 thru '74
935 **Celica Rear Wheel Drive** '71 thru '85
*2038 **Celica Front Wheel Drive** '86 thru '92
1139 **Celica Supra** all models '79 thru '92
361 **Corolla** all models '75 thru '79
961 **Corolla** rear wheel drive models '80 thru '87
*1025 **Corolla** front wheel drive models '84 thru '92
636 **Corolla Tercel** all models '80 thru '82
360 **Corona** all models '74 thru '82
532 **Cressida** all models '78 thru '82
313 **Land Cruiser** all models '68 thru '82
200 **MK II** all 6 cyl models '72 thru '76
*1339 **MR2** all models '85 thru '87
304 **Pick-up** all models '69 thru '78
*656 **Pick-up** all models '79 thru '92
*2048 **Previa** all models '91 thru '93

TRIUMPH
112 **GT6 & Vitesse** all models '62 thru '74
113 **Spitfire** all models '62 thru '81
322 **TR7** all models '75 thru '81

VW
159 **Beetle & Karmann Ghia** '54 thru '79
238 **Dasher** all gasoline models '74 thru '81
*884 **Rabbit, Jetta, Scirocco, & Pick-up** gas models '74 thru '91 & Convertible '80 thru '92
451 **Rabbit, Jetta & Pick-up** diesel models '77 thru '84
082 **Transporter 1600** all models '68 thru '79
226 **Transporter 1700, 1800 & 2000** '72 thru '79
084 **Type 3 1500 & 1600** '63 thru '73
1029 **Vanagon** all air-cooled models '80 thru '83

VOLVO
203 **120, 130 Series & 1800 Sports** '61 thru '73
129 **140 Series** all models '66 thru '74
*270 **240 Series** all models '74 thru '90
400 **260 Series** all models '75 thru '82
*1550 **740 & 760 Series** all models '82 thru '88

SPECIAL MANUALS
1479 **Automotive Body Repair & Painting Manual**
1654 **Automotive Electrical Manual**
1667 **Automotive Emissions Control Manual**
1480 **Automotive Heating & Air Conditioning Manual**
1762 **Chevrolet Engine Overhaul Manual**
1736 **GM & Ford Diesel Engine Repair Manual**
1763 **Ford Engine Overhaul Manual**
482 **Fuel Injection Manual**
2069 **Holley Carburetor Manual**
1666 **Small Engine Repair Manual**
299 **SU Carburetors** thru '88
393 **Weber Carburetors** thru '79
300 **Zenith/Stromberg CD Carburetors** thru '76

HM845
96-TN

** Listings shown with an asterisk (*) indicate model coverage as of this printing. These titles will be periodically updated to include later model years - consult your Haynes dealer for more information.*

Over 100 Haynes motorcycle manuals also available

5-94

Haynes North America, Inc., 861 Lawrence Drive, Newbury Park, CA 91320 • (805) 498-6703